ARTHUR SCHULTZE
FRANK LOUIS SEVENOAK

PLANE AND SOLID GEOMETRY

Elibron Classics
www.elibron.com

Elibron Classics series.

ISBN 1-4021-4484-9 (paperback)
ISBN 1-4212-8120-1 (hardcover)

This Elibron Classics Replica Edition is an unabridged facsimile
of the edition published in 1906 by the Macmillan Company, New York.

PLANE AND SOLID GEOMETRY

PLANE AND SOLID GEOMETRY

BY

ARTHUR SCHULTZE, Ph.D.

ASSISTANT PROFESSOR OF MATHEMATICS, NEW YORK UNIVERSITY
HEAD OF THE DEPARTMENT OF MATHEMATICS, HIGH
SCHOOL OF COMMERCE, NEW YORK

AND

F. L. SEVENOAK, A.M., M.D.

ASSISTANT PRINCIPAL OF THE ACADEMIC DEPARTMENT
STEVENS INSTITUTE OF TECHNOLOGY

New York
THE MACMILLAN COMPANY
LONDON: MACMILLAN & CO. Ltd.
1906

Set up and electrotyped May, 1901. Reprinted March, September, 1902; August, September, 1903; August, 1904; September, November, 1905; January, 1906.

PREFACE

It is generally conceded that the final aim of mathematical teaching should be not only the acquisition of practical knowledge, but that training of the student's mind which gives a distinct gain of mental power. In recognition of this principle nearly all college entrance examinations in geometry require some original work, and most text-books devote considerable space to exercises. Comparatively little, however, has been done to introduce the student *systematically* into original geometrical work. No teacher of physics or chemistry would ask a student to discover a law without so guiding his work as to enable him to reach the desired result; many text-books and teachers expect the pupil to invent geometrical proofs and to solve problems, entirely new to him, without offering any assistance further than a knowledge of the well-established theorems of all text-books. Some writers give a description of the analysis of propositions, which is entirely logical and of great advantage to a person of some mathematical knowledge, but which is usually too abstract to be of any practical value to the beginner. In this book the attempt is made to introduce the student systematically to the solution of geometrical exercises. In the beginning the exercises given in a certain group are of similar kind and related to the preceding proposition; later some general principles are developed which are of fundamental importance for original work, as, for example, the method of proving the equality of lines by means of equal triangles; the method of proving the proportionality of lines by means of similar triangles, etc., and finally the analyses of theorems and problems are introduced, but in a more concrete form than usual.

The propositions are arranged with the view of obtaining a perfect logical and pedagogical order. An unusually large number of exercises, selected with care for the purpose of securing increased mental power, is given.

The general plan and preparation of the greater part of the book are the work of Dr. Schultze, while that of Dr. Sevenoak has been chiefly editorial.

SUGGESTIONS TO TEACHERS

1. Students should be made thoroughly familiar with the definitions, especially in the beginning. Many beginners who are otherwise perfectly capable of understanding logical deductions, fail in geometry because they have not acquired this thorough knowledge of the fundamental concepts. Exercises in geometrical drawing (*e.g.* §§ 113, 114, 115, etc.) and numerical exercises afford an excellent means of familiarizing the student with the definitions.

2. The preliminary propositions (§§ 52–58) on account of their great simplicity are quite often confusing for the beginner and should not be made *the basis of the study of geometrical form.* Later on, however, too much emphasis cannot be laid upon this form in which demonstrations are presented.

3. Do not omit the exercises at the first reading, although it is not necessary at this time to study every exercise given in the text.

4. It is advisable to explain at length to the beginner the meaning of the very first exercises, since many students need some practice before they can understand geometrical language.

5. Many beginners experience considerable difficulty in trying to remember the data of a proposition; they forget the hypothesis. Graphical methods afford an excellent means to overcome this difficulty. Draw equal lines or equal angles in the same color, or if this is impracticable, mark them by equal

crossmarks; denote parallel lines by arrows; draw a greater line in some color, *e.g.* green, a smaller line in another color, *e.g.* yellow, etc. The following diagrams illustrate the method. The diagrams on the left may be used if no colored crayons are available, those on the right should be drawn in the colors indicated. Lines and angles whose relative size is unknown are drawn in white.

I.

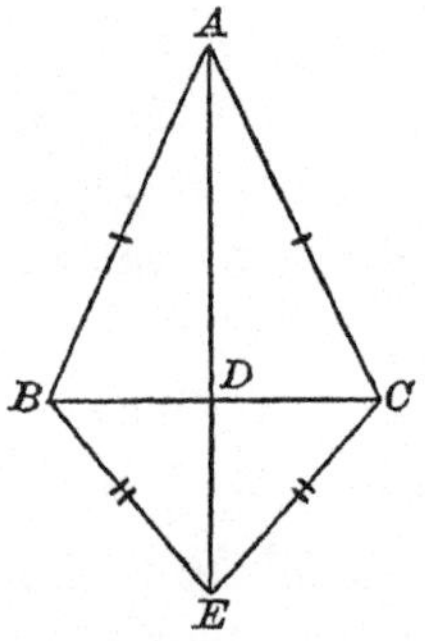

or

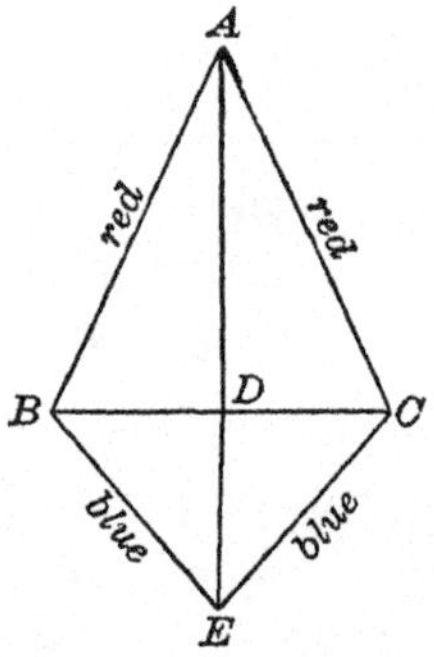

To prove $BD = DC$ and $AD \perp BC$ (Prop. XVII, Bk. I).

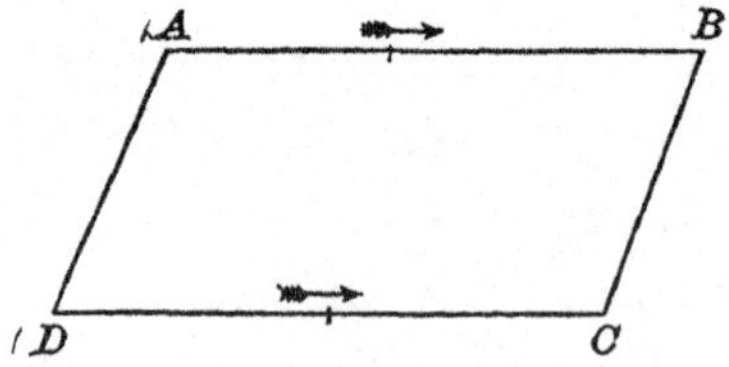

or

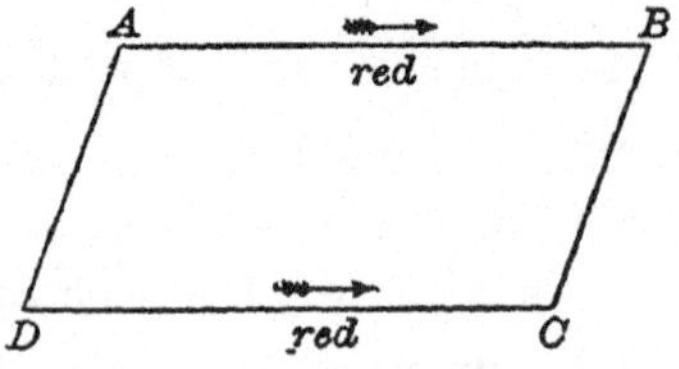

To prove $AD \parallel BC$.

[The hypothesis indicated by either diagram is: In quad rilateral $ABCD$, $AB = CD$, and $AB \parallel CD$.]

III.

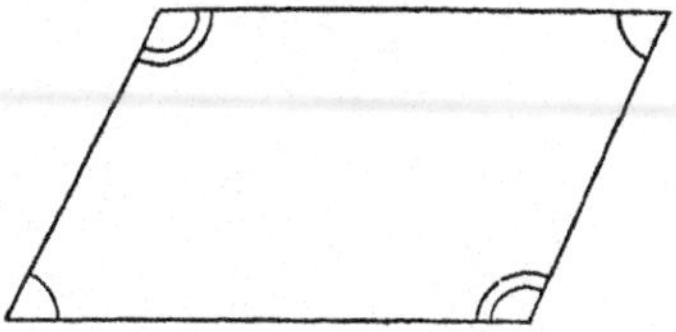

or

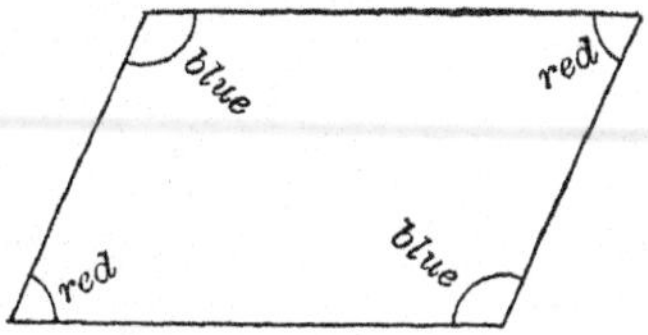

To prove the opposite sides of the quadrilateral are equal.

6. Discourage mere memorizing. Many students form during the first weeks of study the habit of relying too much upon their memories, a habit which often proves disastrous to their future geometrical work.

7. Figures should be drawn as accurately as possible and as general as possible, *i.e.* a figure relating to a triangle in general should not be a right or an isosceles triangle, etc.

8. Give frequent written examinations containing some simple original work.

CONTENTS

SYMBOLS AND ABBREVIATIONS

$+$	. .	plus, or added to.
$-$	. .	minus, or diminished by.
$=$	. .	equals, or is equal to.
$\Leftrightarrow$	. .	is equivalent to.
$>$	. .	is greater than.
$<$	. .	is less than.
$\therefore$	. .	therefore, or hence.
$\perp$	. .	perpendicular, or is perpendicular to.
$\perp$s	. .	perpendiculars.
$\parallel$	. .	parallel, or is parallel to.
$\parallel$s	. .	parallels.
$\sim$	. .	is similar to, or similar.
$\angle$	. .	angle.
$\angle$s	. .	angles.
$\triangle$	. .	triangle.
$\triangle$s	. .	triangles.
▱	. .	parallelogram.
▱s	. .	parallelograms.
$\odot$	. .	circle.
$\odot$s	. .	circles.

adj.	. . .	adjacent.
alt.	. . .	alternate.
ax.	. . .	axiom.
circum.	. .	circumference.
comp.	. .	complement.
con.	. . .	construction.
cor.	. . .	corollary.
corr.	. . .	corresponding.
def.	. . .	definition.
ex.	. . .	exercise.
ext.	. . .	exterior.
hom.	. . .	homologous.
hy.	. . .	hypotenuse.
hyp.	. . .	hypothesis.
int.	. . .	interior.
isos.	. . .	isosceles.
rt.	. . .	right.
st.	. . .	straight.
sup.	. . .	supplementary.

Q.E.D. . . . quod erat demonstrandum (which was to be proved).

Q.E.F. . . . quod erat faciendum (which was to be done).

GEOMETRY

INTRODUCTION

1. A physical body, such as a block of wood or iron, occupies a definite portion of space. The boundary which separates a body from space is called the **surface** of the body.

If the material composing such a body could be removed without altering the surface, there would remain a portion of space called a **geometrical solid** or a **solid**.

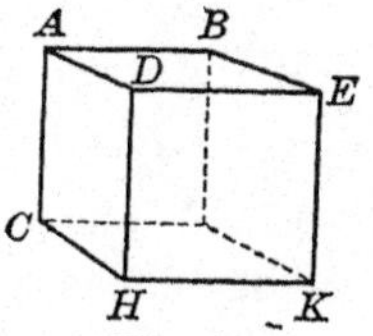

FIG. 1.

DEFINITIONS

2. A **solid** is a limited portion of space. It has three dimensions, **length**, **breadth**, and **thickness**.

3. **Surfaces** are the boundaries of solids, as *ABED* or *ADHC* (Fig. 1). They have two dimensions, **length** and **breadth**.

4. **Lines** are the boundaries of surfaces, as *AB*, *AD* (Fig. 1), etc., and have but one dimension, **length**.

5. **Points** are the boundaries or extremities of lines, and are without dimension, having *position only*.

Surfaces may be conceived as existing independent of the solids whose boundaries they form. In like manner, lines and points may exist independently in space.

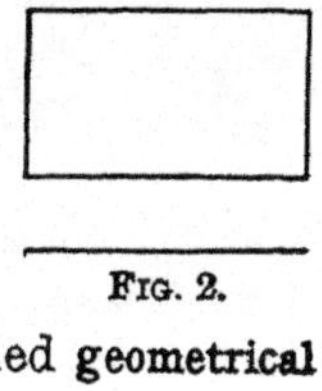

FIG. 2.

6. Solids, surfaces, lines, and points are called **geometrical magnitudes**.

7. **Geometry** is the science that treats of the properties of geometrical magnitudes.

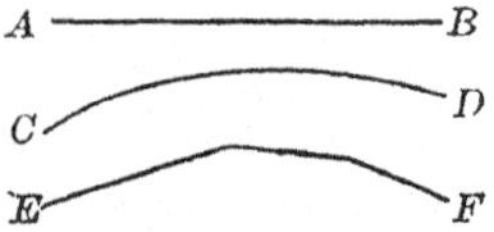

FIG. 3.

8. A **straight line** is a line that has the same direction throughout its length, as *AB*. The word "line" is frequently used to denote a straight line.

9. A **curved line** changes its direction at every point, as *CD*.

10. A **broken line** is composed of several successive straight lines, as *EF*.

11. A **plane surface** or a **plane** is such a surface that a straight line joining any two points in the surface lies entirely in the surface.

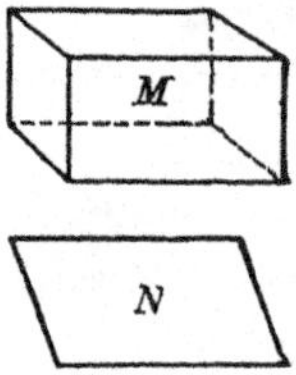

FIG. 4.

12. A **geometrical figure** is any combination of solids, surfaces, lines, or points, as *M* or *N*.

13. A **plane figure** is a geometrical figure, all of whose points lie in the same plane, as *N*.

14. **Plane Geometry** treats of plane figures only.

15. **Solid Geometry** treats of figures which are not plane.

16. When one figure can be placed upon another so that each point of the one lies upon the corresponding point of the other, the figures are said to **coincide**.

17. **Equal magnitudes** are those that can be made to coincide.

18. **Proof by superposition** is the method of proving the equality of two figures by means of coincidence.

QUESTIONS

1. What is the path of a moving point?

2. What geometrical magnitude is, in general, generated by a moving line? by a moving surface?

3. What kind of a surface is represented by the walls of a room?

LINES

19. From the definition of a straight line it appears that

(*a*) two straight lines of unlimited length, coinciding in part, coincide throughout,

(*b*) two straight lines can intersect only once, and

(*c*) two points determine a straight line.

The expression, straight line, is used to denote both an unlimited straight line and a part of such a line.

20. A line of definite length is also called a **segment** and is represented by a line whose ends are marked, as *A* and *B* (Fig. 5).

FIG. 5.

21. The length of this line is called the **distance** from *A* to *B*.

A line whose ends are not marked represents a line of indefinite length, as *CD*.

22. The **direction** of the line *AB* means the direction from *A* toward *B*; of *BA*, from *B* toward *A*.

FIG. 6.

23. To **produce** the line *AB* means to prolong it through *B*; to **produce** *BA* means to prolong it through *A*.

24. To **bisect** a geometrical magnitude means to divide it into two equal parts. Thus, *AC* is bisected if *AD* equals *DC*.

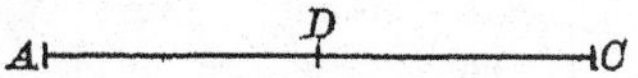

FIG. 7.

25. Two points, *A* and *B*, are **equidistant** from a third point, *C*, if $CA = CB$.

ANGLES

26. An **angle** is the inclination of two intersecting lines to each other.

27. The **vertex** is the point of intersection, and the lines are the **sides** of the angle. The lines *BA* and *BC*, meeting in *B*, form the angle *ABC*. *B* is the vertex, and *BA* and *BC* are the sides.

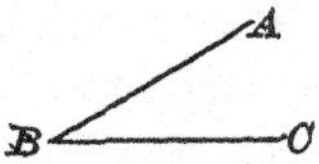

FIG. 8.

28. If there be only one angle at a vertex, it may be designated by one letter, as the angle *B*; but if there be two or more, three letters are necessary, as angle *ABD*. An angle may be denoted also by a number or letter placed within, as angle 1, angle 2, angle *a*.

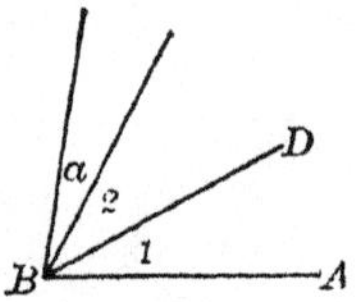

Fig. 9.

29. To **bisect** an angle means to divide it into two equal parts. Thus, *BC* bisects the angle *ABD*, if angle *ABC* equals angle *CBD*. *BC* is called the **bisector** of angle *ABD*.

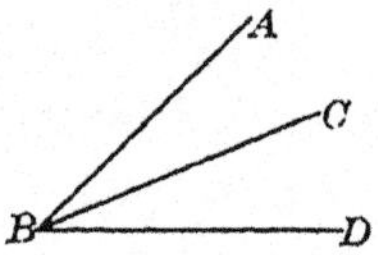

Fig. 10.

30. A **straight angle** is an angle whose sides lie in the same straight line but extend in opposite directions, as *ACB*.

31. When two straight lines intersect so as to form four equal angles, each angle is called a **right angle**, as *EOD*, *DOF*, etc. A right angle is half a straight angle.

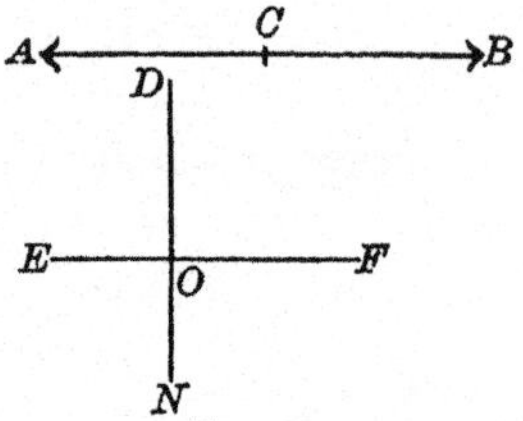

Fig. 11.

32. An **acute angle** is an angle less than a right angle, as angle *IKH*.

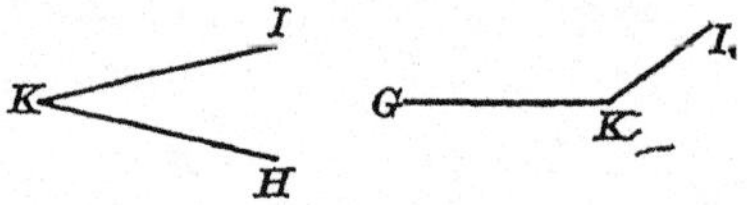

Fig. 12.

33. An **obtuse angle** is greater than a right angle, but less than a straight angle, as *GKI*.

34. A **reflex angle** is one greater than a straight angle, but less than two straight angles, as *ABC*.

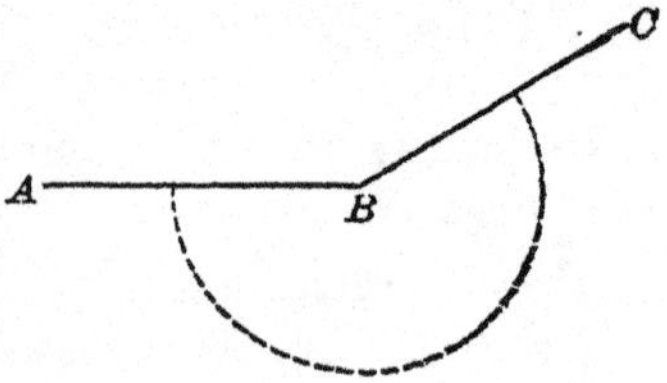

Fig. 13.

35. Two lines are **perpendicular** to each other if they meet at right angles, as *DN* and *EF* (Fig. 11). The point *O* is called the **foot** of the perpendicular *DO*.

36. **Oblique angles** are acute, obtuse, or reflex.

37. An angle is **measured** by finding how many times it contains a certain unit. The usual unit is the *degree,* or one-ninetieth ($\frac{1}{90}$) of a right angle. A degree is divided into sixty equal parts called *minutes,* and a minute into sixty equal parts called *seconds.* Degrees, minutes, and seconds are expressed by symbols, as in $6° 50' 12''$. Read six degrees, fifty minutes, and twelve seconds.

38. **Adjacent angles** are those that have a common vertex and a common side between them, as the angles *ABC* and *CBD.*

39. **Vertical angles** have a common vertex, and the sides of the one are prolongations of the sides of the other, as the angles *EKH,* and *GKF.*

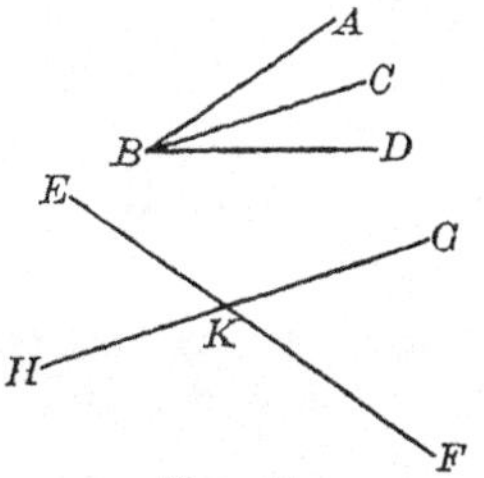

FIG. 14.

40. Two angles whose sum equals a right angle are **complementary angles.** Each is called the *complement* of the other. The angles *LMN* and *NMO* are complementary.

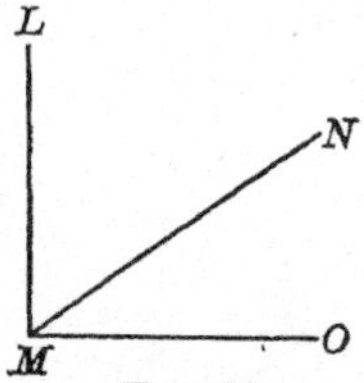

FIG. 15.

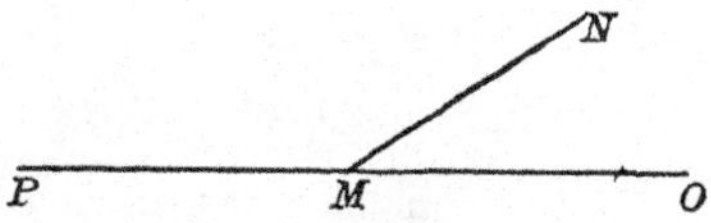

FIG. 16.

41. Two angles whose sum equals a straight angle are **supplementary angles.** Each angle is called the *supplement* of the other. The angles *PMN* and *NMO* are supplementary.

EXERCISES

4. What is the supplement of an angle of 23°?

5. What is the complement of an angle of 45°?

6. How many degrees are there in an angle which is twice its complement?

7. If an angle is equal to twice its supplement, what part of a straight angle is it?

8. What angle is formed by the hands of a clock at one o'clock? at 2:30? at 2:45?

9. How many minutes does it take the minute hand of a clock to describe (*a*) a right angle? (*b*) an angle of 60°? (*c*) an angle of 45°?

10. What is the complement of an angle of a degrees?

11. What is the supplement of an angle of n right angles?

12. If six lines radiate from a point, forming equal angles, find one angle (*a*) in degrees, (*b*) in right angles, and (*c*) in straight angles.

13. What kind of an angle is less than its supplement? equal to its supplement? greater than its supplement?

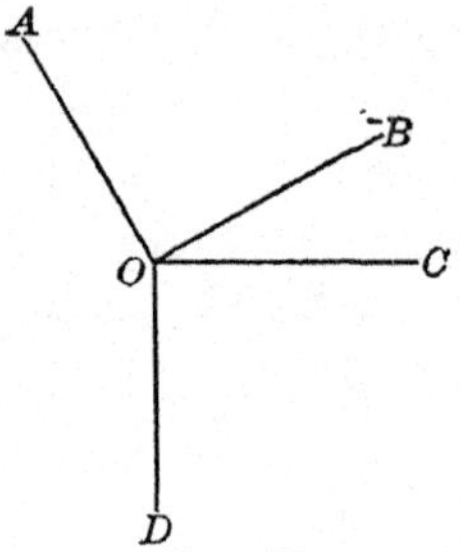

FIG. 17.

14. Four lines, AO, BO, CO, and DO, meet in a point. Which angle is the sum of AOB and BOC? the sum of BOC and COD? the difference between BOD and COD?

15. In the same diagram, if

$AOB = 60°$, $BOC = 30°$, and $COD = 90°$,

find AOC, BOD, and AOD.

16. If the four lines meet in a point so that

$AOB = 120°$, and $BOC = COD = DOA$, find BOC.

17. In Fig. 17 name four pairs of adjacent angles.

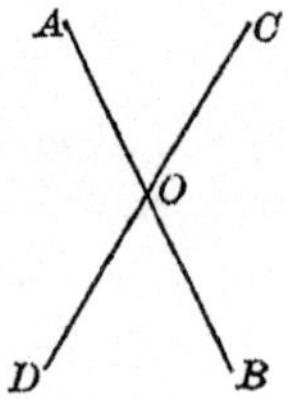

FIG. 18.

18. If two lines, AB and CD, intersect in O, making $AOC = 60°$, find the other angles.

19. In the same diagram, if $AOC = m$ degrees, how many degrees are in DOB? in BOC?

20. If the four lines mentioned in Ex. 14 be drawn so that AO is perpendicular to BO, and CO to DO, find AOD (*a*) if $BOC = 60°$, (*b*) if $BOC = m$ degrees.

21. What relation exists between the angles BOC and AOD in the preceding exercise?

22. If, in the annexed diagram, AO is perpendicular to CO, and BO is perpendicular to DO, find AOD. (*a*) if $COB = 40°$, (*b*) if $COB = m°$.

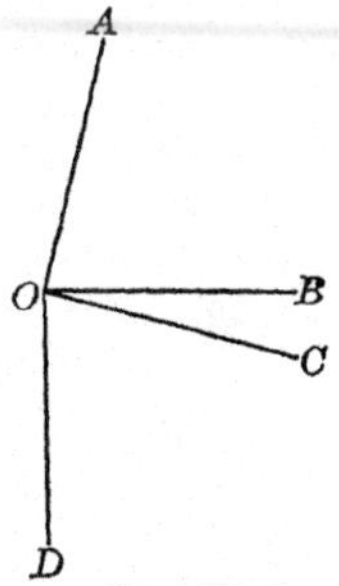

FIG. 19.

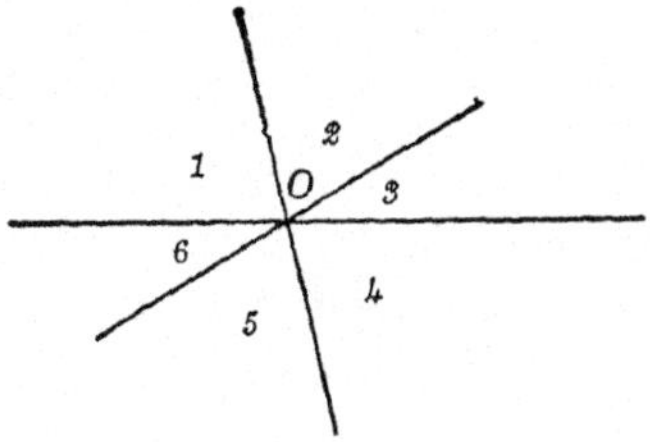

FIG. 20.

23. What relation exists between AOD and BOC in the preceding exercise?

24. What angle is formed by the bisectors of a pair of adjacent supplementary angles?

25. Three lines meet in a point, O, forming six angles, 1, 2, 3, 4, 5, and 6. Find angle 3, if angle $1 = 80°$ and angle $5 = 60°$.

GENERAL TERMS

42. A **theorem** is a geometrical truth requiring demonstration.

43. An **axiom** is a geometrical truth assumed as self-evident.

44. A **problem** is a question to be solved.

45. A **proposition** is a general term for a theorem or a problem.

46. A **corollary** is a theorem easily derived from another theorem.

47. A **scholium** is a remark.

48. A **postulate** is a construction so simple that its possibility is admitted without further demonstration.

49. The **hypothesis** is that which is assumed in the statement of a theorem.

50. The **conclusion** is that which follows from the hypothesis.

51. A proposition is the **converse** of another, when the *hypothesis and the conclusion of the one are respectively the conclusion and the hypothesis of the other.*

Axioms

1. Things that are equal to the same or equal things are equal to each other.
2. If equals be added to equals, the sums are equal.
3. If equals be subtracted from equals, the remainders are equal.
4. If equals be added to unequals, the sums are unequal.
5. If equals be subtracted from unequals, or unequals from equals, the remainders are unequal.
6. Doubles of equals are equal.
7. Halves of equals are equal.
8. The whole equals the sum of its parts.
9. The whole is greater than any of its parts.
10. A straight line is the shortest distance between two points.

(For Axiom 11, see page 17 on parallel lines.)

Postulates

1. A straight line can be drawn between any two points.
2. A straight line can be produced indefinitely.

PLANE GEOMETRY

BOOK I

LINES AND RECTILINEAR FIGURES *

PRELIMINARY THEOREMS

52. *All straight angles are equal.*

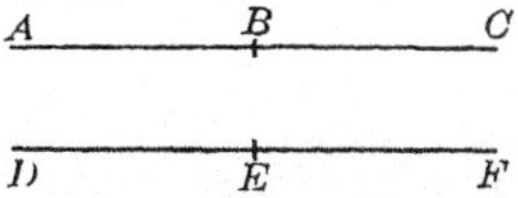

Hyp. Angles ABC and DEF are straight angles.

To prove $\angle ABC = \angle DEF$.

Proof. Apply $\angle DEF$ to $\angle ABC$ so that the vertex E coincides with the vertex B, and ED coincides with BA. Then EF will fall on BC (straight lines coinciding in part coincide throughout).

Hence $\angle DEF = \angle ABC$. Q.E.D.

53. *All right angles are equal.* (Ax. 7.)

54. *At a given point in a given line there can be but one perpendicular to the line.* (53)

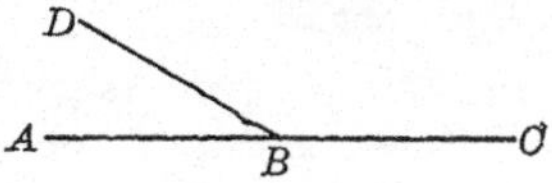

55. *The complements of the same or of equal angles are equal.* (Ax. 3.)

56. *The supplements of the same or of equal angles are equal.* (Ax. 3.)

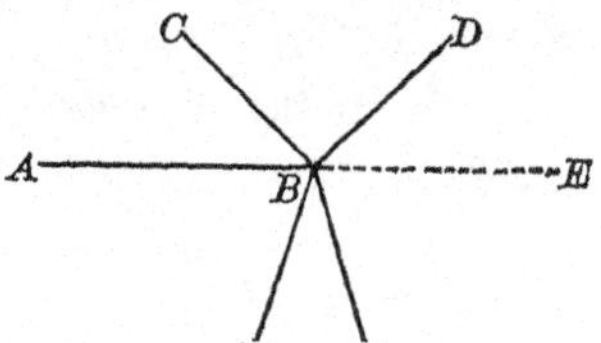

57. *If two adjacent angles have their exterior sides in a straight line, these angles are supplementary.* (Ax. 8.)

58. *The sum of all the angles formed at a point in a plane is equal to two straight angles.*

* A figure formed by straight lines only is called a rectilinear figure.

PROPOSITION I. THEOREM

59. *Vertical angles are equal.*

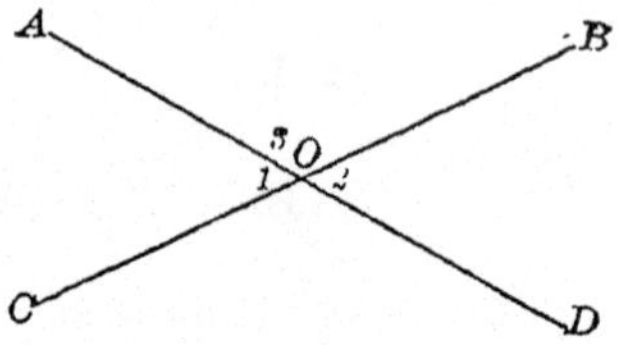

Hyp. Angles 1 and 2 are vertical angles.

To prove $\angle 1 = \angle 2$.

Proof. $\angle 1$ is a supplement of $\angle 3$,

$\angle 2$ is a supplement of $\angle 3$,

(*two adjacent angles whose exterior sides are in a straight line are supplementary*).

$\therefore \angle 1 = \angle 2$,

(*supplements of equal* ∠s *are equal*). Q.E.D.

Ex. 26. If, in the diagram for above proposition, $\angle AOC$ is 80°, find the other angles.

Ex. 27. If, in the same figure, $\angle AOB$ be bisected, and the bisector be produced through O, prove that $\angle COD$ is also bisected.

Ex. 28. If three lines, AB, CD, and EF, meet in a point, O, prove $\angle AOE - \angle FOD = \angle AOC$.

Ex. 29. In the same diagram, prove:

$\angle AOE + \angle EOD - \angle AOD = 2\angle AOC.$

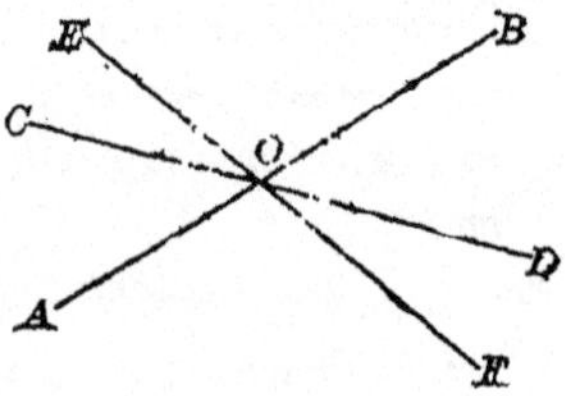

Ex. 30. In the same diagram, prove:

$\angle AOC + \angle EOB + \angle DOF = 2$ rt. ∠s.

Ex. 31. In the same diagram, prove:

$\angle AOE + \angle COB + \angle EOD =$ 4 rt. ∠s.

60. DEF. A **polygon** is a portion of a plane bounded by three or more straight lines, which are termed **sides**, and whose sum is the **perimeter** of the polygon.

The angles included by the adjacent sides are the **angles** of the polygon, and their vertices are the **vertices** of the polygon. An **exterior angle** is formed by a side and the prolongation of an adjacent one. A **diagonal** is a straight line joining the vertices of two non-adjacent angles.

61. A polygon of three sides is called a **triangle**; one of four sides, a **quadrilateral**.

TRIANGLES

62. A triangle having three unequal sides is a **scalene triangle**. An **isosceles triangle** has two of its sides equal. An **equilateral triangle** has its three sides equal.

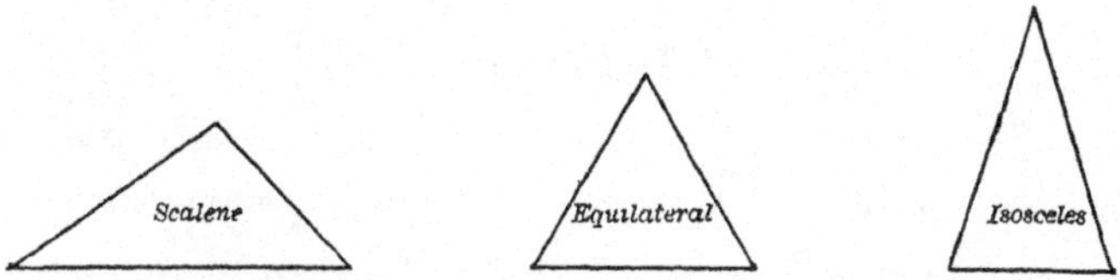

63. A triangle is called **acute** if all its angles are acute; **right**, if one of its angles is a right angle; **obtuse**, if one of its angles is obtuse. A triangle is called **equiangular** if all its angles are equal.

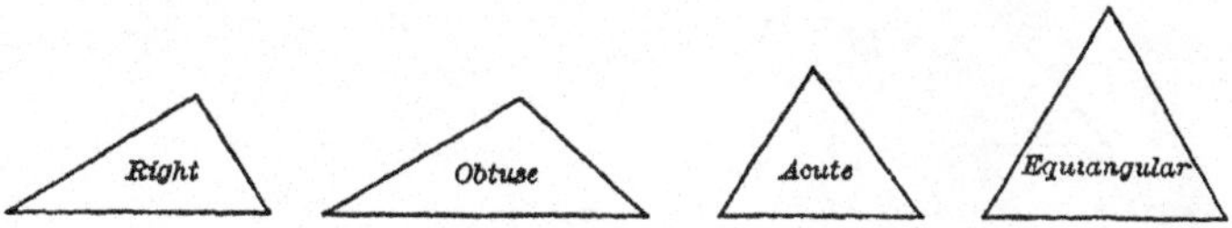

64. The **base** of a triangle is the side on which the figure is supposed to stand. The base of an isosceles triangle is that side which is equal to no other; the two equal sides are called the **arms**.

65. The **vertical angle** of a triangle is the angle opposite the **base**.

66. In a right triangle the side opposite the right angle is called the **hypotenuse**, and the other two sides, the **arms.**

67. The three perpendiculars from the vertices of a triangle to the opposite sides (produced if necessary) are called the **altitudes** of the triangle; the three bisectors of the angles are

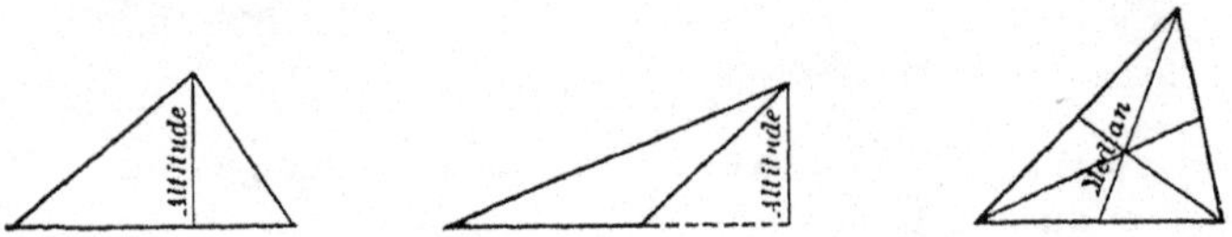

called the **bisectors** of the triangle; and the three lines from the vertices to the midpoints of the opposite sides are called the **medians** of the triangle.

PROPOSITION II. THEOREM

68. *Two triangles are equal when a side and two adjacent angles of the one are equal respectively to a side and two adjacent angles of the other.*

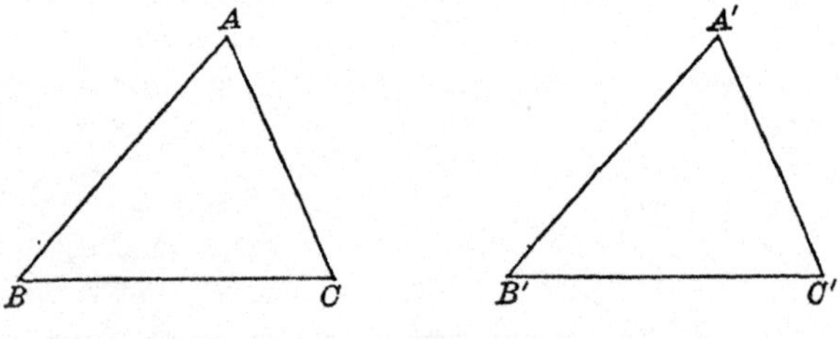

Hyp. In triangles ABC and $A'B'C'$,

$$AB = A'B', \angle A = \angle A', \text{ and } \angle B = \angle B'.$$

To prove $\triangle ABC = \triangle A'B'C'$.

Proof. Apply $\triangle ABC$ to $\triangle A'B'C'$ so that AB shall coincide with $A'B'$.

BC will take the direction of $B'C'$,

$(\angle B = \angle B'$ *by hyp.*).

C will fall upon $B'C'$ or its prolongation.

AC will take the direction of $A'C'$,

$(\angle A = \angle A'$ *by hyp.*).

C will fall upon $A'C'$ or its prolongation.

$\therefore$ the point C falling upon both the lines $B'C'$ and $A'C'$, must fall upon the point common to both lines, namely, C'.

$\therefore$ $\triangle$ ABC and $A'B'C'$ coincide

$\therefore \triangle ABC = \triangle A'B'C'$. Q.E.D.

69. NOTE. — This method of proof (**superposition**) is employed in fundamental propositions only. The student should place those parts upon each other whose equality is known, and, by successive steps, trace the position of the rest of the figure.

70. NOTE. — In order to facilitate the citing of propositions, the following abbreviation is suggested for the above proposition : *a. s. a.* = *a. s. a.* Similar abbreviations will be suggested for other propositions.

71. DEF. Polygons are **mutually equiangular** if their angles are respectively equal, and **mutually equilateral** if their sides are respectively equal.

If two polygons are mutually equiangular, lines or angles similarly situated are called **homologous lines or angles.** Thus AB and $A'B'$ (Prop. II) are homologous sides, C and C' homologous angles, the medians drawn from A and A' respectively homologous medians, etc.

Ex. 32. Hyp. BD bisects $\angle B$,

$EF \perp BD$.

To prove $\triangle EBD = \triangle FBD$.

(See diagram.)

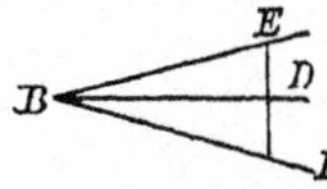

Ex. 33. Hyp. I is the midpoint of GH, $KG \perp GH$, $HL \perp GH$, KL is a straight line.

To prove $\triangle KGI = \triangle HLI$.

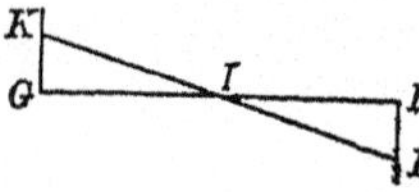

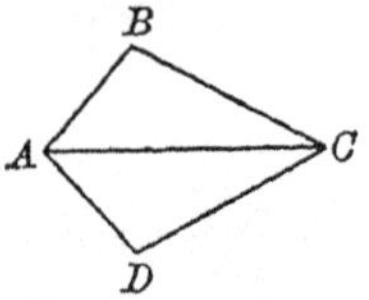

Ex. 34. If a diagonal of a quadrilateral bisects those angles whose vertices it joins, the diagonal divides the figure into two equal triangles.

PROPOSITION III. THEOREM

72. *Two triangles are equal if two sides and the included angle of the one are equal respectively to two sides and the included angle of the other* (*s.a.s.* = *s.a.s.*).

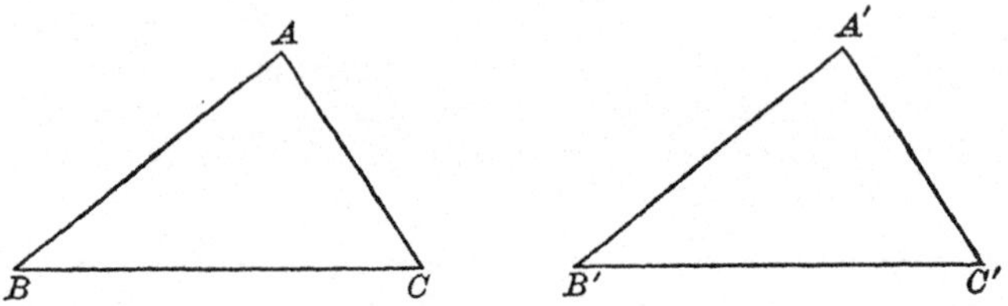

Hyp. In $\triangle ABC$ and $A'B'C'$,

$$AB = A'B',\ BC = B'C',\ \text{and } \angle B = \angle B'.$$

To prove $\triangle ABC = \triangle A'B'C'$.

Proof. Apply $\triangle ABC$ to $\triangle A'B'C'$ so that BC shall coincide with $B'C'$.

BA will take the direction of $B'A'$,

($\angle B = \angle B'$ *by hyp.*).

The point A will fall upon the point A',

($AB = A'B'$ *by hyp.*).

$\therefore AC$ will coincide with $A'C'$.

$\therefore$ $\triangle ABC$ and $A'B'C'$ coincide.

$\therefore \triangle ABC = \triangle A'B'C'$. Q.E.D.

Ex. 35. Hyp. Two lines AB and CD bisect each other in E.

To prove $\triangle ADE = \triangle CEB$.

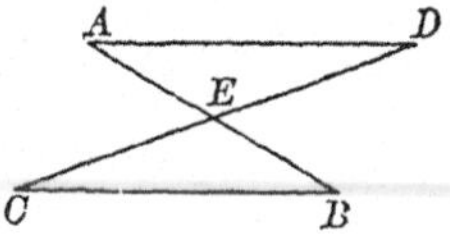

73. Remark. — **The equality of lines and angles is usually proved by means of equal triangles.**

Ex. 36. Hyp. DC is a perpendicular bisector of AB.

To prove $AD = BD$.

Ex. 37. The bisector of the vertical angle of an isosceles triangle bisects the base.

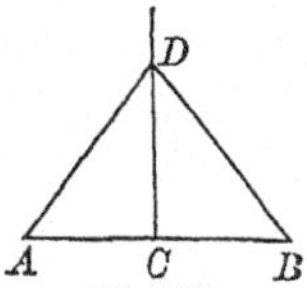

(Ex. 36)

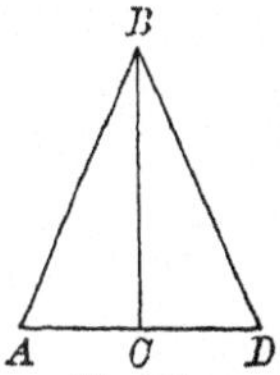

(Ex. 37)

Ex. 38. Hyp. BF bisects $\angle DBE$.

$BD = BE$.

To prove $\angle FDB = \angle FEB$.

Ex. 39. Hyp. In $\triangle ABC$, $\angle CAB = \angle ABC$, and $\angle ABD = \angle BAE$.

To prove $BD = AE$.

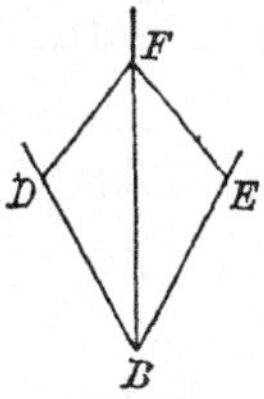

(Ex. 38)

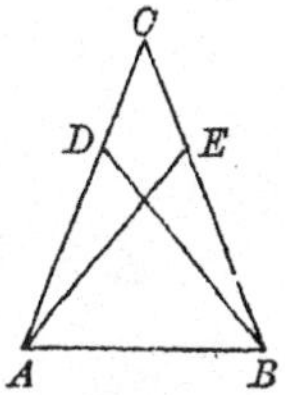

(Ex. 39)

Ex. 40. Hyp. In $\triangle ABC$, $\angle A = \angle C$, and AD and CE are bisectors.

To prove $AD = CE$.

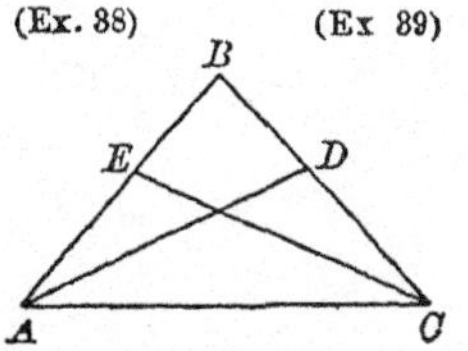

Ex. 41. Hyp. In $\triangle HIK$,

$HI = KI$, and $HL = KM$.

To prove $HM = LK$.

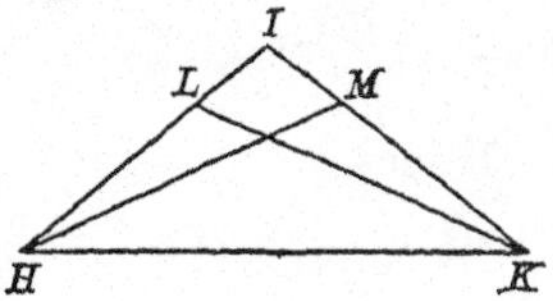

74. Remark. — **If the lines and angles whose equality is to be proved are not parts of triangles, try to construct such triangles.**

Ex. 42. If in triangle ABC, $AB = BC$, then $\angle A = \angle C$.

Proposition IV. Theorem

75. *An exterior angle of a triangle is greater than either remote interior angle.*

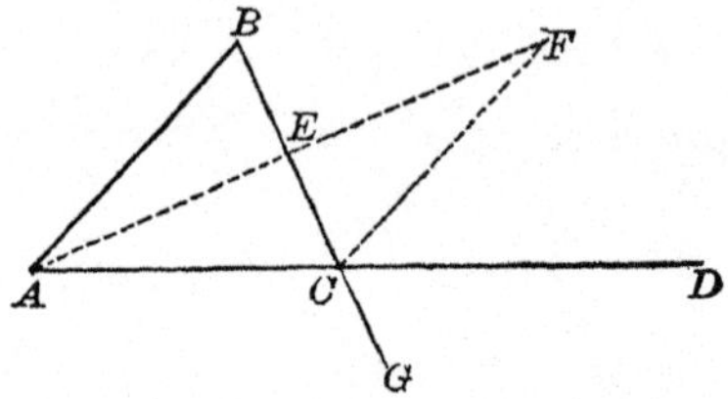

Hyp. $\angle BCD$ is an ext. $\angle$ of $\triangle ABC$.

To prove $\angle BCD > \angle A$ or $\angle B$.

Proof. Let E be the midpoint of BC. Draw AE and produce it its own length to F. Draw FC.

In $\triangle$s ABE and FCE, $AE = EF$ and $BE = EC$. (Con.)

$$\angle BEA = \angle FEC,$$

(*vertical* $\angle$s).

$$\therefore \triangle ABE = \triangle FCE,$$

(*s. a. s.* = *s. a. s.*).

$$\therefore \angle ECF = \angle B,$$

(*hom. parts of equal* $\triangle$s).

But $\angle BCD > \angle ECF$. (Ax. 9)

Hence $\angle BCD > \angle B$.

By joining the midpoint of AC to B, it follows in the same manner that

$$\angle ACG > \angle A.$$

But $\angle ACG = \angle BCD$,

(*vertical* $\angle$s).

Hence $\angle BCD > \angle A$. Q.E.D.

Ex. 43. If four points, A, B, C, D, in a straight line be joined to a point, E, without, then

(1) $\angle ABE > \angle ACE$, (2) $\angle ABE > \angle CED$,
(3) $\angle ABE > \angle ADE$.

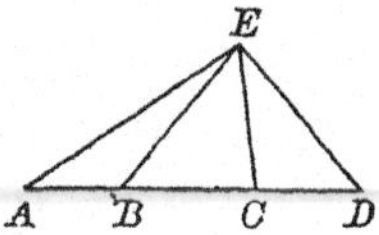

Ex. 44. If in quadrilateral $ABCD$ the diagonals AC and BD meet in E, find four angles smaller than angle BEA.

Ex. 45. If from any point D in $\triangle ABC$ DA and DB are drawn, then $\angle D > \angle C$.

HINT. — Draw DC.

76. DEF. When two straight lines, AB and CD, are cut by a third straight line, EF, called **a transversal**, then the angles a, b, g, and h are called **exterior angles**, the angles c, d, e, and f are called **interior angles**, the angles a and e, b and f, c and g, and d and h are called **corresponding angles**, the angles c and f, and d and e are called **alternate interior angles**, and the angles b and g, a and h, are called **alternate exterior angles**.

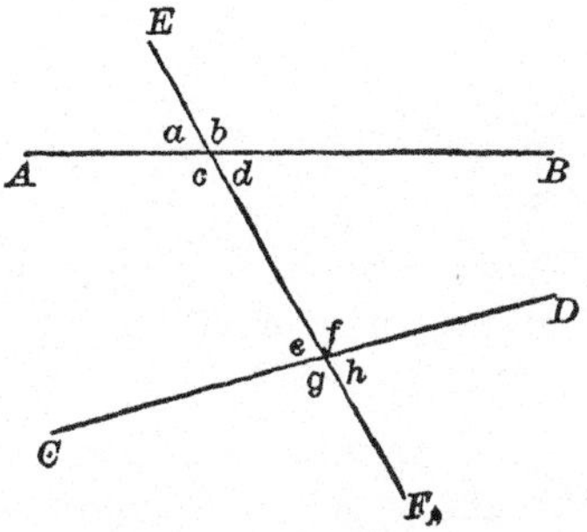

PARALLEL LINES

77. DEF. **Parallel lines** are lines which lie in the same plane and do not meet if produced indefinitely. AB and CD are parallel.

78. AXIOM 11. Two intersecting lines cannot be both parallel to a third straight line.

79. THEOREM. *Two straight lines which are parallel to a third straight line are parallel to each other.*

For if the two lines should meet, we would have two intersecting straight lines parallel to a third straight line, which contradicts Axiom 11.

PROPOSITION V. THEOREM

80. *Two lines are parallel if a transversal to these lines makes the alternate interior angles equal.*

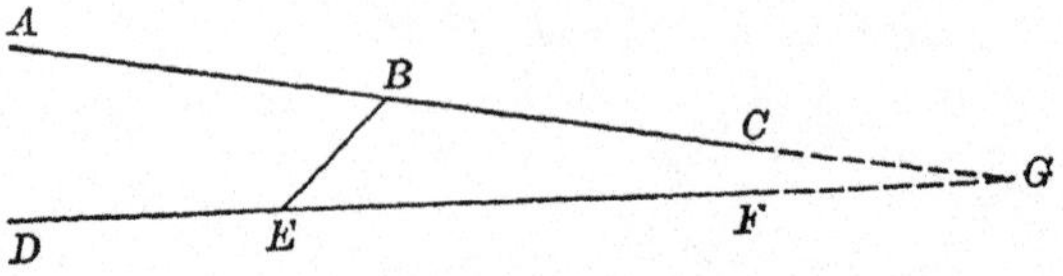

Hyp. AC and DF are intersected by BE so that

$$\angle ABE = \angle BEF.$$

To prove $AC \parallel DF$.

Proof. AC and DF either meet or are parallel.

Suppose they meet in G.

Then BEG is a triangle whose exterior $\angle ABE$ is equal to a remote interior $\angle BEG$, which is impossible.

Hence AC and DF are parallel. Q.E.D.

81. SCHOLIUM. That the lines AC and DF cannot meet on the side of A and D can be proved by the same method.

82. REMARK. — **In order to demonstrate that two lines are parallel, prove the equality of a pair of alternate interior angles**

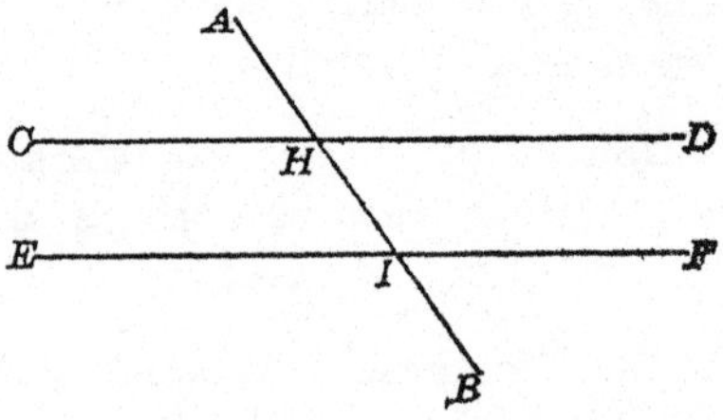

Ex. 46. In the annexed diagram, if

$\angle IHC = 120°$,

and $\angle FIH = 120°$,

is CD parallel to EF?

Ex. 47. In the same diagram, if $\angle DHI = 60°$,

and $\angle FIH = 120°$,

CD is parallel to EF.

Ex. 48. In the same diagram, if $\angle AHC = \angle EIH$, CD is parallel to EF.

Ex. 49. Lines are parallel if a transversal makes the alternate exterior angles equal.

Ex. 50. If two sides of a triangle are produced their own lengths through the common vertex, a line joining their ends is parallel to the third side of the triangle.

Ex. 51. If in the diagram for Prop. V $\angle ABE = \angle BEF$, prove that the bisectors of $\angle CBE$ and DEB are parallel.

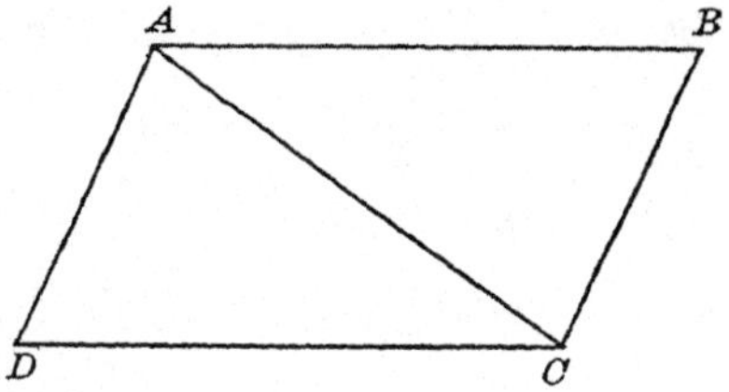

Ex. 52. If in the annexed diagram

$$AB = CD,$$

and $\angle BAC = \angle ACD$,

then AD is parallel to BC.

Proposition VI. Theorem

83. *Two lines are parallel if a transversal to these lines makes the corresponding angles equal.*

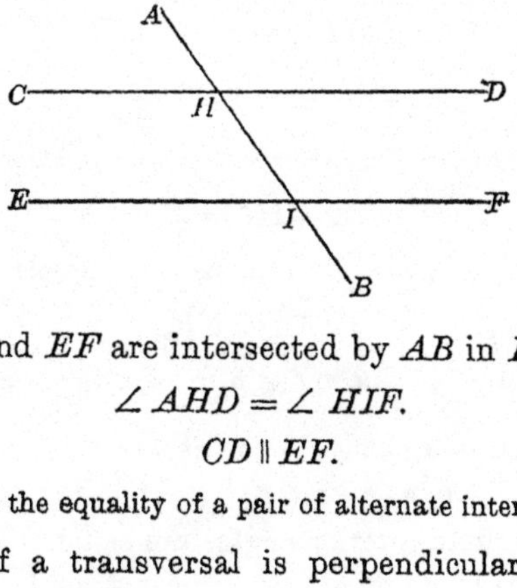

Hyp. CD and EF are intersected by AB in H and I respectively, and $\angle AHD = \angle HIF$.

To prove $CD \parallel EF$.

Hint. — Prove the equality of a pair of alternate interior angles.

84. Cor. If a transversal is perpendicular to two lines, these lines are parallel.

Ex. 53. If in the diagram for Prop. VI $\angle AHC = 60°$, and $\angle HIE = 60°$, is CD parallel to EF?

Ex. 54. In the same diagram, if $\angle AHD$ is the supplement of $\angle EIH$, CD is parallel to EF.

PROPOSITION VII. THEOREM

85. *Two lines are parallel if a transversal to these lines makes the interior angles on the same side of the transversal supplementary.*

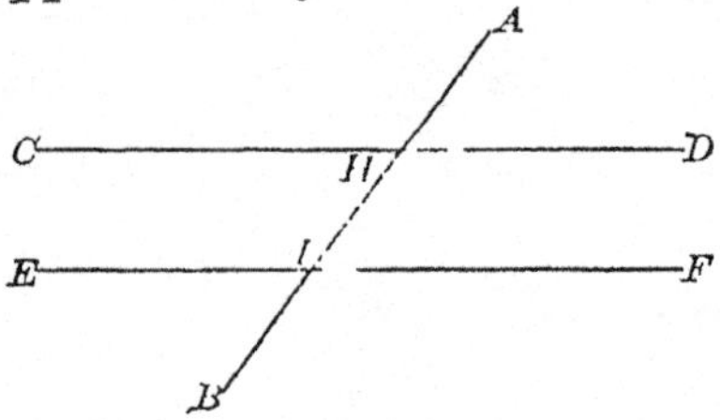

Hyp. CD and EF are intersected by a transversal AB in H and I, respectively, and $\angle DHI + \angle HIF = 2$ rt. $\angle$s.

To prove $CD \parallel EF$.

Proof. $\angle DHI$ is sup. to $\angle HIF$, (Hyp.)

$\angle EIH$ is sup. to $\angle HIF$,

(*if two adjacent angles have their ext. sides in a st. line, they are sup.*).

$\therefore \angle DHI = \angle EIH$,

(*supplements of same $\angle$ are equal*).

Hence $CD \parallel EF$, Q.E.D.

(*alt. int. $\angle$s being equal*).

86. REMARK. — **Lines are demonstrated to be parallel by proving that**

(*a*) **Two alternate interior angles are equal,**

(*b*) **Two corresponding angles are equal, or**

(*c*) **Two interior angles on the same side of a transversal are supplementary.**

Ex. 55. Two lines are parallel if a transversal to these lines makes the exterior angles on the same side of the transversal supplementary.

Ex. 56. If in the diagram for Prop. VII $\angle AHD = \angle HIF$, and perpendiculars be erected upon CD and EF at H and I, respectively, the perpendiculars are parallel.

Ex. 57. In the same diagram, if $\angle AHD$ is the supplement of $\angle EIH$, CD is parallel to EF.

Proposition VIII. Theorem

87. *If two parallels are cut by a transversal, the alternate interior angles are equal.*

[Converse of Prop. V.]

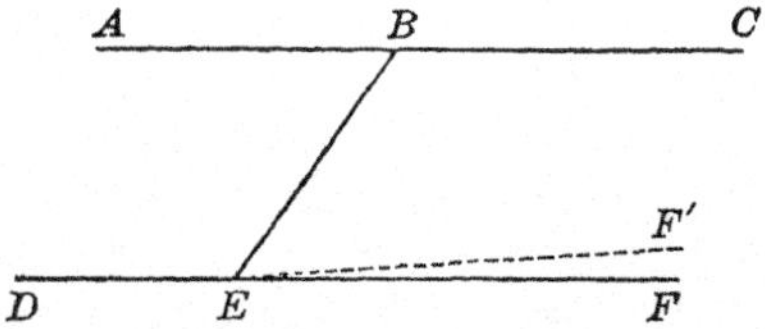

Hyp. The parallel lines AC and DF are intersected by a transversal in B and E.

To prove $\angle ABE = \angle BEF$.

Proof. $\measuredangle s\ ABE$ and BEF are either equal or unequal. Suppose they are unequal, and let EF' be drawn so that $\angle BEF' = \angle ABE$.

Then $EF' \parallel AB$,

(*two lines are* $\parallel$ *if a transversal makes the alt. int.* $\measuredangle s$ *equal*).

But $EF \parallel AB$. (Hyp.)

Therefore two intersecting lines EF and EF' would be parallel to AC, which is impossible. (Ax. 11.)

Hence $\angle ABE = \angle BEF$. Q.E.D.

88. Cor. If a transversal is perpendicular to one of two parallel lines, it is perpendicular to the other also.

Ex. 58. If two parallels are cut by a transversal, the alternate exterior angles are equal.

Ex. 59. In the diagram for Prop. VII, if $\angle AHD = 40°$, how many degrees are in $\measuredangle s\ HIF$, HIE, and EIB?

Ex. 60. If in the diagram for Prop. VIII the transversal be produced through B and E, the figure will contain three angles equal to $\angle ABE$. Find these angles.

Ex. 61. If the opposite sides of a quadrilateral are parallel, they must be equal.

PROPOSITION IX. THEOREM

89. *If two parallel lines are cut by a transversal, the corresponding angles are equal.*

[Converse of Prop. VI.]

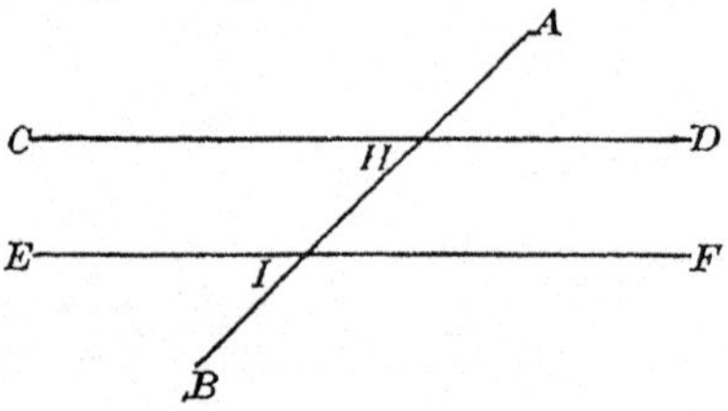

Hyp. Two parallel lines CD and EF are intersected by AB in H and I, respectively.

To prove $\angle AHD = \angle HIF$.

Proof. $\angle AHD = \angle CHI$,

(*vertical* $\angle$s).

$\angle CHI = \angle HIF$,

(*alt. int.* $\angle$s *of* $\parallel$ *lines*).

$\angle AHD = \angle HIF$. (Ax. 1.) Q.E.D.

Ex. 62. In the diagram for Prop. IX, if $\angle AHD = 50°$, how many degrees are in $\angle$s EIB, CHI, AIF, and EIA?

Ex. 63. In the same diagram, prove that $\angle CHI + \angle EIH = 2$ rt. $\angle$s.

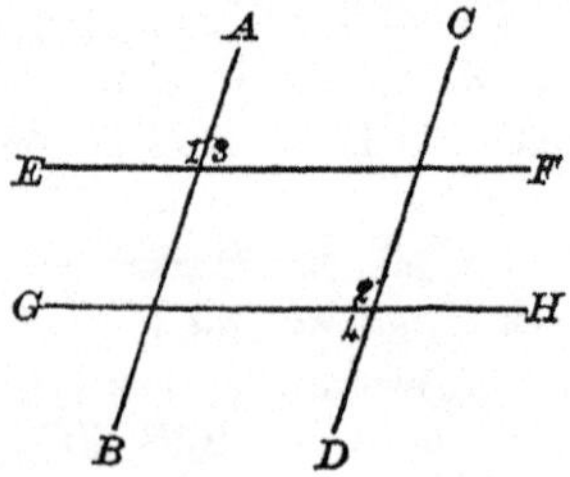

Ex. 64. If two parallel lines AB and CD are intersected by parallel transversals HG and EF, prove (*a*) $\angle 1 = \angle 2$, (*b*) $\angle 3 = \angle 4$. (See annexed diagram.)

Ex. 65. In the same diagram, prove that $\angle 1 + \angle 4 = 2$ rt. $\angle$s.

Ex. 66. If three points A, B, and C be joined, and BC be produced to D,

$$\angle ACD = \angle A + \angle B.$$

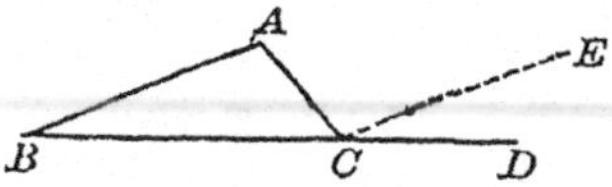

HINT. — Draw $EC \parallel BA$ (annexed diagram).

Ex. 67. In the same diagram, if CE bisects $\angle ACD$, then $\angle A = \angle B$.

Ex. 68. The bisectors of supplementary adjacent angles are perpendicular to each other.

PROPOSITION X. THEOREM

90. *If two parallel lines are cut by a transversal, the interior angles on the same side of the transversal are supplementary.*

[Converse of Prop. VII.]

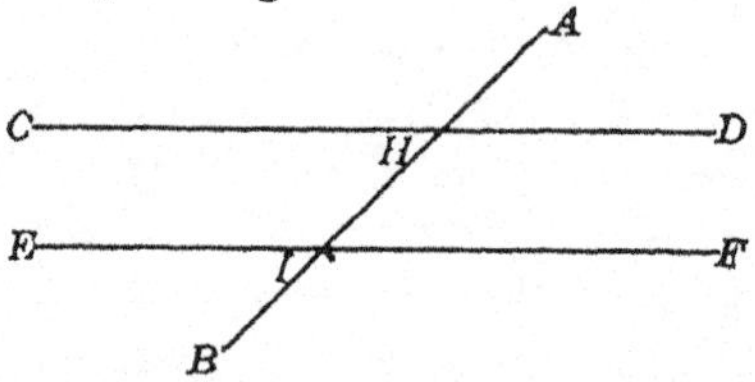

Hyp. Two parallel lines CD and EF are intersected by AB in H and I, respectively.

To prove $\angle DHI + \angle HIF = 2$ rt. $\angle$s.

Proof. $\angle DHI = \angle HIE$,

(*alt. int. $\angle$s of $\parallel$ lines*).

But $\angle HIE + \angle HIF = 2$ rt. $\angle$s,

(*two adj. $\angle$s whose ext. sides are in a st. line are supplementary*).

$\therefore \angle DHI + \angle HIF = 2$ rt. $\angle$s. Q.E.D.

Ex. 69. In the diagram for Prop. IX, if CD is parallel to EF, prove that

$$\angle AHD + \angle HIE = 2 \text{ rt. } \angle s.$$

Ex. 70. If two parallel lines are cut by a transversal, the exterior angles on the same side of the transversal are supplementary.

Ex. 71. The bisectors of a pair of corresponding angles, formed by parallel lines, are parallel.

Ex. 72. If upon each of two parallel lines a perpendicular be erected at any point, these perpendiculars either coincide or are parallel.

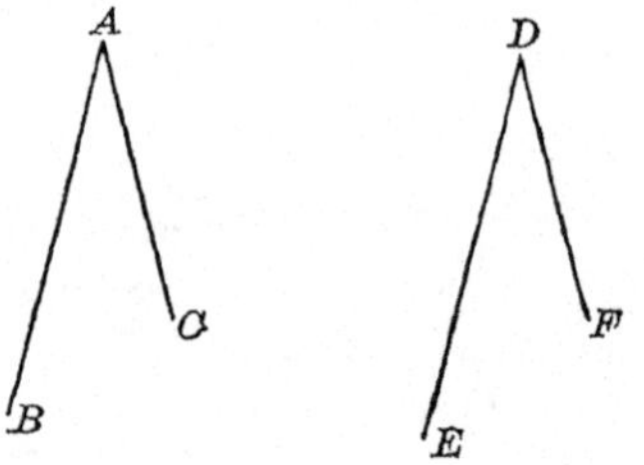

Ex. 73. In the annexed diagram, if AB is parallel to ED, and $\angle A = \angle D$, prove that AC is parallel to DF.

Ex. 74. If in the annexed diagram ∠s A, B, and C are right angles, $\angle D$ is also a right angle.

Ex. 75. If in the annexed diagram AB is parallel to CD, prove that

$$\angle 1 = \angle 2 + \angle 3.$$

Ex. 76. State and prove the converse of the preceding exercise.

Ex. 77. The bisectors of two interior angles on the same side of a transversal to two parallel lines are perpendicular to each other.

Ex. 78. State and prove the converse of Ex. 71.

Proposition XI. Theorem

91. *Angles whose corresponding sides are parallel are either equal or supplementary.*

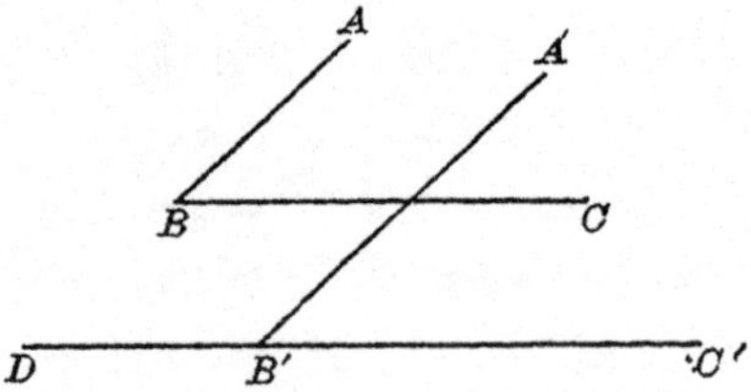

Hyp. $AB \parallel A'B'$ and $BC \parallel B'C'$.

To prove $\angle B = \angle A'B'C'$

$\angle B + \angle A'B'D = 2$ rt. ∠s.

[The proof is left to the student.]

92. Scholium. The angles are equal if both the corresponding pairs of sides lie in the same or in opposite directions from the vertex.

Ex. 79. If two sides of a quadrilateral are equal and parallel, the other two sides must be parallel.

Ex. 80. If two sides of a triangle are respectively parallel to two homologous sides of an equal triangle, the third side of the first must be parallel to the third side of the second.

Proposition XII. Theorem

93. *The sum of the angles of a triangle is equal to a straight angle.*

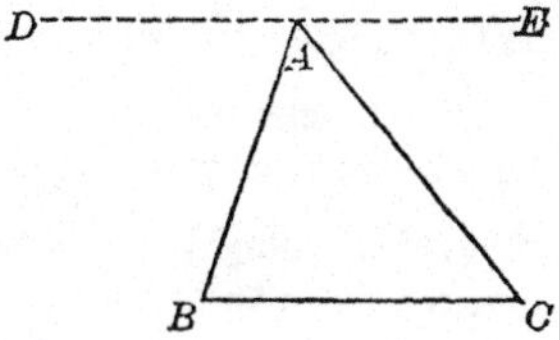

Hyp. ABC is a triangle.

To prove $\angle A + \angle B + \angle C =$ a st. $\angle$.

Proof. Draw line $DE \parallel BC$ through A.

$\angle B = \angle DAB$, $\angle C = \angle CAE$ (*alt. int.* $\angle$*s of* $\parallel$ *lines.*)

$\therefore \angle A + \angle B + \angle C = \angle A + \angle DAB + \angle CAE$. (Ax. 2.)

But $\angle A + \angle DAB + \angle CAE = \angle DAE =$ a st. $\angle$.

$\therefore \angle A + \angle B + \angle C =$ a st. $\angle$. (Ax. 1.) Q.E.D.

94. Cor. 1. In a triangle there can be only one obtuse or one right angle.

95. Cor. 2. The acute angles of a right triangle are complementary.

96. Cor. 3. If two triangles have two angles of the one respectively equal to two angles of the other, the third angles are equal.

97. Cor. 4. Two triangles are equal, if a side, the opposite angle, and any other angle of the one are equal respectively to a side, the opposite angle, and any other angle of the other triangle (*s. a. a.* = *s. a. a.*).

98. Cor. 5. From a point without a line there can be only one perpendicular to that line.

99. Cor. 6. Each angle of an equiangular triangle is equal to sixty degrees.

Ex. 81. If an angle of a triangle is (1) 40°, (2) $m°$, what is the sum of the other two angles?

Ex. 82. If one angle of a triangle is equal to the sum of the other two, (1) how many degrees has that angle? (2) what is such a triangle called?

Ex. 83. If two angles of a triangle are 60° and 40°, respectively, what is the angle formed by the bisectors of these angles?

Ex. 84. Find each angle of a triangle if the second equals twice the first, and the third equals three times the first.

Ex. 85. If two angles of a triangle are (1) 40° and 60°, (2) $m°$ and $n°$, find the other interior and the exterior angles of the triangle.

Ex. 86. If an exterior angle of a triangle is three-fourths of a right angle, a remote interior angle one-half of a right angle, find the other interior angles.

Ex. 87. If two lines are intersected by a transversal, and the bisectors of the interior angles on the same side of the transversal are perpendicular to each other, these lines are parallel.

Ex. 88. The altitude upon the hypotenuse of a right triangle divides the right angle into two parts, which are respectively equal to the two acute angles of the right triangle.

Ex. 89. Find the sum of the four angles of a quadrilateral.

Ex. 90. If two angles of a triangle are equal, the bisector of the third angle divides the figure into two equal triangles.

Ex. 91. Two right triangles are equal if the hypotenuse and an acute angle of the one are equal respectively to the hypotenuse and an acute angle of the other.

Ex. 92. The altitudes upon the arms of an isosceles triangle are equal.

Proposition XIII. Theorem

100. *An exterior angle of a triangle is equal to the sum of the two remote interior angles.*

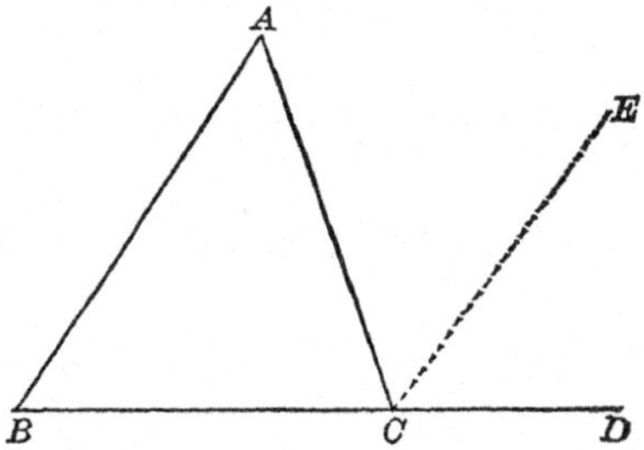

Hyp. $\angle ACD$ is an exterior angle of $\triangle ABC$.

To prove $\angle ACD = \angle A + \angle B$.

Hint. — Draw $CE \parallel BA$.

Ex. 93. If two angles of a triangle are equal, the bisector of the remote exterior angle is parallel to the opposite side of the triangle.

Ex. 94. If in the diagram for Prop. XIII DA is drawn, then

$$\angle ADB < \angle ACB.$$

Ex. 95. Prove Prop. XIII by means of (56).

Ex. 96. The sum of the three exterior angles of a triangle is equal to four right angles.

Ex. 97. If the sum of two exterior angles of a triangle is equal to three right angles, the triangle is a right one.

Ex. 98. If from a point without an acute angle, perpendiculars are drawn upon the sides of the angle, these perpendiculars include an angle equal to the original angle.

* **Ex. 99.** The bisectors of two exterior angles of a triangle include an angle equal to half the third exterior angle.

* Exercises denoted by (*) are difficult and may be omitted at a first readin

PROPOSITION XIV. THEOREM

101. *The base angles of an isosceles triangle are equal.*

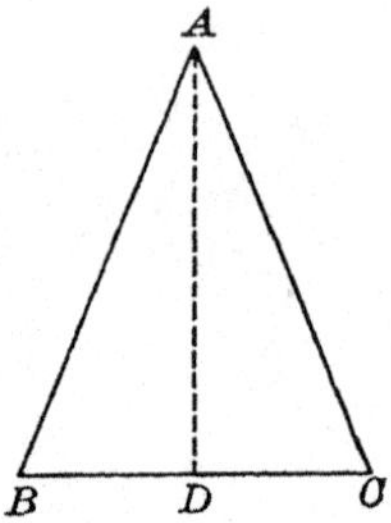

Hyp. In $\triangle ABC$, $AB = AC$.

To prove $\angle B = \angle C$.

HINT. — Let AD be the bisector of $\angle BAC$, and prove the equality of the two triangles.

102. COR. An equilateral triangle is equiangular.

Ex. 100. If the base of an isosceles triangle is trisected, the lines joining the points of division with the vertex are equal.

Ex. 101. The vertical angle of an isosceles triangle is: (1) 40°; (2) $m°$. Find the base angles.

Ex. 102. How many degrees are in each angle of an isosceles right triangle?

Ex. 103. A base angle of an isosceles triangle is: (1) two-fifths of a right angle; (2) m right angle. Find the other angles.

Ex. 104. Each base angle of an isosceles triangle is one-half the remote exterior angle.

Ex. 105. The perpendiculars from the midpoint of the base upon the arms of an isosceles triangle are equal.

Ex. 106. The vertical angle of an isosceles triangle is: (1) 40°; (2) $m°$. Find the angle formed by the base, and an altitude upon an arm.

Ex. 107. If an arm of an isosceles triangle is produced by its own length through the vertex, and the end of the prolongation is joined to the nearest end of the base, the line joining is perpendicular to the base.

Ex. 108. The bisectors of the base angles of an isosceles triangle are equal.

PROPOSITION XV. THEOREM

103. *If two angles of a triangle are equal, the sides opposite these angles are equal.*

[Converse of XIV.]

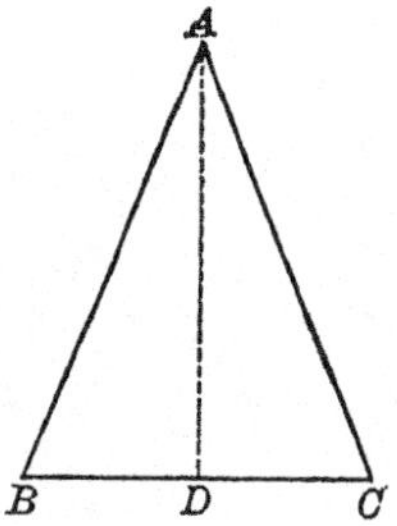

Hyp. In $\triangle ABC$, $\angle B = \angle C$.

To prove $AB = AC$.

HINT. — Let AD be the bisector of angle BAC, and prove the equality of the triangles.

104. COR. An equiangular triangle is also equilateral.

105. REMARK. — **The above two propositions may also be used to prove the equality of lines and angles.**

Ex. 109. The bisectors of the base angles of an isosceles triangle form, if they meet, an isosceles triangle.

Ex. 110. If at the ends of the base of an isosceles triangle perpendiculars are erected upon the arms, another isosceles triangle is formed.

Ex. 111. A line parallel to the base of an isosceles triangle, and intersecting the two arms, cuts off another isosceles triangle.

Ex. 112. If two exterior angles, formed by producing one side of a triangle at both ends, are equal, the other sides are equal.

Ex. 113. If one angle of an isosceles triangle is 60°, the triangle is equilateral.

Ex. 114. If the bisector of an exterior angle of a triangle is parallel to one side, the triangle is isosceles.

Ex. 115. If in quadrilateral $ABCD$ $BA = BC$, and $DA = DC$, then $\angle A = \angle C$.

Ex. 116. If two triangles ADC and BEF have their three sides equal, respectively, then the homologous angles, *e.g.* $\triangle$ A and E, are equal.

HINT. — Place the $\triangle$ together, so as to form a quadrilateral.

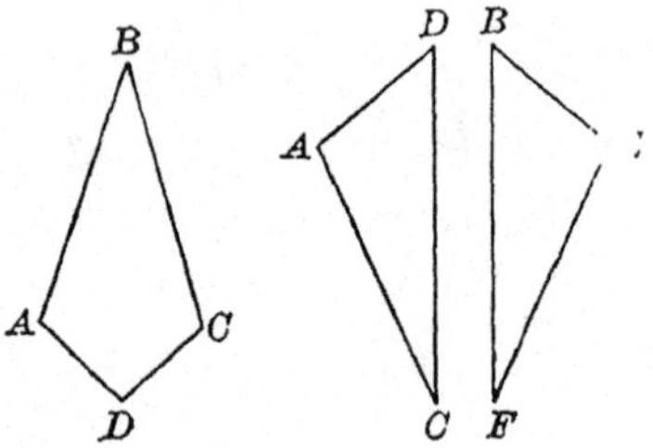

PROPOSITION XVI. THEOREM

106. *Two triangles are equal if three sides of the one are respectively equal to three sides of the other.* (s. s. s = s. s. s.)

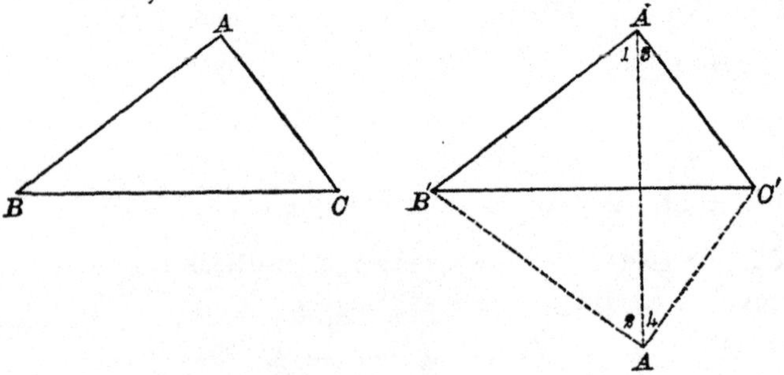

Hyp. In $\triangle$ ABC and $A'B'C'$,

$$AB = A'B',\ BC = B'C',\ AC = A'C'.$$

To prove $\triangle ABC = \triangle A'B'C'$.

Proof. Apply $\triangle ABC$ to $\triangle A'B'C'$ so that its greatest side BC coincides with $B'C'$, and A and A' lie on opposite sides of $B'C'$.

Join AA'.

$$AB' = A'B', \text{ and } AC' = A'C'. \qquad \text{(Hyp.)}$$

$$\therefore \angle 1 = \angle 2, \text{ and } \angle 3 = \angle 4,$$

(*base* $\angle$s *of an isos.* $\triangle$).

$$\therefore \angle 1 + \angle 3 = \angle 2 + \angle 4, \qquad \text{(Ax. 2.)}$$

i.e.

$$\angle BAC = \angle B'A'C'.$$

$$\therefore \triangle A'B'C' = \triangle AB'C',$$

(*s. a. s.* = *s. a. s.*).

$$\therefore \triangle ABC = \triangle A'B'C'.$$

Q E.D.

Ex. 117. If the opposite sides of a quadrilateral are equal, the opposite angles are equal.

Ex. 118. The median to the base of an isosceles triangle is perpendicular to the base.

Ex. 119. If a quadrilateral has two pairs of equal adjacent sides, the angles included by the unequal sides are equal.

Ex. 120. The medians to the arms of an isosceles triangle are equal.

Ex. 121. If the opposite sides of a quadrilateral are equal, the sides are parallel.

Ex. 122. If in the sides of an equilateral $\triangle ABC$, the points D, E, and F be taken so that

$$AD = BE = CF,$$

then $\triangle DFE$ is equilateral.

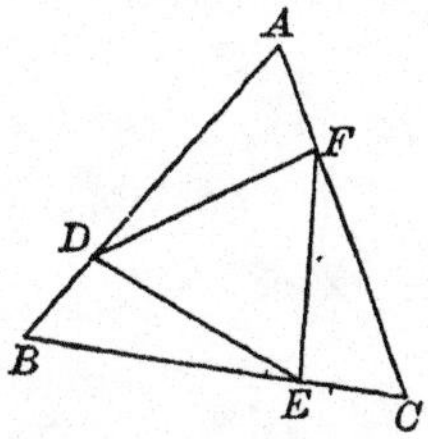

Ex. 123. If the sides of an equilateral $\triangle ABC$ are produced so that

$$AD = BE = CF,$$

then $\triangle DFE$ is equilateral.

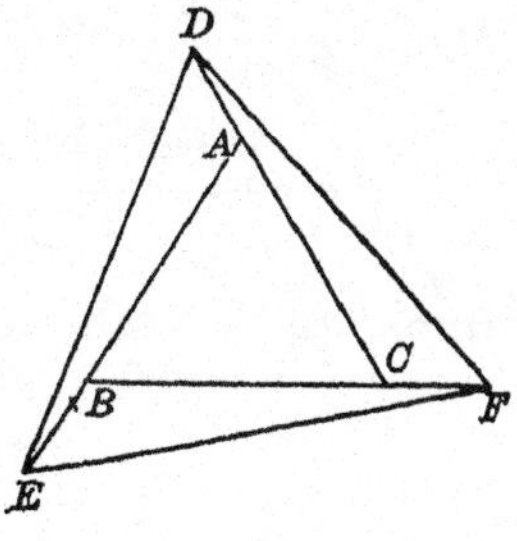

107. Remark. — **If it is impossible to prove the equality of the required pair of triangles, prove first the equality of some other pair, whose homologous parts will aid in proving the equality of the original pair**

Ex. 124. If the opposite sides of a quadrilateral are equal, and a line be drawn through the midpoint of the diagonal and terminating in two sides, this line is bisected.

Ex. 125. Two triangles are equal if two sides and the median to one of these sides are equal respectively to two sides and the homologous median of the other.

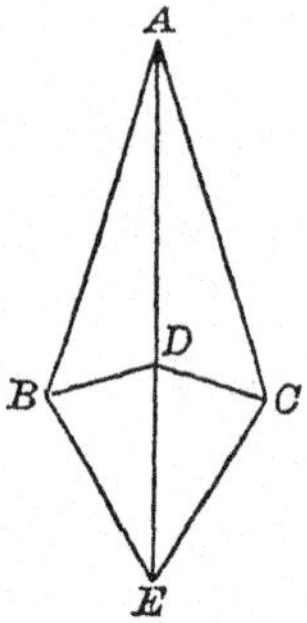

Ex. 126. Two isosceles triangles are equal if one angle and the altitude upon one arm of one triangle are equal respectively to one angle and the altitude upon one arm of the other.

Ex. 127. If the opposite sides of a quadrilateral are equal, the diagonals bisect each other.

Ex. 128. Two triangles are equal if the base, an adjacent angle, and the altitude upon the base of one triangle are equal respectively to the base, an adjacent angle, and the altitude of the other.

Ex. 129. In the annexed diagram if $AB = AC$, and AD bisects $\angle BAC$, then $\angle DBE = \angle DCE$.

PROPOSITION XVII. THEOREM

108. *Two right triangles are equal if the hypotenuse and an arm of the one are respectively equal to the hypotenuse and an arm of the other.* (hy. arm = hy. arm.)

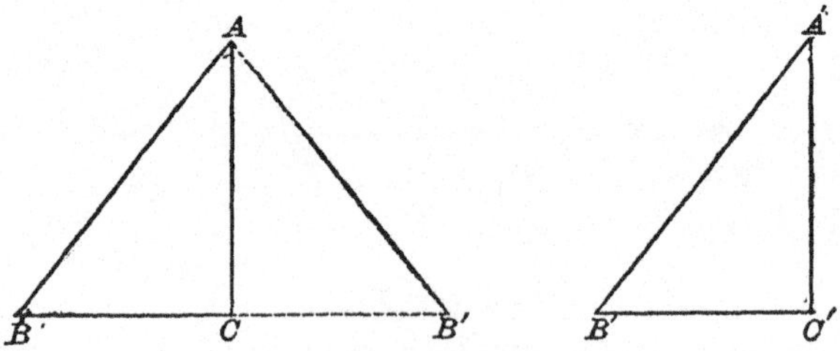

Hyp. In the right $\triangle\triangle\ ABC$ and $A'B'C'$,

$$AB = A'B';\ AC = A'C'.$$

To prove $\triangle ABC = \triangle A'B'C'$.

Proof. Apply $\triangle A'B'C'$ to $\triangle ABC$, so that $A'C'$ coincides with AC, and B' falls remote from B.

BCB' is a straight line,

($\angle BCB' = 2$ *rt.* $\angle$s).

$AB = AB'$. (Hyp.)

$\therefore \angle B = \angle B'$,

(*base* $\angle$s *of an isos.* $\triangle$).

$\therefore \triangle ABC = \triangle AB'C$,

(*s. a. a.* = *s. a. a.*).

$\therefore \triangle ABC = \triangle A'B'C'$. Q.E.D.

109. Def. A **circumference** is a curved line, all of whose points are equidistant from a point within called the center, as *ABC*, the center being *D*.

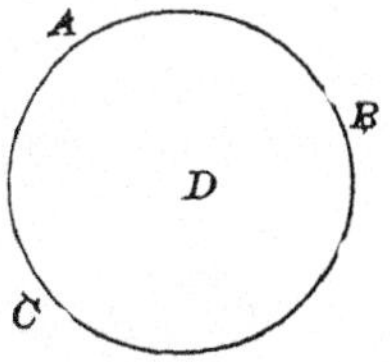

A **circle** is the portion of a plane bounded by a circumference, and is usually read "the circle *D*" or "the $\odot ABC$."

A **radius** is any straight line drawn from the center to the circumference.

An **arc** is any portion of a circumference.

Ex. 130. If two altitudes of a triangle are equal, the corresponding sides are equal, and the triangle is isosceles.

Ex. 131. Two triangles are equal if two sides and the altitude upon one of them of one triangle are respectively equal to two sides and the homologous altitude of the other.

Ex. 132. If the perpendiculars from the midpoint of one side of a triangle upon the two other sides are equal, the triangle is isosceles.

Ex. 133. Two triangles are equal if the base, the median, and altitude to the base of one triangle are equal respectively to the base, the median, and altitude to the base of the other.

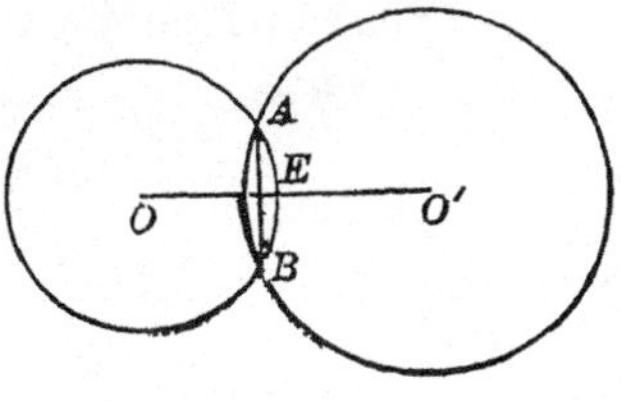

Ex. 134. If two circles intersect in *A* and *B*, a line joining their centers bisects *AB* in *E*.

Ex. 135. If on the sides of an equilateral triangle, ABC, the points, D, E, F, are taken, so that $AD = BE = CF$, and E, D, and F are joined to the opposite vertices, then $\triangle A'B'C'$ is equilateral.

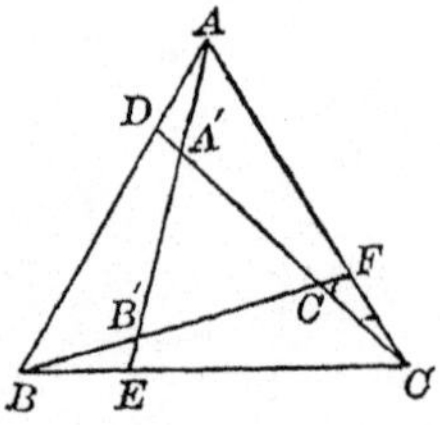

Ex. 136. If the opposite sides of a polygon of six sides are equal, and one pair of opposite sides are parallel, then the opposite angles are equal.

Ex. 137. Two triangles are equal if one side and the altitudes upon the other sides of one triangle are equal respectively to one side and the altitudes upon the other sides of the other triangle.

PROPOSITION XVIII. THEOREM

110. *The line that joins the vertices of two isosceles triangles on the same base bisects the common base at right angles.*

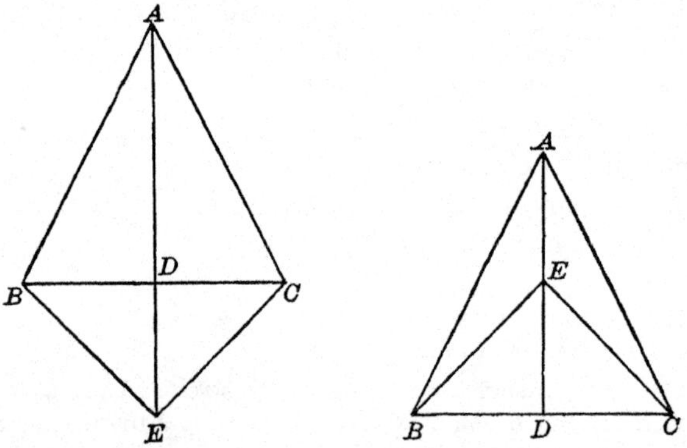

Hyp. $\triangle ABC$ and BCE are isosceles, and have the common base BC.

To prove AE is the perpendicular-bisector of BC.

Proof. $AB = AC$, $BE = EC$ (Hyp.), and AE is common

$$\therefore \triangle BAE = \triangle CAE,$$

(*s. s. s.* = *s. s. s.*).

$$\therefore \angle BAE = \angle EAC,$$

(*hom. parts of equal* $\triangle$s).

AD is common.

$AB = AC$. (Hyp.)

$\therefore \triangle BAD = \triangle DAC$,

(*s. a. s.* = *s. a. s.*).

$\therefore BD = CD$, and $\angle BDA = \angle CDA$,

(*hom. parts of equal* $\triangle$s).

$\therefore \angle BDA =$ rt. $\angle$ (being one-half of a st. $\angle$).

$\therefore AE$ is the perpendicular-bisector of BC. Q.E.D.

111. Cor. 1. Two points equidistant from the ends of a line determine the perpendicular-bisector to that line.

112. Cor. 2. Every point equidistant from the ends of a line lies in the perpendicular-bisector to that line.

Ex. 138. If the four sides of a quadrilateral are equal, the diagonals bisect each other.

CONSTRUCTIONS

Proposition XIX. Problem

113. *To bisect a given straight line.*

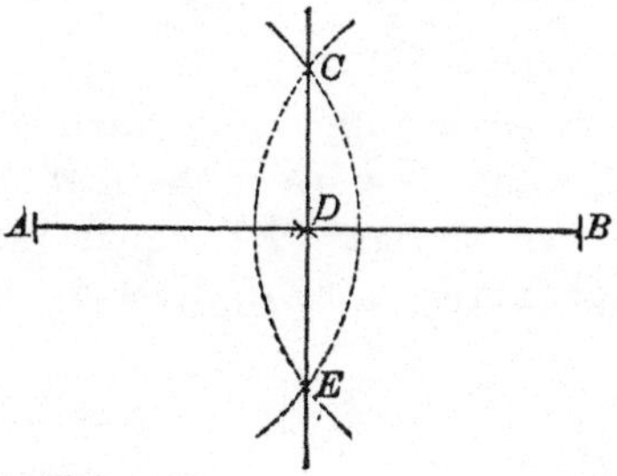

Given a straight line AB.

Required to bisect AB.

Construction. From A and B as centers, with equal radii greater than $\frac{1}{2}$ AB, describe arcs intersecting at C and E.

Join CE.

Then the line CE bisects AB at D. Q.E.F.

Proof. Two points equidistant from the ends of a line determine the perpendicular-bisector to that line.

SCHOLIUM. *CE* is the perpendicular-bisector of *AB*. (111)

Ex. 139. Divide a given line into four equal parts.

Ex. 140. Construct the three medians of a triangle.

PROPOSITION XX. PROBLEM

114. *To bisect a given angle.*

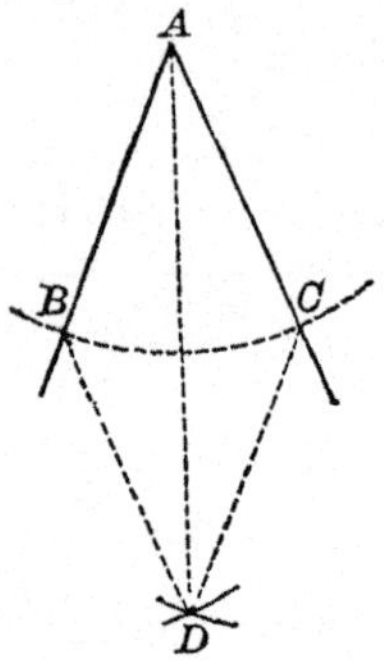

Given. ∠ *CAB*.

Required. To bisect ∠ *CAB*.

Construction. From *A* as a center, with any radius, as *AB*, describe an arc cutting the sides of the ∠ *A* at *B* and *C*.

From *B* and *C* as centers, with equal radii greater than one-half the distance from *B* to *C*, describe two arcs intersecting at *D*.

Join *AD*.

AD bisects ∠ *CAB*. Q.E.F

HINT. — What is the usual means of proving the equality of angles?

Ex. 141. Divide an angle into four equal parts.

Ex. 142. To bisect a straight angle.

Ex. 143. Construct an angle of 90°; of 135°.

Ex. 144. Draw the three bisectors of a triangle.

PROPOSITION XXI. PROBLEM

115. *At a given point in a given straight line, to erect a perpendicular to that line.*

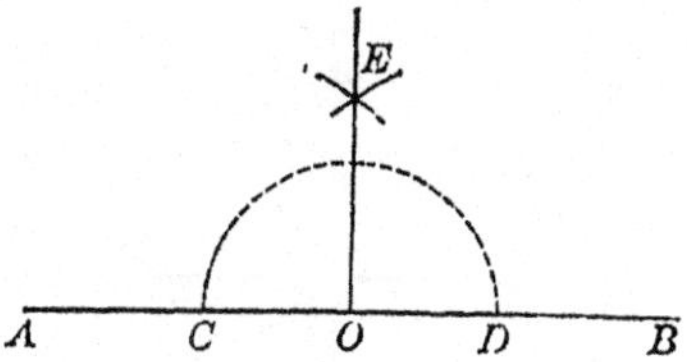

Given. Point O in line AB.

Required. A perpendicular to the line AB at O.

Construction. From O as a center, with any radius OC, describe an arc intersecting AB in C and D.

From C and D as centers, with equal radii greater than OC, describe two arcs intersecting at E.

Join OE.

OE is the perpendicular to the line AB at O. Q.E.F.

[The proof is left to the student.]

Ex. 145. Construct the complement of a given angle. What kind of an angle must the given angle be?

Ex. 146. By what angle is the supplement of an angle greater than its complement?

Ex. 147. Construct the supplement of a given angle.

Ex. 148. To bisect a given reflex angle.

Ex. 149. Construct an angle of 67° 30′.

Ex. 150. Construct an angle of 60°.

Ex. 151. Construct an angle of 30°.

Ex. 152. Construct an angle of 120°; of 75°.

Ex. 153. Construct the three perpendicular bisectors of the three sides of any given triangle.

Proposition XXII. Problem

116. *From a point without a straight line, to let fall a perpendicular upon that line.*

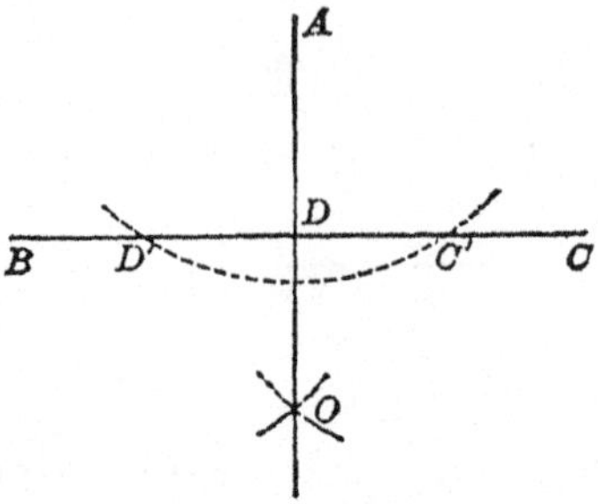

Given. A straight line BC, and a point A without the line.

Required. A perpendicular from the point A to the line BC.

Construction. From A as a center, with a radius sufficiently great, describe an arc cutting BC in C' and D'.

From D' and C' as centers, with equal radii greater than $\frac{1}{2}$ $D'C'$, describe two arcs intersecting at O.

Draw AO intersecting BC in D.

AD is a perpendicular from the point A to the line BC. Q.E.F.

[The proof is left to the student.] (111)

Ex. 154. Construct the three altitudes of an acute triangle; of an obtuse triangle.

Ex. 155. From a given point without a line, to draw a line forming with the given line an angle equal to half a right angle.

Ex. 156. From a given point without a line, to draw a line forming with the given line an angle of 60°. How many such lines can be drawn?

Ex. 157. From a given point without a line, to draw a line forming with the given line an angle of 30°.

PROPOSITION XXIII. PROBLEM

117. *At a point C, in a given straight line AB, to construct an angle equal to a given angle O.*

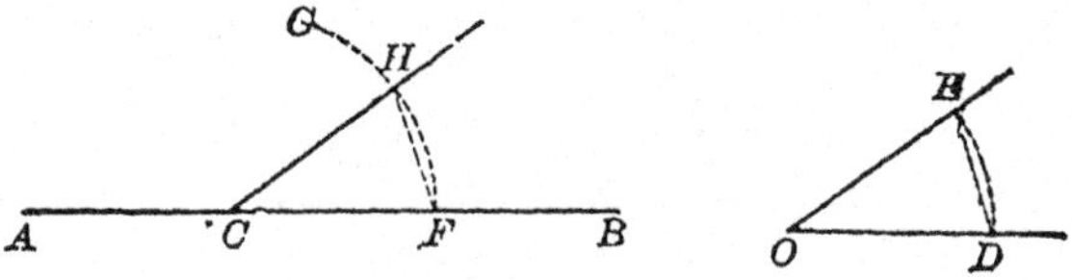

Given. Point C in line AB and $\angle O$.

Required. An $\angle$ at $C = \angle O$.

Construction. From O as a center, with any radius, describe an arc cutting the sides of $\angle O$ in D and E.

From C as a center, with the same radius, describe arc FG, intersecting CB in F. From F as a center, with a radius equal to DE, draw an arc intersecting FG in H.

Join CH.

$\angle HCF$ is an $\angle$ at $C = \angle O$.

HINT. — What is the usual means of proving the equality of angles?

Q.E.F.

NOTE. — Two letters, *e.g.* DE, used as above, denote a straight line.

Ex. 158. Given a polygon of five sides (pentagon) $ABCDE$. Required a pentagon having four of its sides equal respectively to AB, BC, CD, and DE, and the included angles equal to $\angle s\ B$, C, and D.

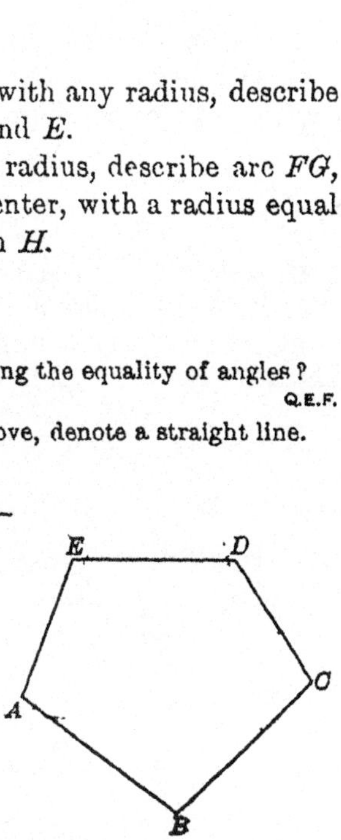

PROPOSITION XXIV. PROBLEM

118. *Through a given point to draw a parallel to a given line.*

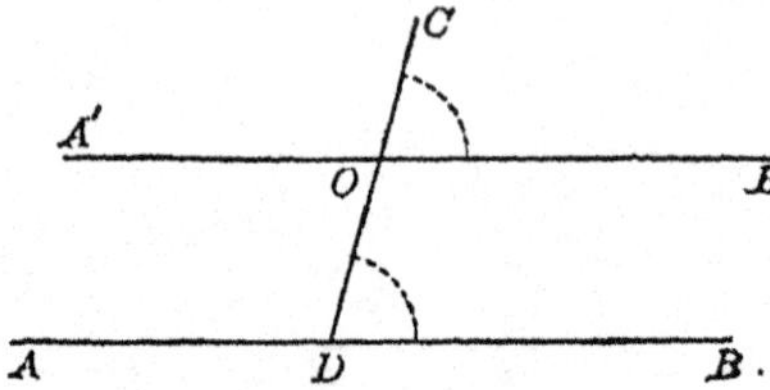

Given. Line AB and point O.

Required. A line through $O \parallel AB$.

Construction. Through O draw DOC, forming $\angle D$ with AB.
At O construct $\angle COB' = ODB$. (117)
$A'B'$ is a line through $O \parallel AB$. Q.E.F.

HINT. — What is the usual means of proving that two lines are parallel? (86)

Ex. 159. From a given point without a line to draw a line, making a given angle with the given line.

Ex. 160. Through a given point to draw a parallel to a given line by means of (*a*) Prop. V, (*b*) Ex. 50.

PROPOSITION XXV. THEOREM

119. ***Every point in the perpendicular-bisector of a line is equidistant from the extremities of that line.***

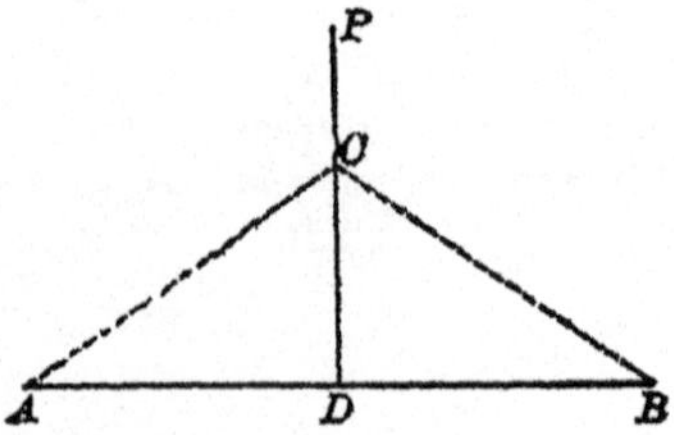

Hyp. *PD* is the perpendicular-bisector of *AB*.

C is a point in *PD*.

To prove C is equidistant from *A* and *B*.

HINT. — What is the usual means of proving the equality of lines?

Ex. 161. In a given line *AB* find a point equidistant from two given points *P* and *Q*.

Ex. 162. In a given circumference find a point equidistant from two given points.

Ex. 163. Find a point equidistant from three given points.

120. DEF. The **distance** of a point from a line is the length of the perpendicular from the point to the line.

PROPOSITION XXVI. THEOREM

121. *Every point in the bisector of an angle is equidistant from the sides of the angle.*

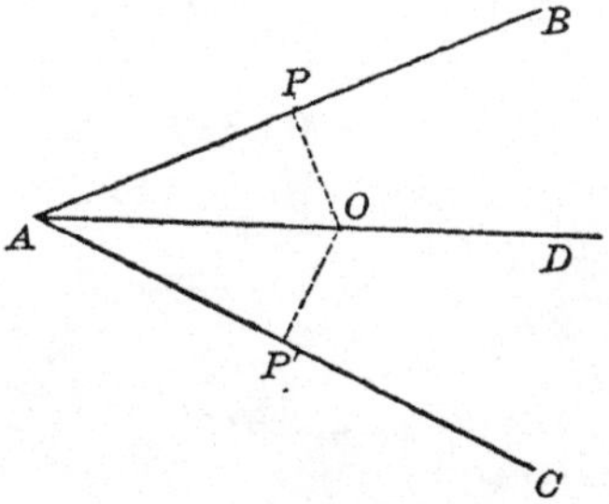

Hyp. $\angle DAB = \angle DAC$ and *O* is any point in *AD*.

To prove O is equidistant from *AB* and *AC*.

Draw $OP \perp AB$ and $OP' \perp AC$, and prove the equality of the two triangles.

122. COR. A point equidistant from the sides of an angle is in the bisector of that angle.

Ex. 164. In a given circumference find a point equidistant from two given lines.

Ex. 165. Find a point equidistant from three given lines.

UNEQUAL LINES AND UNEQUAL ANGLES

PROPOSITION XXVII. THEOREM

123. *The sum of two sides of a triangle is greater, and their difference is less, than the third side.*

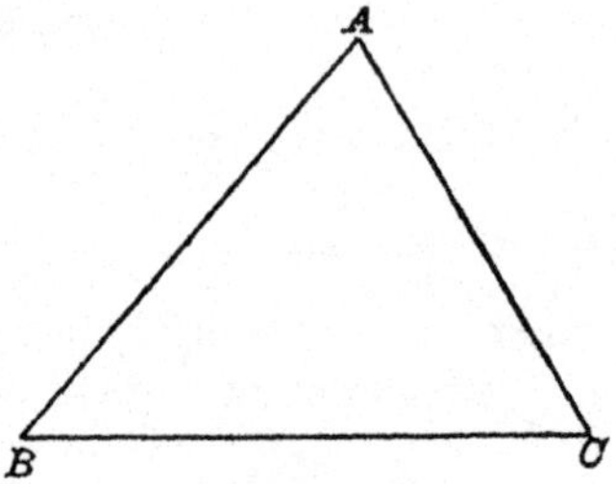

Hyp. ACB is a $\triangle$.

To prove $AB + AC > BC$; and $BA > CB - AC$.

Proof. 1°. A straight line is the shortest distance between two points. (Ax. 10.)

$\therefore AB + AC > BC$. Q.E.D.

2°. $AB + AC > BC$. (above)

$AB > BC - AC$. (Ax. 5.) Q.E.D.

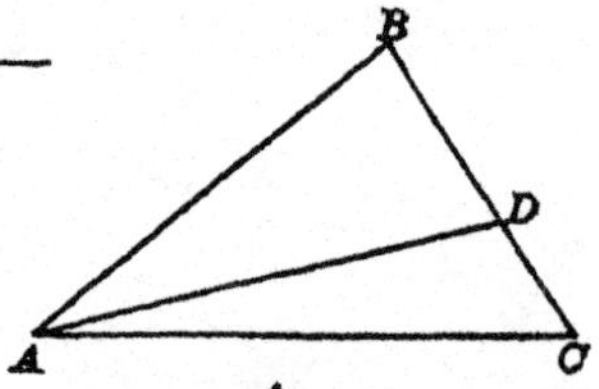

Ex. 166. If in the side BC of $\triangle ABC$ any point D be taken, then

$$AB + BC > AD + DC.$$

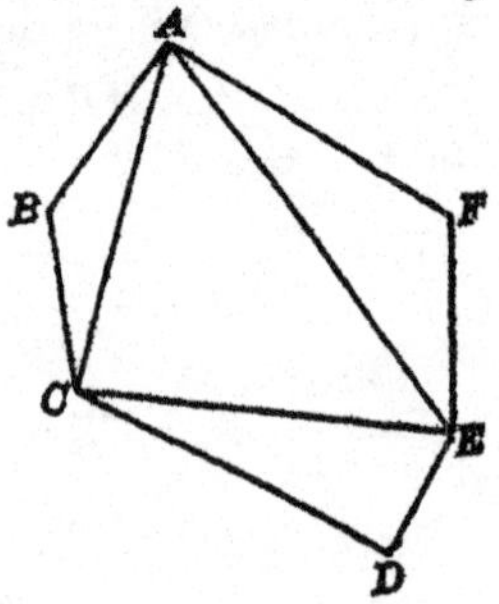

Ex. 167. Prove that the perimeter of polygon $ABCDEF >$ perimeter of $\triangle ACE$.

***Ex. 168.** If from any point D in $\triangle ABC$ DB and DC are drawn, then

$$AB + AC > DB + DC.$$

PROPOSITION XXVIII. THEOREM

124. *If two sides of a triangle are unequal, the angles opposite are unequal, and the greater angle is opposite the greater side.*

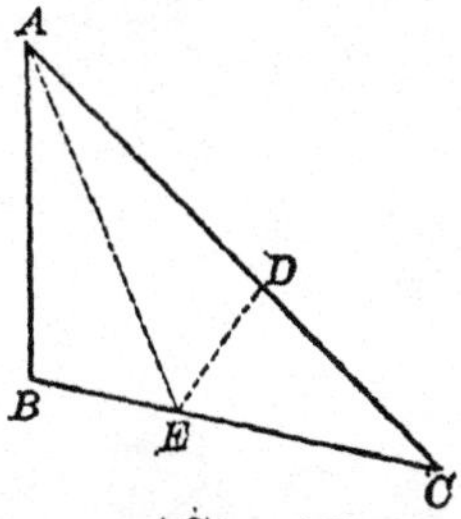

Hyp. In $\triangle ABC$ $AC > AB$.

To prove $\angle B > \angle C$.

Proof. Draw AE, bisecting $\angle BAC$. On AC lay off $AD = AB$ and draw DE.

$$\triangle ABE = \triangle ADE,$$

(*s. a. s. = s. a. s.*).

$$\therefore \angle EDA = \angle B,$$

(*hom. parts of equal* ▲).

But $\angle EDA > \angle C$,

(*exterior* $\angle$ *of a* $\triangle$ *is* > *either remote interior* $\angle$).

$$\therefore \angle B > \angle C.$$

Q.E.D

Ex. 169. If one arm AB of an isosceles triangle ABC is produced to D, then

$$\angle ACD > \angle ADC.$$

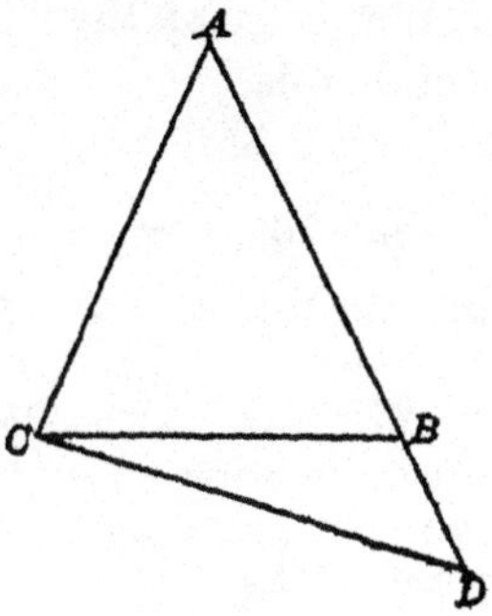

PROPOSITION XXIX. THEOREM

125. *If two angles of a triangle are unequal, the sides opposite are unequal, and the greater side is opposite the greater angle.*

[Converse of Prop. XXVIII.]

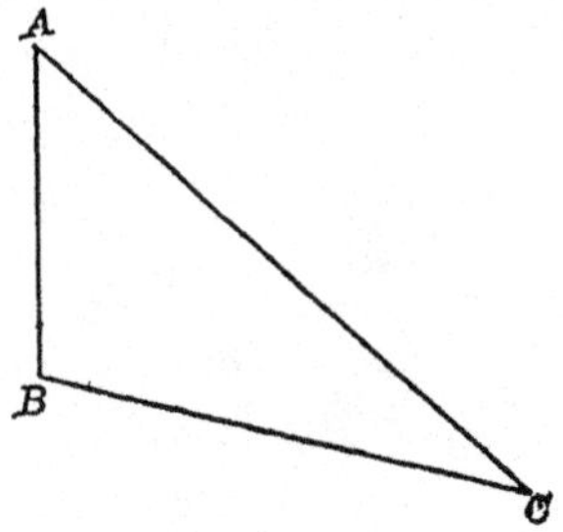

Hyp. In $\triangle ABC \angle B > \angle C$.

To prove $AC > AB$.

Proof. AB must be either $>$, $=$, or $<$ AC.

If $AB > AC$, then $\angle C > \angle B$. (124)

If $AB = AC$, $\angle C = \angle B$. (101)

But both of the above consequences would contradict the hypothesis.

$\therefore AC > AB$. Q.E.D.

126. COR. The perpendicular is the shortest line that can be drawn from a point to a given line.

127. NOTE.—The method used in the above proof is known as the "indirect method" or "reductio ad absurdum." We examine every case, and prove the impossibility of all except one. (Compare Props. V and VIII.)

PROPOSITION XXX. THEOREM

128. *If two triangles have two sides of the one equal respectively to two sides of the other, but the included angle of the first greater than the included angle of the second, then the third side of the first is greater than the third side of the second.*

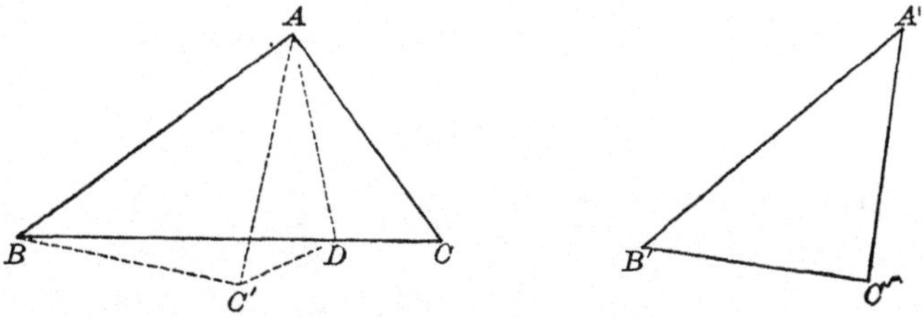

Hyp. In ▲ ABC and $A'B'C'$

$$AB = A'B'; \; AC = A'C'; \; \angle A > \angle A'.$$

To prove $BC > B'C'$.

Proof. Apply $\triangle A'B'C'$ upon $\triangle ABC$ so that $A'B'$ coincides with AB.

Since $\angle A > \angle A'$ (Hyp.), $A'C'$ will fall between AB and AC as AC'.

Draw AD bisecting $\angle CAC'$ to meet BC in D, and join DC'.

$$\triangle ADC' = \triangle ADC,$$

(*s. a. s.* = *s. a. s*).

$$\therefore DC' = DC,$$

(*hom. parts of equal* ▲).

But $$BD + DC' > BC',$$

(*the sum of 2 sides of a* $\triangle >$ *third side*).

$$\therefore BD + DC > BC',$$

or $$BC > BC' \text{ or } B'C'.$$ Q.E.D.

Ex. 170. Which is the greatest side of a right triangle? of an obtuse triangle?

Ex. 171. If two angles of a triangle are 50° and 60°, respectively, which is the greater of the two opposite sides? which is the greatest side of the triangle? which is the smallest?

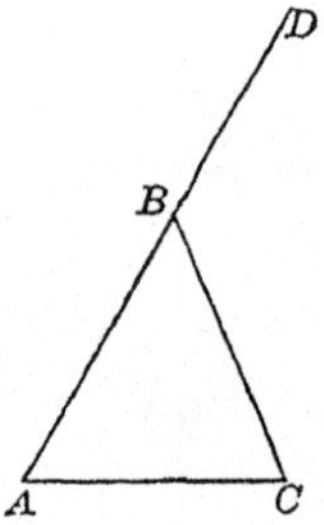

Ex. 172. If $\angle A$ of $\triangle ABC$ equals two-thirds of a right angle, and the exterior angle DBC equals five-fourths of a right angle, which is the greatest side of the triangle? which is the smallest?

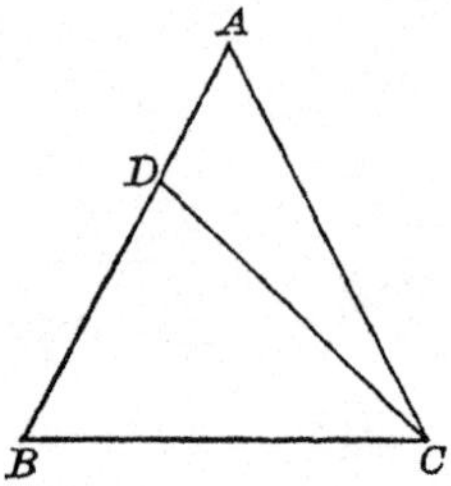

Ex. 173. If $AB = AC$ in $\triangle ABC$, prove that

$$DC > DB.$$

Ex. 174. If O and O' are 2 equal circles, OA, OB, $O'A'$, and $O'B'$ radii including,

$$\angle A'O'B' > \angle AOB, \text{ then } A'B' > AB.$$

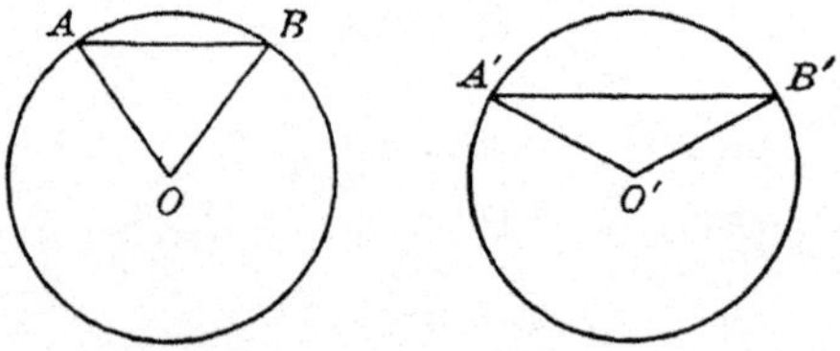

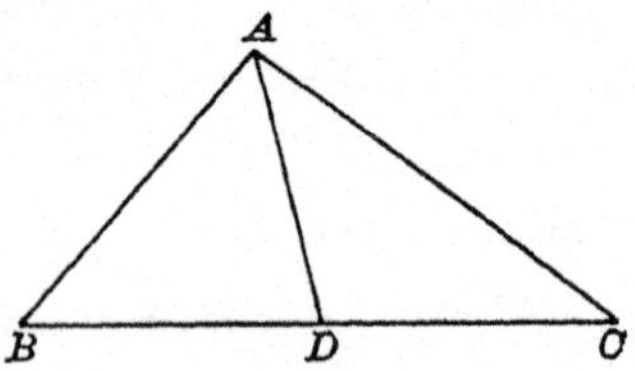

Ex. 175. If in triangle ACB the median AD be drawn forming the acute angle ADB, which is the greater side, AB or AC?

PROPOSITION XXXI. THEOREM

129. *If two triangles have two sides of the one equal respectively to two sides of the other, but the third side of the first greater than the third side of the second, then the included angle of the first is greater than the included angle of the second.*

[Converse of Prop. XXX.]

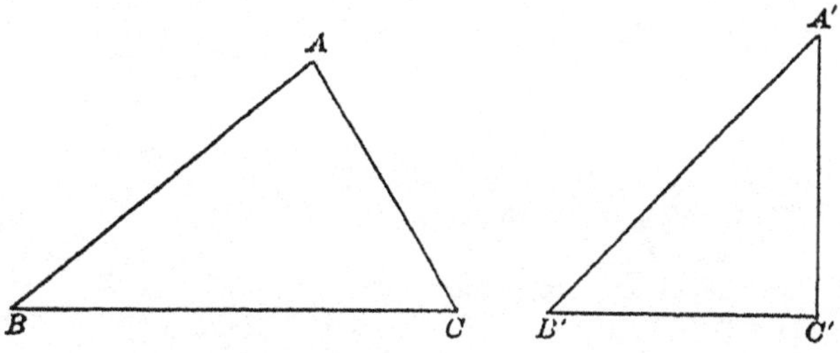

Hyp. In $\triangle s$ ABC and $A'B'C'$

$$AB = A'B';\ AC = A'C';\ BC > B'C'.$$

To prove $\angle A > \angle A'$.

Proof. $\angle A$ must be either $>$, $=$, or $< \angle A'$.

If $\angle A < \angle A'$, $BC < B'C'$. (128)

If $\angle A = \angle A'$, $BC = B'C'$. (72)

But both of the above consequences would contradict the hypothesis.

$$\therefore \angle A > \angle A'.$$ Q.E.D.

Ex. 176. If in $\triangle ABC$, having $AB > BC$, the median BE is drawn, then $\angle AEB$ is an obtuse angle.

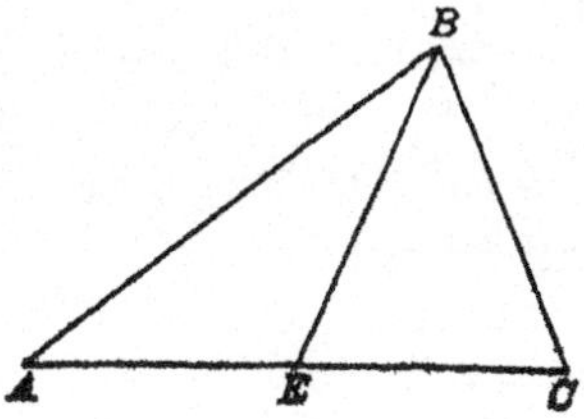

130. REMARK. — Inequality of lines is generally proved by means of Props. XXVII, XXIX, XXX. If it is impossible to discover any relation of the angles, Prop. XXVII is used; if the two lines are parts of the same triangle, and the opposite angles can be proved unequal, Prop. XXIX is used; but if the sides or angles are parts of different triangles, Prop. XXX is used. In some cases several methods will lead to the desired result.

The inequality of angles is proved in a similar manner by Props. IV, XXVIII and Prop. XXXI.

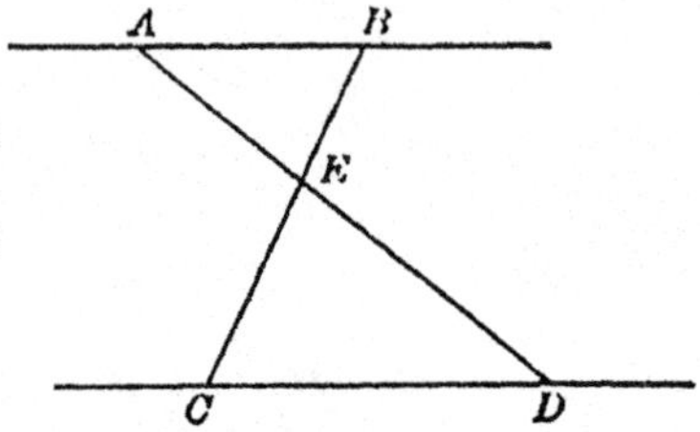

Ex. 177. If AD and BC intersect in E, prove $AD + BC > AB + CD$. (Which of the three propositions must be used?)

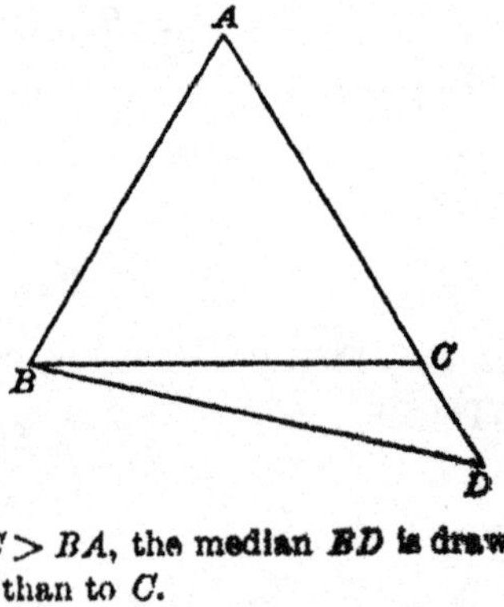

Ex. 178. If AC, a side of an equilateral triangle, is produced to D, and BD is joined,

then $AD > BD > AB$.

* Ex. 179. If in $\triangle ABC$, having $BC > BA$, the median BD is drawn, then any point E in BD is nearer to A than to C.

* Ex. 180. If in $\triangle ABC$, having $AB < BC$, the bisector BD is drawn, meeting AC in D, then $AD < DC$.

Ex. 181. A point P, not in the perpendicular-bisector of AB, is unequally distant from A and B. (Find 3 proofs corresponding with the methods of 130.)

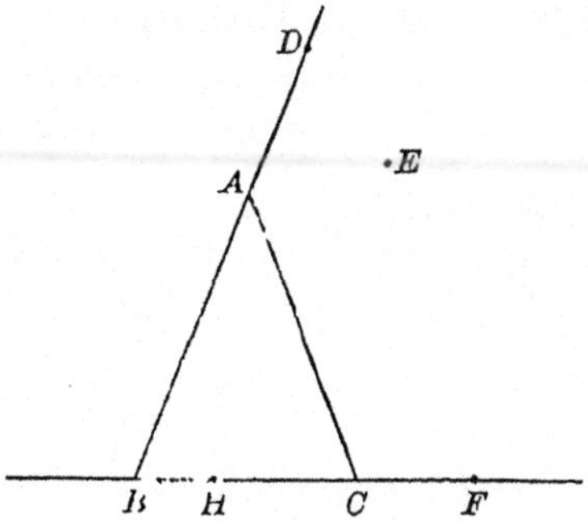

Ex. 182. In the annexed diagram, if $AB = AC$, prove that $BD > DC$.

Ex. 183. In the diagram of Ex. 182, prove that $BE > EC$.

Ex. 184. In the diagram of Ex. 182, prove that $AF > AB$.

Ex. 185. In the diagram of Ex. 182, prove that $AB > AH$.

* **Ex. 186.** If the opposite sides of a quadrilateral $ABCD$ are equal, but $AB > AD$, prove that $\angle AOB > \angle AOD$.

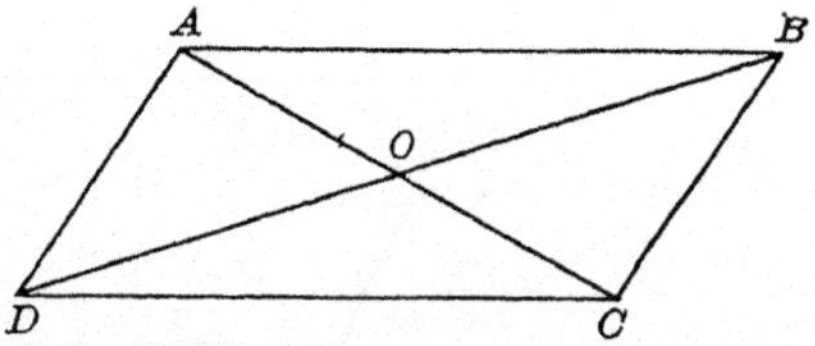

* Ex. 187. The sum of the lines drawn from any point in a triangle to its vertices is less than the perimeter, but greater than the semiperimeter of the triangle.

Ex. 188. The sum of the diagonals of any quadrilateral is less than the perimeter, but greater than the semiperimeter of the quadrilateral.

QUADRILATERALS

131. Def. A **trapezoid** is a quadrilateral that has one pair of sides parallel. A **parallelogram** has its opposite sides parallel.

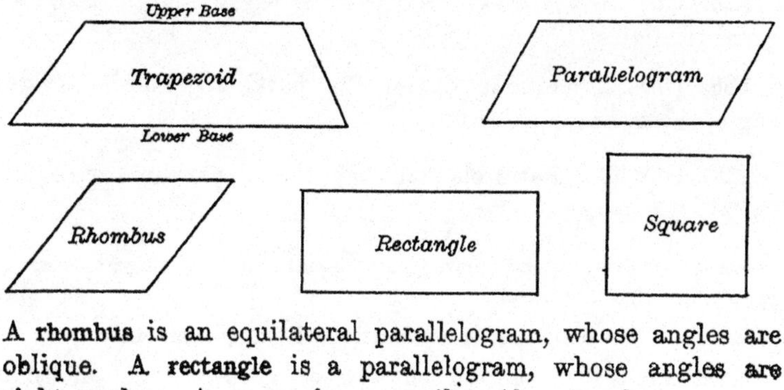

A **rhombus** is an equilateral parallelogram, whose angles are oblique. A **rectangle** is a parallelogram, whose angles are right angles. A **square** is an equilateral rectangle.

132. Def. An **isosceles trapezoid** is one whose non-parallel sides are equal. The parallel sides of a trapezoid are called its **bases**, and are distinguished as **upper** and **lower**.

133. Def. A **diagonal** of a quadrilateral is a straight line joining opposite vertices. The **altitude** of a parallelogram or trapezoid is the perpendicular distance between the two bases.

Proposition XXXII. Theorem

134. *The opposite sides and angles of a parallelogram are equal.*

Hyp. $ABCD$ is a parallelogram.

To prove $AD = BC$; $AB = CD$; $\angle A = \angle C$; $\angle B = \angle D$.

Hint. — What is the usual means of proving the equality of lines and angles?

135. Cor. 1. A diagonal divides a parallelogram into two equal triangles.

136. Cor. 2. If one angle of a parallelogram is a right angle, the figure is a rectangle.

137. Cor. 3. Parallels included between parallels are equal.

Ex. 189. The perpendiculars to a diagonal of a parallelogram from the opposite vertices are equal.

Ex. 190. The diagonals of a parallelogram bisect each other.

* Ex. 191. A line bisecting one side of a triangle and parallel to another side, bisects the third side also.

PROPOSITION XXXIII. THEOREM

138. *If the opposite sides of a quadrilateral are equal, the figure is a parallelogram.*

Hyp. In quadrilateral $ABDC$

$AB = CD$; $AD = BC$.

To prove $AD \parallel BC$; $AB \parallel CD$.

HINT. — Prove the equality of a pair of alternate interior angles by means of equal triangles.

Ex. 192. If the diagonals of a quadrilateral bisect each other, the figure is a parallelogram.

PROPOSITION XXXIV. THEOREM

139. *If two sides of a quadrilateral are equal and parallel, the figure is a parallelogram.*

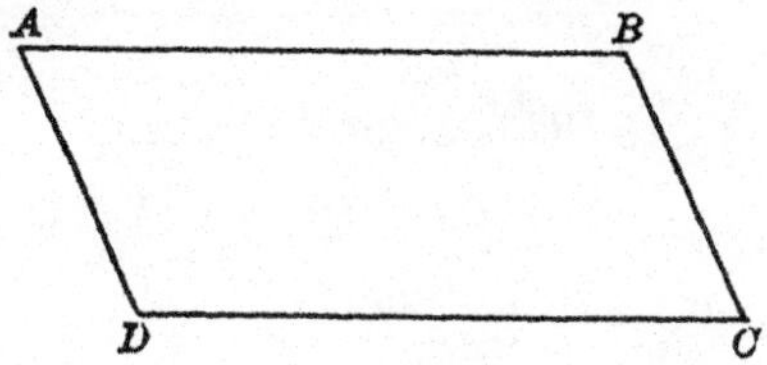

Hyp. In quadrilateral $ABCD$

$AB \parallel CD$; and $AB = CD$.

To prove $ABCD$ is a parallelogram.

[The proof is similar to that in Prop. XXXIII.]

Ex. 193. The bisectors of two opposite angles of a parallelogram are parallel.

Ex. 194. If two opposite sides of a parallelogram are produced by the same length in opposite directions and their ends joined to the nearest vertices, another parallelogram is formed.

140. Remark. — **Lines may be shown to be parallel by proving them to be opposite sides of a parallelogram.**

Ex. 195. If two opposite sides of a parallelogram are divided into three equal parts, and the respective points of division are joined, the lines are parallel.

Ex. 196. If two opposite sides of a parallelogram are produced by the same length in the same direction, a line joining the ends is parallel to the other sides of the parallelogram.

Ex. 197. If from two opposite vertices of a parallelogram lines be drawn bisecting two opposite sides, respectively, the lines are parallel.

Proposition XXXV. Theorem

141. *The diagonals of a parallelogram bisect each other.*

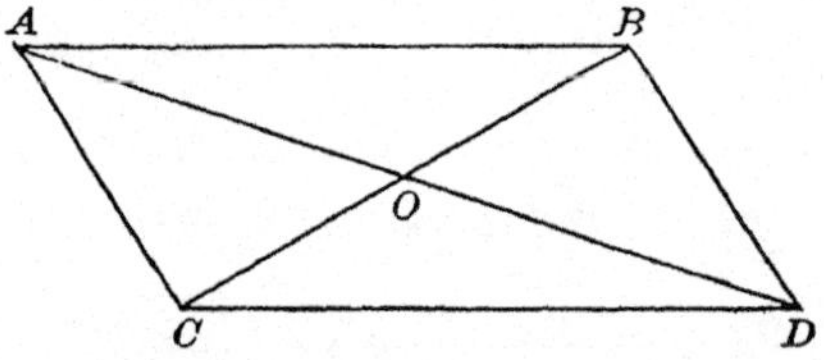

Hyp. $ABDC$ is a parallelogram.

To prove $AO = OD$; $BO = OC$.

[The proof is left to the student.]

Ex. 198. The diagonals of a rhombus are perpendicular to each other.

Ex. 199. If the diagonals of a parallelogram are perpendicular to each other, the figure is a rhombus, or a square.

Ex. 200. If each half of the diagonals of a parallelogram is bisected and the midpoints are joined in order, another parallelogram is formed.

Ex. 201. If a diagonal bisects an angle of a parallelogram, the figure is equilateral.

Ex. 202. The diagonals of a rectangle are equal.

Ex. 203. State and prove the converse of the preceding exercise.

Ex. 204. If the ends of two diameters of a circle be joined in succession a rectangle is formed.

Ex. 205. The base angles of an isosceles trapezoid are equal.

Ex. 206. State and prove the converse of the preceding exercise.

Proposition XXXVI. Theorem

142. *Two parallelograms are equal if two adjacent sides and the included angle of one are equal, respectively, to two adjacent sides and the included angle of the other.*

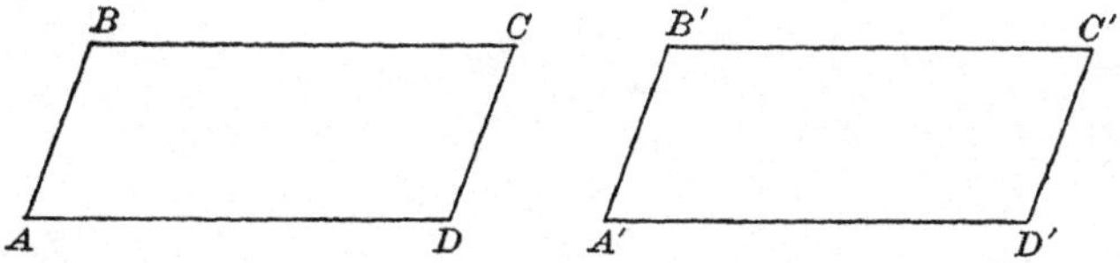

Hyp. In ▱s $ABCD$ and $A'B'C'D'$, $AB = A'B'$, $AD = A'D'$, $\angle A = \angle A'$.

To prove ▱ $ABCD =$ ▱ $A'B'C'D'$.

Proof. Apply ▱ $ABCD$ to $A'B'C'D'$, so that AB coincides with $A'B'$. Then AD takes the direction $A'D'$,

($\angle A = \angle A'$).

And D coincides with D',

(*for* $AD = A'D'$).

BC takes the direction of $B'C'$, (Ax. 11.)
and C must lie in $B'C'$ or in $B'C'$ produced.

For a similar reason, C must lie in $D'C'$ or in $D'C'$ produced.

Hence C coincides with C',

and ▱ $ABCD$ coincides with ▱ $A'B'C'D'$.

$\therefore$ ▱ $ABCD =$ ▱ $A'B'C'D'$ Q.E.D.

143. REMARK.—**In order to demonstrate that one line (or angle) is twice as large as another line (or angle), either double the smaller, and prove its double equal the greater, or bisect the greater, and prove its half equal the smaller.**

Ex. 207. If two angles of a triangle are 30° and 60°, respectively, the side opposite the angle of 30° is one-half the side included by the two angles.

Ex. 208. The median to one side of a triangle is less than one-half the sum of the other two sides.

Ex. 209. The median to the hypotenuse of a right triangle is one-half the hypotenuse.

Ex. 210. The angle formed by the base of an isosceles triangle and the altitude upon one arm is one-half the vertical angle.

PROPOSITION XXXVII. THEOREM

144. *If three or more parallels intercept equal lengths on one transversal, they intercept equal lengths on every transversal.*

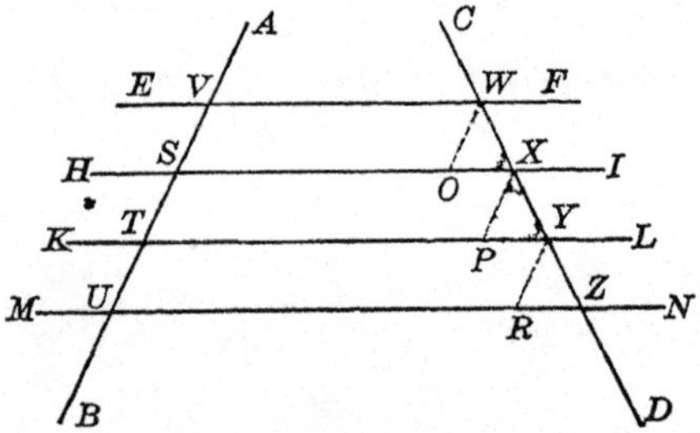

Hyp. $EF \parallel HI \parallel KL \parallel MN$, $VS = ST = TU$.

To prove $WX = XY = YZ$.

Proof. Draw WO, XP, $YR \parallel AB$.

Then $WVSO$, $XSTP$, $YTUR$ are **parallelograms**,

(*hyp. and con.*).

$\therefore$ $VS = WO$, $ST = XP$, $TU = YR$,

(*opposite sides of parallelogram*).

But as $VS = ST = TU,$ (Hyp.)

$\therefore\ WO = XP = YR.$ (Ax. 1.)

$\angle OWX = \angle PXY = \angle RYZ,$

(*corr.* $\angle$s *of* $\parallel$ *lines*).

$\angle WXO = \angle XYP = \angle YZR,$

(*corr.* $\angle$s *of* $\parallel$ *lines*).

$\therefore\ \triangle WOX = \triangle XPY = \triangle YRZ,$

(*s. a. a.* = *s. a. a.*).

$\therefore\ WX = XY = YZ,$

(*hom. parts of equal* $\triangle$s). Q.E.D.

145. Cor. A parallel to one side of a triangle, bisecting another side, bisects the third side also.

Proposition XXXVIII. Problem

146. *To divide a given line into any number of equal parts.*

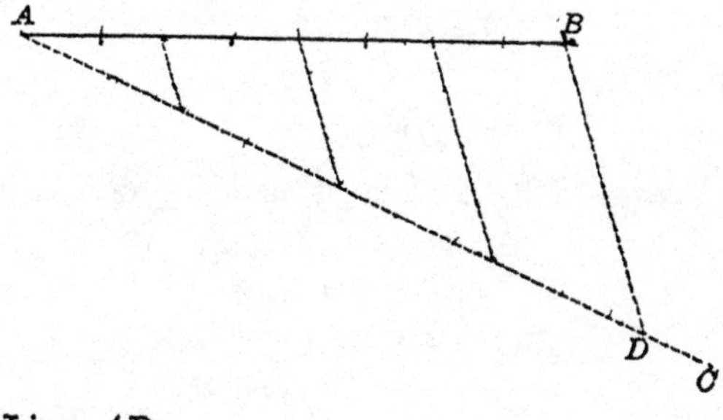

Given. Line AB.

Required. To divide AB into a number of equal parts.

Construction. From A draw any line AC.

On AC lay off any convenient length as often as the number of required parts, and from last point D, draw DB.

Through the points of division on AC, draw lines $\parallel$ BD.

These lines divide AB into the required number of equal parts. Q.E.F.

[The proof is left to the student.]

Ex. 211. A line bisecting one non-parallel side of a trapezoid, and parallel to the base, bisects the other non-parallel side.

PROPOSITION XXXIX. THEOREM

147. *A line which joins the midpoints of two sides of a triangle is parallel to the third side and equal to half of it.*

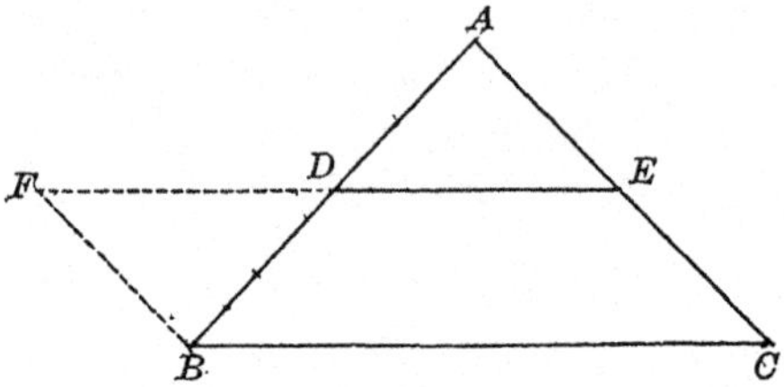

Hyp. In $\triangle ABC$:

$$AD = DB, AE = EC.$$

To prove 1°. $DE \parallel BC$.

2°. $DE = \frac{1}{2}BC$.

Proof. Draw $FB \parallel AC$, meeting ED produced to F.

$$\triangle ADE = \triangle BDF,$$

(*a. s. a. = a. s. a.*).

$$\therefore DE = DF, \text{ and } AE = BF,$$

(*hom. parts of equal* $\triangle$s).

But $AE = EC$.

$$\therefore FB = EC. \quad \text{(Ax. 1.)}$$

$\therefore BCEF$ is a parallelogram,

(*two opposite sides are equal and* $\parallel$).

$$\therefore DE \parallel BC, \text{ and } DE (= \tfrac{1}{2}FE) = \tfrac{1}{2}BC,$$

(*opposite sides of a parallelogram are parallel and equal*). Q.E.D.

Ex. 212. A line joining the midpoints of two adjacent sides of a quadrilateral is equal and parallel to the line joining the midpoints of the other two.

Ex. 213. If the midpoints of the three sides of a triangle are joined, four equal triangles are formed.

***Ex. 214.** A line joining the midpoints of the non-parallel sides of a trapezoid is parallel to the base, and equal to half the sum of the bases.

148. DEF. An **equiangular polygon** is one whose angles are all equal; an **equilateral polygon** one whose sides are equal. All the polygons discussed are understood to be **convex**, *i.e.* such that no side produced will fall within the polygon. A polygon of five sides is known as a **pentagon**; one of six sides, a **hexagon**; seven sides, a **heptagon**; eight sides, an **octagon**; ten sides, a **decagon**; twelve sides, a **dodecagon.**

149. Polygons that are mutually equiangular and equilateral are equal, for they can be made to coincide.

PROPOSITION XL. THEOREM

150. *The sum of all angles of a polygon is equal to as many straight angles as the figure has sides, less two.*

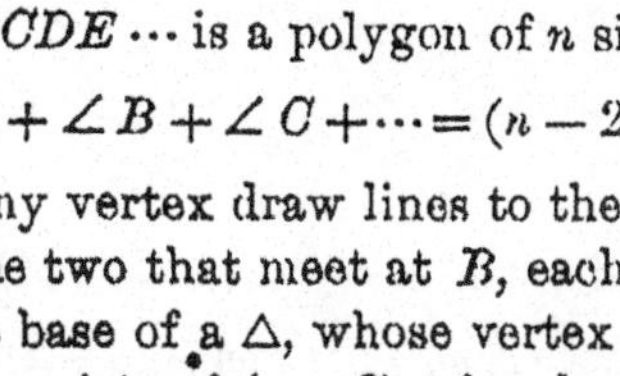

Hyp. *ABCDE* ⋯ is a polygon of n sides

To prove $\angle A + \angle B + \angle C + \cdots = (n-2)$ straight angles.

Proof. From any vertex draw lines to the other vertices.

Since, except the two that meet at B, each of the n sides of the polygon is the base of a △, whose vertex is B,

∴ the polygon consists of $(n-2)$ triangles.

∴ the sum of the ∠s of the polygon $= (n-2)$ straight angles. (93) Q.E.D.

151. Cor. Each angle of an equiangular polygon of n sides equals $\frac{n-2}{n}$ straight angles.

Ex. 215. Show that the sum of the interior angles of a hexagon equals eight right angles.

Ex. 216. How many degrees are in each angle of an equiangular quadrilateral? pentagon? hexagon? decagon?

Ex. 217. How many sides has a polygon, the sum of whose angles equals sixteen right angles? 360°? 4 straight angles? n right angles?

Proposition XLI. Theorem

152. *If the sides of any polygon be successively produced, forming one exterior angle at each vertex, the sum of these angles is equal to two straight angles.*

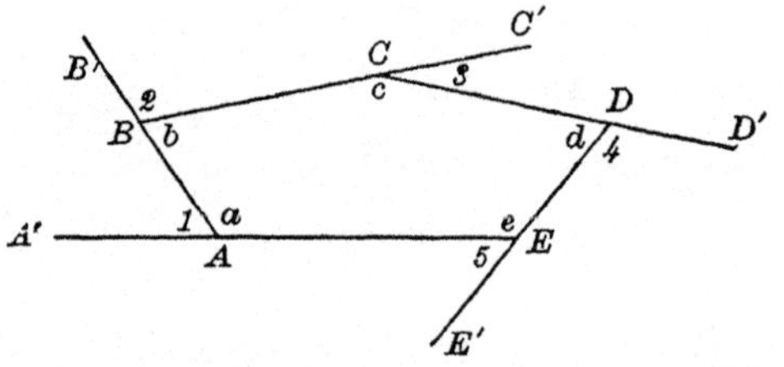

Hyp. $ABCDE \cdots$ is a polygon of n sides.

To prove

$$\angle A'AB + \angle B'BC + \angle C'CD + \cdots = \text{two straight } \angle s.$$

Proof. $(\angle a + \angle 1) + (\angle b + \angle 2) + \cdots = n$ straight $\angle s$

(con.).

But $\angle a + \angle b + \angle c + \cdots = (n-2)$ straight $\angle s$,

(sum of interior ∠s of a polygon = (n − 2) straight ∠s).

$\therefore \angle 1 + \angle 2 + \angle 3 + \cdots = (n - n + 2)$ straight $\angle s$, (Ax. 3.)

or $\angle A'AB + \angle B'BC + \angle C'CD + \cdots =$ two straight $\angle s$. Q.E.D.

153. Cor. Each exterior angle of an equiangular polygon equals $\frac{2}{n}$ straight angles.

Ex. 218. How many sides has a polygon, the sum of whose interior angles equals twice the sum of its exterior angles?

Ex. 219. How many sides has an equiangular polygon, three of whose exterior angles are together equal to 90°?

Ex. 220. How many sides has an equiangular polygon, four of whose angles are together equal to seven right angles?

Proposition XLII. Theorem

154. *The bisectors of the angles of a triangle meet in a common point, which is equidistant from the sides of the triangle.*

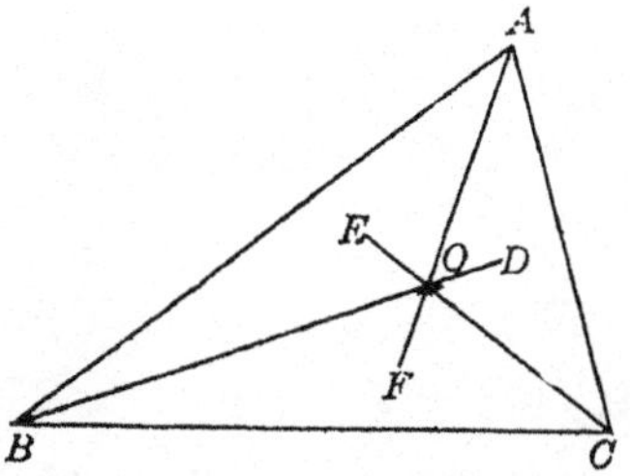

Hyp. In $\triangle ABC$:

$\angle ABD = \angle DBC$; $\angle ECB = \angle ECA$; $\angle CAF = \angle FAB$.

To prove 1°. Point O is a common point of BD, CE, and AF; and

2°. Point O is equidistant from BC, AB, and AC.

Proof. Let BD and AF meet in O.

O is equidistant from BC and AB.
Also O is equidistant from AB and AC. } Every point in the bisector of an angle is equidistant from the sides of the angle.

$\therefore$ O is equidistant from AC and BC.

$\therefore$ O is in the bisector CE,

(*point equidistant from the sides of an angle is in the bisector of the angle*).

∴ O is a common point of BD, CE, and AF; and is equidistant from BC, AB, and AC. Q.E.D.

Ex. 221. If in the diagram for Prop. XLII (1) $\angle A = 60°$; $\angle B = 50°$; (2) $\angle A = m°$; $\angle B = n°$. Find all other angles of the figure.

PROPOSITION XLIII. THEOREM

155. *The perpendiculars erected at the midpoints of the sides of a triangle meet in a common point which is equidistant from the vertices of the triangle.*

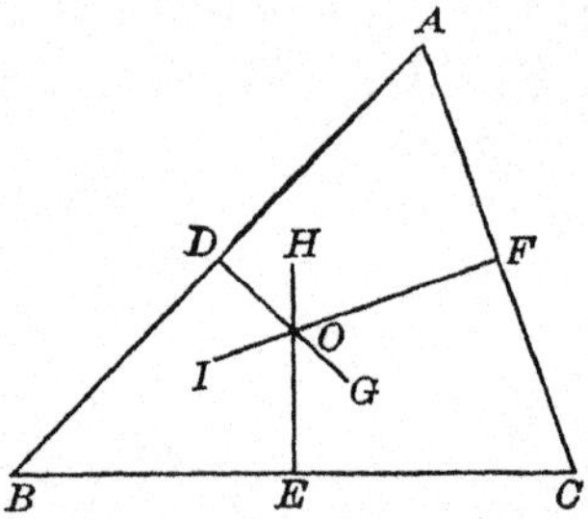

Hyp. In $\triangle ABC$,

DG, EH, and FI are perpendicular-bisectors of AB, BC, and CA, respectively.

To prove 1°. O is common to DG, EH, and FI; and

2°. O is equidistant from A, B, and C.

Proof. Let DG and EH intersect at O.

∴ O is equidistant from A and B, (119)

also O is equidistant from B and C. (119)

∴ O is equidistant from A and C.

∴ O lies in the perpendicular-bisector FI. (112)

∴ O is common to DG, EH, and FI; and is equidistant from A, B, and C. Q.E.D.

Ex. 222. Construct a circumference passing through the vertices of a given triangle.

Ex. 223. In what kind of a triangle will the point of intersection of the perpendicular-bisectors be within the triangle? without? in one side?

ANALYSIS OF THEOREMS

156. An **analysis of a theorem** is the course of reasoning by which a proof is discovered. We inquire what means we have to prove a conclusion, as illustrated in the following example.

157. THEOREM. *The bisectors of the opposite angles of a parallelogram are parallel.*

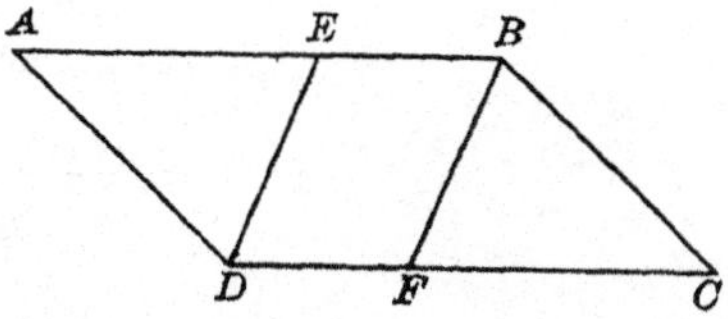

Hyp. The angles B and D of the parallelogram $ABCD$ are bisected.

To prove $ED \parallel BF$.

Analysis. 1. The means for proving that two lines are parallel are:

(*a*) A parallelogram.

(*b*) Equal alternate interior angles.

(*c*) Corresponding angles.

(*d*) Supplementary interior angles, etc.

Let us select any of these methods, *e.g.* 1 (*a*), *i.e.*

to prove $EBFD$ is a parallelogram.

2. The means for proving that a quadrilateral is a parallelogram are:

(*a*) Opposite sides are equal.

(*b*) Two opposite sides are equal and parallel, etc.

Again we select any one, *e.g.* 2 (*a*), *i.e.*

to prove $ED = BF$, $EB = DF$.

3. The means for proving the equality of lines is usually a pair of equal triangles, *i.e.*

to prove $\triangle AED = \triangle BCF.$

4. The equality of the two triangles is easily established.

Hence, **proof.** $AD = BC,$

(*opposite sides of a parallelogram*).

$\angle A = \angle C,$

(*opposite* ∠s *of a parallelogram*).

$\angle ADE = \angle FBC,$

(*halves of equal* ∠s).

$\therefore \triangle ADE = \triangle FBC,$

(*a. s. a.* = *a. s. a.*).

$\therefore ED = BF$, and $AE = FC,$

(*hom. parts of equal* △s).

But $AB = DC,$

$\therefore DF = EB.$ Ax. 3)

$\therefore DEBF$ is a parallelogram,

(*opposite sides are equal*).

$\therefore ED \parallel BF.$ Q.E.D.

The above demonstration is not the only one nor the shortest one. Each of the possibilities indicated under 1 (*a*), 1 (*b*), 1 (*c*), 1 (*d*), 2 (*a*), 2 (*b*), will furnish one or more proofs.

Ex. 224. Prove the above proposition by means of 2 (*b*).

Ex. 225. Prove the above proposition by means of 1 (*b*).

Ex. 226. Prove the above proposition by means of 1 (*c*).

Ex. 227. Prove the above proposition by means of 1 (*d*).

158. THEOREM. *If a median is intersected by another one, the segment of the median between the point of intersection and vertex is twice its other segment.*

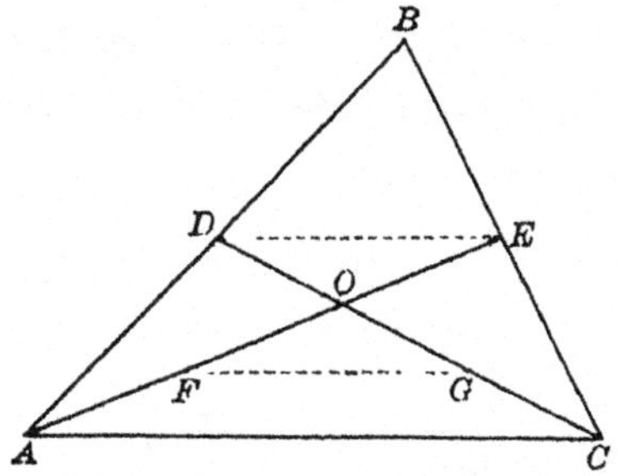

Hyp. DC and AE are medians of $\triangle ABC$.

To prove $AO = 2\,(OE)$; and $CO = 2\,(OD)$.

Analysis. 1. The means of proving that one line equals twice another is:

(a) Bisect the greater
(b) Double the smaller } (143)

Each method furnishes a proof. Select 1 (a). Bisect AO and CO in F and G, respectively, and prove $FO = OE$, $GO = DO$.

2. The means of proving the equality of lines is usually a pair of equal triangles, *i.e.*

to prove (a) $\triangle DOE = \triangle FOG$,

or (b) $\triangle DOF = \triangle OGE$.

Either pair may be used, *e.g.* 2 (a).

3. Draw DE and FG.

$DE = \frac{1}{2} AC$ and $\parallel AC$
$FG = \frac{1}{2} AC$ and $\parallel AC$ } why?

The equality of the triangles may now be easily established. Hence,

Proof. DE is parallel and equal to $\frac{1}{2}AC$. (147)

FG is parallel and equal to $\frac{1}{2}AC$. (147)

$\therefore DE$ is parallel and equal to FG. (Axs. 10 and 1)

$$\therefore \angle EDO = \angle OGF,$$

(*alternate interior* ∠s *of parallel lines*).

and $$\angle DOE = \angle FOG,$$

(*vertical* ∠s).

$$\therefore \triangle DOE = \triangle FOG,$$

(*s. a. a.* = *s. a. a.*).

$$\therefore FO = OE;\ GO = OD.$$

$$\therefore AO = 2(OE);\ OC = 2(OD).$$ Q.E.D.

Ex. 228. Demonstrate the proposition by proving the equality of $\triangle DOF$ and $\triangle EOG$. (Draw OB.)

Ex. 229. Prove the proposition by showing that $DFGE$ is a parallelogram.

Ex. 230. Prove the proposition by means of 1 (b) of analysis (three methods).

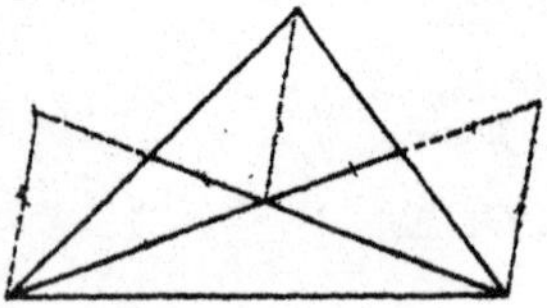

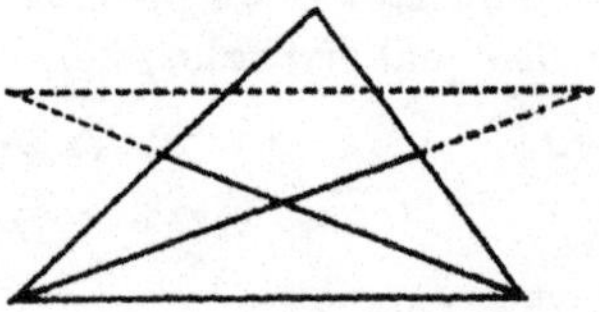

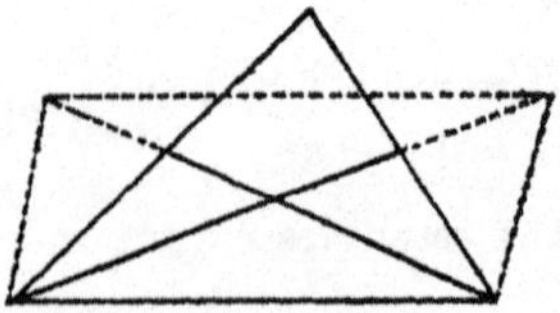

PROPOSITION XLIV. THEOREM

159. *The three altitudes of a triangle (produced if necessary) meet in a common point.*

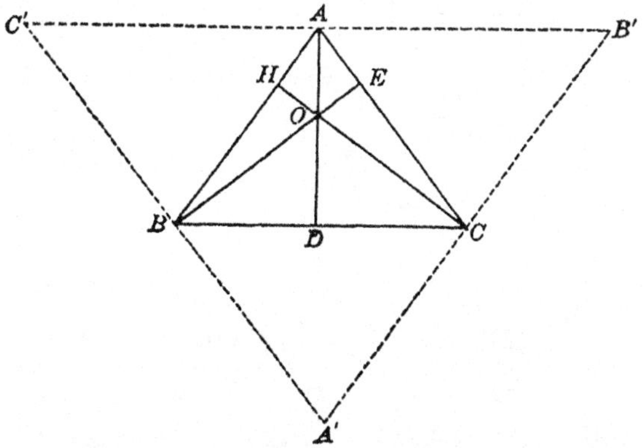

Hyp. In $\triangle ABC$

$$AD \perp BC,\ BE \perp AC,\ CH \perp AB.$$

To prove O is common to AD, BE, CH.

Proof. Draw $A'B' \parallel AB$ through C; $B'C' \parallel BC$ through A; $A'C' \parallel AC$ through B.

$AD \perp BC$ (Hyp.), $\therefore AD \perp B'C'$. (88)

But $ABCB'$ and $ACBC'$ are parallelograms. (Con.)

$\therefore AB' = BC$, and $AC' = BC$. (134)

$\therefore AB' = AC'$ (Ax. 1)

$\therefore AD$ is the perpendicular-bisector of $C'B'$.

[The rest of the proof is left to the student.]

Ex. 231. In what kind of a triangle is the point of intersection of the altitudes of a triangle within the triangle? without?

Ex. 232. In what kind of a triangle do altitudes and perpendicular-bisectors coincide?

F

PROPOSITION XLV. THEOREM

160. *The medians of a triangle meet in a common point.*

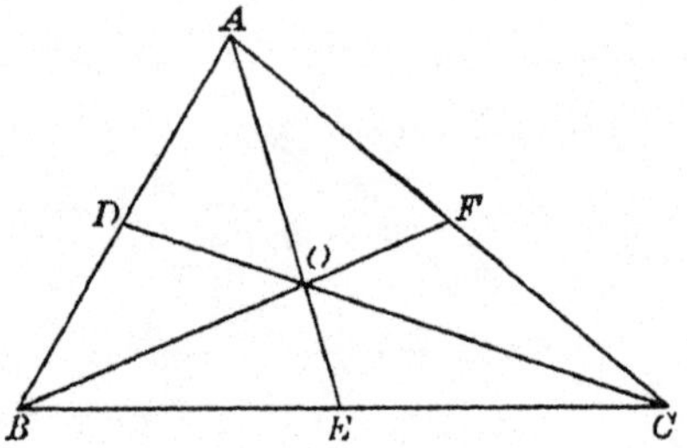

Hyp. AE, BF, CD, are the medians of $\triangle ABC$.

To prove AE, BF, CD, meet in a point.

Proof. Both BF and AE intersect CD so that

$$CO = 2\,(DO). \qquad (158)$$

$\therefore$ AE, BF, CD, meet in a common point. Q.E.D.

Ex. 233. If two medians of a triangle are equal, the triangle is isosceles.

Ex. 234. Given the vertical angle B of an isosceles triangle, required $\angle DAC$, formed by the base and the altitude upon one arm, in terms of B.

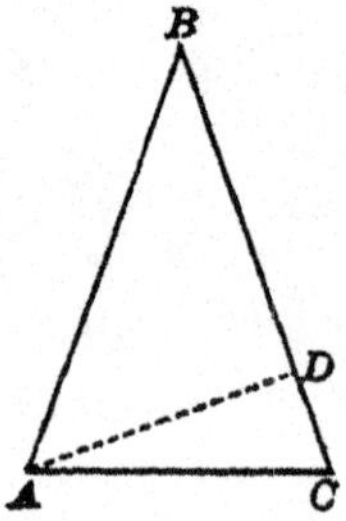

$\angle C + \angle CAB = 2$ rt. $\angle s - \angle B$, (Why?)

but $\angle C = \angle CAB$. (Why?)

$\therefore 2\angle C = 2$ rt. $\angle s - \angle B$.

$\therefore \angle C = \text{rt.}\angle - \frac{B}{2}$,

but $\angle DAC$ is the complement of $\angle C$. (Why?)

$\therefore \angle DAC = \text{rt.}\angle - \left(\text{rt.}\angle - \frac{B}{2}\right)$.

$\therefore \angle DAC = \frac{B}{2}$.

Therefore the angle formed by the base and the altitude upon an arm of an isosceles triangle equals one-half the vertical angle.

161. REMARK. — **Geometrical propositions may be demonstrated by algebraical computation.**

Although a direct geometrical proof is always preferable, the above method is very useful, especially for establishing relations between angles.

The propositions to be used are mainly the proposition of the sum of the angles of a polygon (and triangle), and the proposition of the exterior angles of a polygon (and triangle).

If the algebraical solution should be too difficult for the beginner, numerical examples ought to be solved at first.

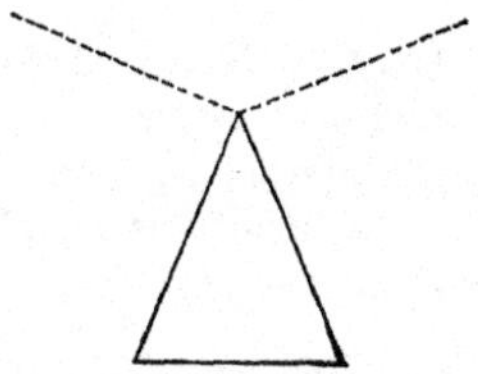

Ex. 235. Perpendiculars drawn at the vertex of an isosceles triangle upon the arms enclose an angle equal to twice the base angle.

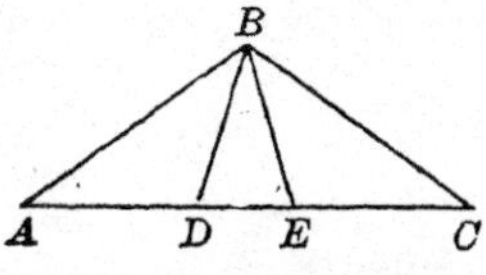

Ex. 236. If in the base AC of an isosceles $\triangle ABC$, 2 points D and E be taken such that

$$AB = AE \text{ and } BC = CD,$$

prove $$\angle DBE = \angle A.$$

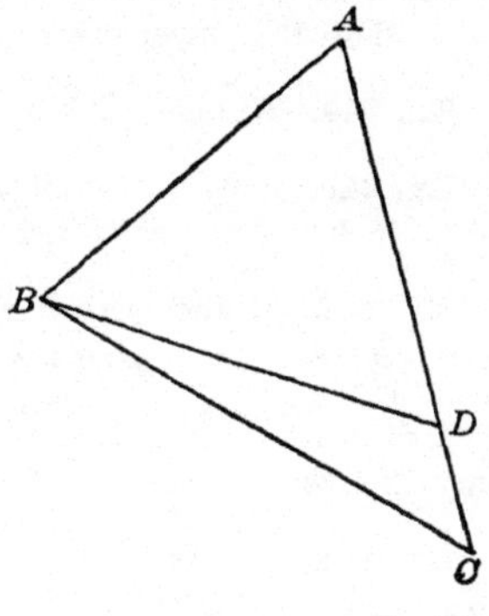

Ex. 237. Given $\triangle ABC$, and point D in AC, so that $AD = AB$.

Prove $\angle DBC = \frac{1}{2}(\angle ABC - \angle C)$.

(HINT. — Express successively all the angles of the figure in terms of ABC and C.)

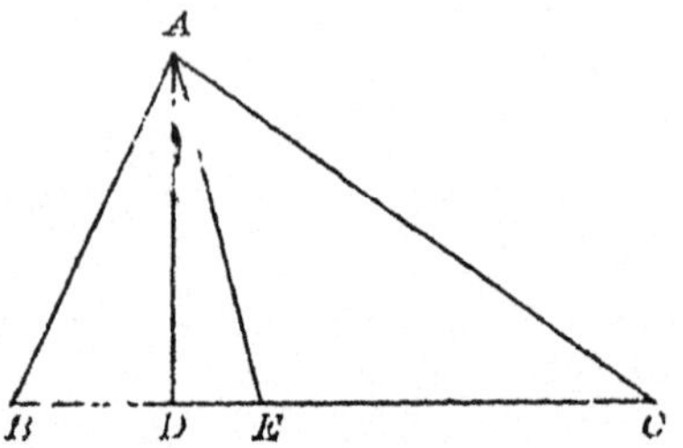

Ex. 238. If AD is the altitude and AE the bisector of the angle BAC of the triangle ABC,

prove $\angle DAE = \frac{1}{2}(\angle B - \angle C)$.

Ex. 239. The sum of three angles of a quadrilateral, diminished by the fourth exterior angle, is equal to a straight angle.

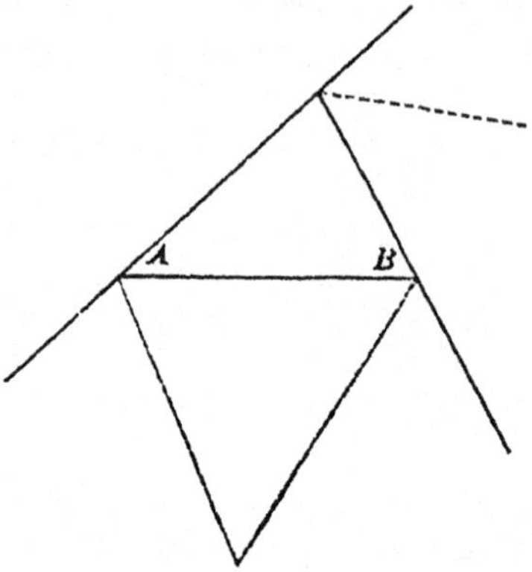

Ex. 240. The bisectors of two exterior angles of a triangle include an angle equal to one-half the third exterior angle.

(HINT. — Express every angle in terms of $\angle A$ and $\angle B$.)

Ex. 241. The bisectors of the angles of a quadrilateral form a quadrilateral the sum of whose opposite angles is equal to two right angles.

* **Ex. 242.** Prove Exs. 237 and 238 by a purely geometrical method.

MISCELLANEOUS EXERCISES

Ex. 243. If the bisectors of two adjacent angles are perpendicular to each other, the angles are supplementary.

Ex. 244. The bisectors of vertical angles are in a straight line.

Ex. 245. Perpendiculars drawn from a point within an angle, upon the sides, include an angle which is the supplement of the given angle.

Ex. 246. If the vertical angles of two isosceles triangles are supplementary, the base angles are complementary.

Ex. 247. If the base angles of two isosceles triangles are complementary, the vertical angles are supplementary.

Ex. 248. The exterior angle at the base of an isosceles triangle is equal to a right angle increased by one-half the vertical angle.

Ex. 249. If each side of an equilateral triangle be trisected, and the points of division nearest to each vertex be joined respectively, a hexagon is formed which is equiangular and equilateral.

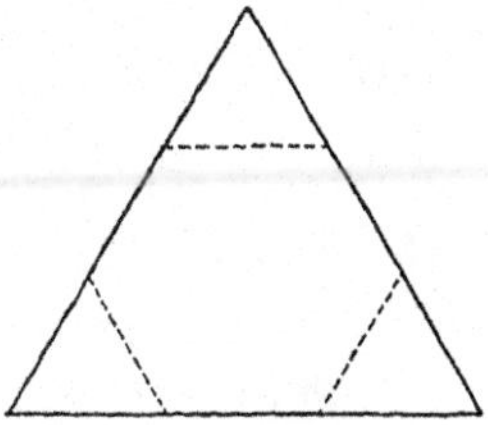

Ex. 250. Homologous medians of equal triangles are equal.

Ex. 251. Homologous altitudes of equal triangles are equal.

Ex. 252. Two isosceles triangles are equal if the vertical angle and the altitude upon an arm of the one are respectively equal to the vertical angle and the homologous altitude of the other.

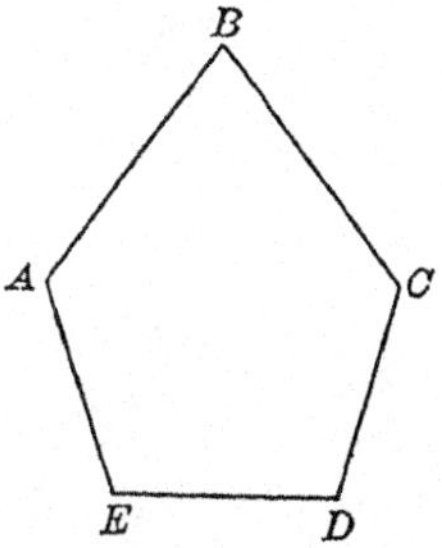

Ex. 253. If a median of a triangle is perpendicular to the base, the triangle is isosceles.

Ex. 254. If in the pentagon $ABCDE$ $AB = BC$, $AE = CD$, and $\angle A = \angle C$, then $BE = BD$, and $\angle E = \angle D$.

Ex. 255. If the opposite sides of a hexagon are parallel, and one pair of opposite sides are equal, all opposite sides are equal.

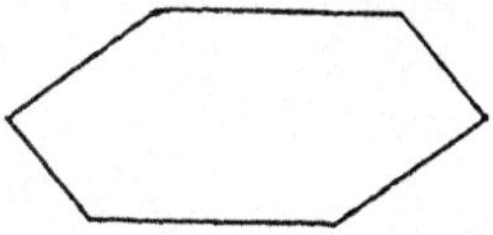

Ex. 256. If from the ends of the base, BC, of an isosceles triangle, ABC, equal parts, BD and CE, be laid off on one arm and the prolongation of the other, the line joining D and E is bisected by the base.

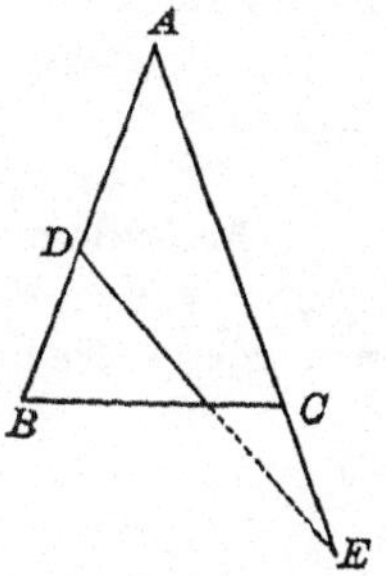

Ex. 257. If through any point, D, on the bisector of an angle, A, a parallel be drawn to one of the sides, to meet the other side in B, then $AB = BD$.

Ex. 258. Angles whose corresponding sides are perpendicular to each other, are either equal or supplementary.

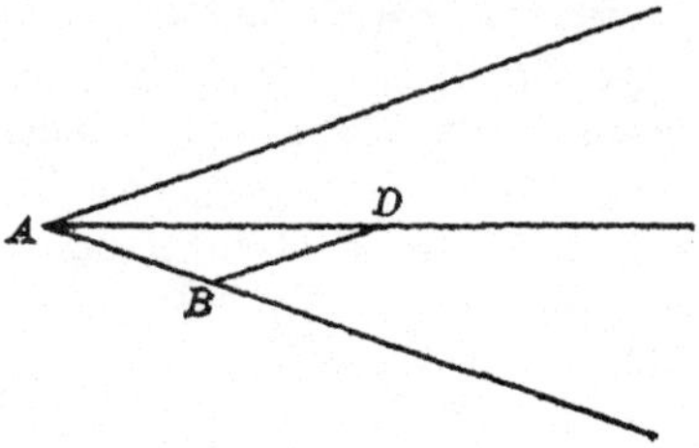

Ex. 259. Two equilateral triangles are equal if the altitude of one equals the altitude of the other.

Ex. 260. Any straight line that passes through the midpoint of one of the diagonals of a parallelogram, bisects the parallelogram.

Ex. 261. The number of all diagonals of a polygon of n sides is $\frac{n(n-3)}{2}$.

Ex. 262. If a perpendicular be dropped from the vertex to the base of a triangle, each segment of the base will be smaller than the adjacent side of the triangle.

Ex. 263. How many sides has a polygon, the sum of whose interior angles is equal to three times the sum of the angles of a hexagon?

Ex. 264. How many sides has an equiangular polygon, whose exterior angle equals the interior angle of an equilateral triangle?

Ex. 265. Prove the proposition of the sum of the interior angles of a polygon by joining any point within to the vertices of the polygon.

Ex. 266. If the vertices of a triangle lie in the sides of another triangle, the perimeter of the first is less than the perimeter of the second.

Ex. 267. The perpendiculars from two vertices of a triangle upon the median drawn from the third vertex are equal.

Ex. 268. The altitude upon the hypotenuse of a right triangle divides the figure into two triangles which are mutually equiangular.

Ex. 269. If the upper base of an isosceles trapezoid is equal to the arms, the diagonals bisect the angles at the lower base.

Ex. 270. The bisectors of the four angles of a parallelogram enclose a rectangle.

Ex. 271. The lines joining the midpoints of the sides of a rectangle, taken in order, enclose a rhombus.

Ex. 272. The lines joining the midpoints of the sides of a rhombus, taken in order, enclose a rectangle.

Ex. 273. The lines joining the midpoints of the sides of any quadrilateral, taken in order, enclose a parallelogram. (Ex. 212.)

Ex. 274. The lines joining the midpoints of opposite sides of any quadrilateral, bisect each other. (Ex. 273.)

Ex. 275. If the vertical angle of an isosceles triangle is one-half of a base angle, a bisector of a base angle divides the figure into two isosceles triangles.

***Ex. 276.** If a line from one end of the base of an isosceles triangle to the opposite side divides the figure into two isosceles triangles, then the line is a bisector of the base angle, and each base angle equals the double of the vertical angle.

***Ex. 277.** The midpoints of two opposite sides of a quadrilateral and the midpoints of the diagonals determine the vertices of a parallelogram.

***Ex. 278.** The lines joining the midpoints of the opposite sides of a quadrilateral and the line joining the midpoints of the diagonals meet in a point. (Exs. 277, 274.)

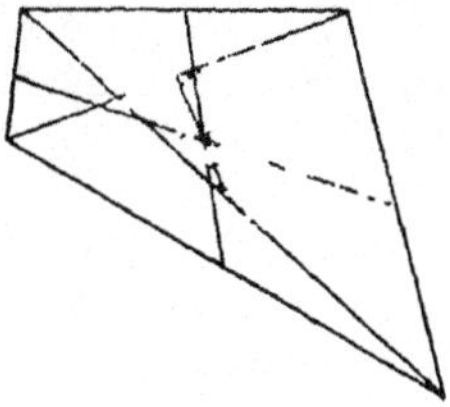

Ex. 279. The bisectors of the exterior angles of a quadrilateral form a quadrilateral, the sum of whose opposite angles is equal to one straight angle.

Ex. 280. A line from the vertex of an isosceles triangle to any point in the base is smaller than the arms.

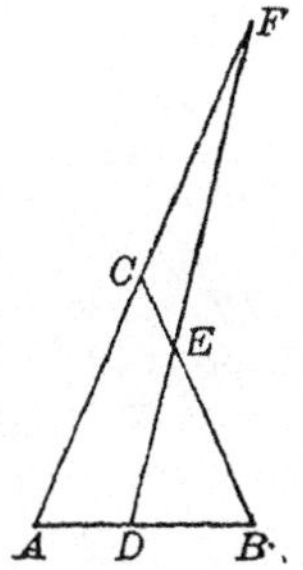

***Ex. 281.** If from the midpoint, D, of the base of an isosceles triangle, ABC, a line be drawn to meet one arm in E, the prolongation of the other in F, then

$$CE < CF.$$

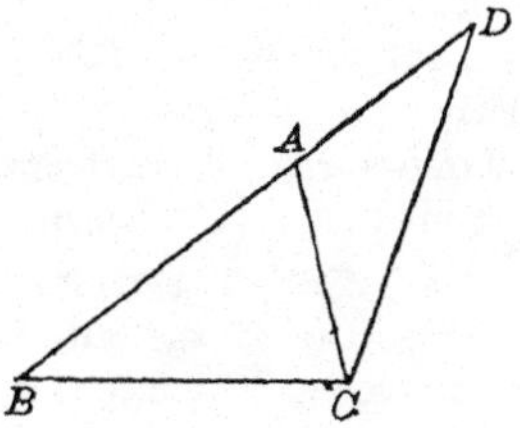

Ex. 282. If in the triangle ABC

$$AB > AC,$$

and D is a point in the prolongation of BA, then

$$DB > DC.$$

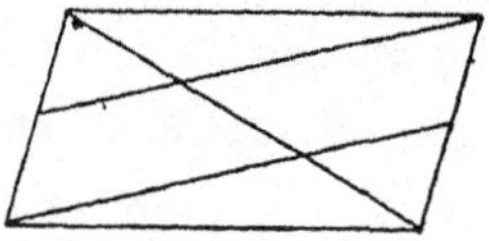

***Ex. 283.** Lines joining the midpoints of two opposite sides of a parallelogram to the ends of a diagonal trisect the other diagonal.

162. Remark. — **In order to prove that the sum of two lines, *a* and *b*, equals a third line, *c*, either**

(*a*) Construct the sum of a and b, and prove the line so obtained is equal to c, or (*b*) Lay off *a* (or *b*) on *c*, and prove that the line representing the difference equals *b* or (*a*).

Ex. 284. The sum of the perpendiculars dropped from any point in the base of an isosceles triangle to the arms, is equal to the altitude upon one of the arms.

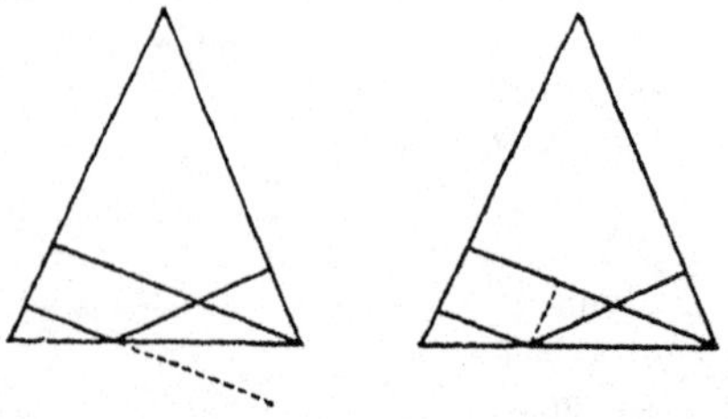

Ex. 285. The sum of the three perpendiculars dropped from any point of an equilateral triangle upon the sides is constant, and equal to the altitude of the triangle. (Ex. 284.)

Ex. 286. If the altitude *BD* of $\triangle ABC$ is intersected by another altitude in *G*, and *EH* and *HF* are perpendicular-bisectors, prove $BG = 2\,(HE)$, and $AG = 2\,(HF)$. (143.)

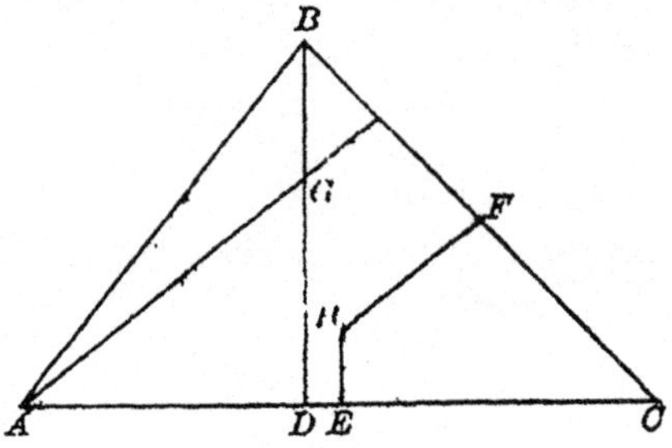

Ex. 287. The line joining the point of intersection of the altitudes of a triangle and the point of intersection of the three perpendicular-bisectors, cuts off one-third of the corresponding median. (Ex. 286.)

Ex. 288. The points of intersection of the altitudes, medians, and perpendicular bisectors of a triangle lie in a straight line.

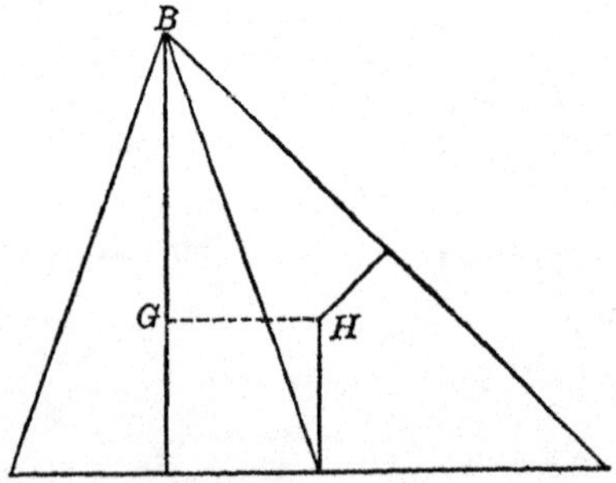

Ex. 289. The diagonals of an isosceles trapezoid are equal.

Ex. 290. If the opposite angles of a quadrilateral are equal, the figure is a parallelogram.

* **Ex. 291.** If the diagonals of a trapezoid are equal, the trapezoid is isosceles.

Ex. 292. The median drawn to any side of a triangle is less than half the sum of the other two sides.

BOOK II

THE CIRCLE. CONSTRUCTIONS

DEFINITIONS

163. A **circumference** is a curved line all points of which are equidistant from a point within called the **center**. A **circle** is a portion of a plane bounded by a circumference and is usually

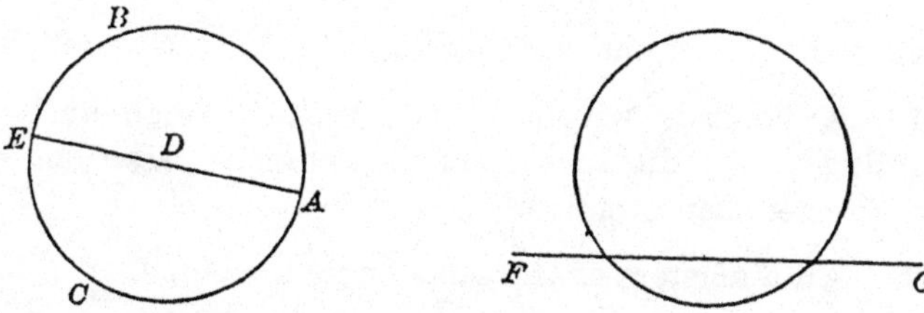

read $\odot ABC$ or $\odot D$. A **radius** is any straight line drawn from the center to the circumference, as DA. A **diameter** is a straight line passing through the center, and terminating in the circumference; as AE.

164. An **arc** is a part of the circumference. A **semicircumference** is half of the circumference. A **minor arc** is an arc less than a semicircumference; a **major arc** is an arc greater than a semicircumference.

165. A **secant** is a straight line intersecting the circumference in two points; as FG. A **tangent** is a straight line, which touches the circumference at one point only, and does not intersect it if produced; as *MN*.

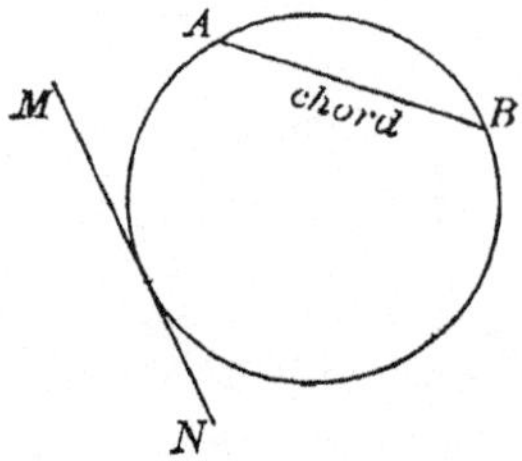

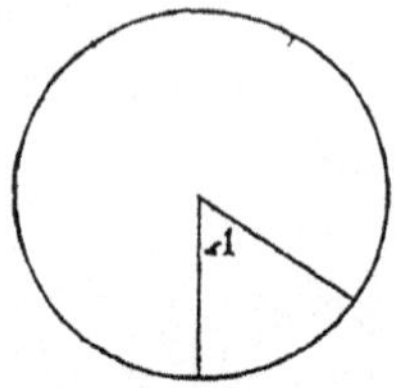

166. A **chord** is a straight line joining any two points in the circumference; as *AB*.

167. A **central angle** is an angle formed by two radii; as $\angle A$.

168. Circles having the *same center* are called **concentric**.

PRELIMINARY THEOREMS

169. All radii of the same circle are equal. (By definition.)

170. A point is within, on, or without a circumference, according as the distance from the center is less than, equal to, or greater than a radius.

171. All diameters of the same circle are equal.

172. Two circles are equal if their radii are equal. (Prove by superposition.)

173. A diameter bisects a circle and its circumference. (Prove by superposition.)

Ex. 293. The radii of two concentric circles are 6 inches and 9 inches respectively. If a point is 7 inches from the common center, does it lie within the larger circle? Within the smaller?

Ex. 294. The distance between the centers of two circles is 4, the radii are 6 and 9 respectively. Does every point of the smaller circle lie within the greater?

PROPOSITION I. THEOREM

174. *In the same circle or in equal circles, equal central angles intercept equal arcs; and, conversely, equal arcs subtend equal central angles.*

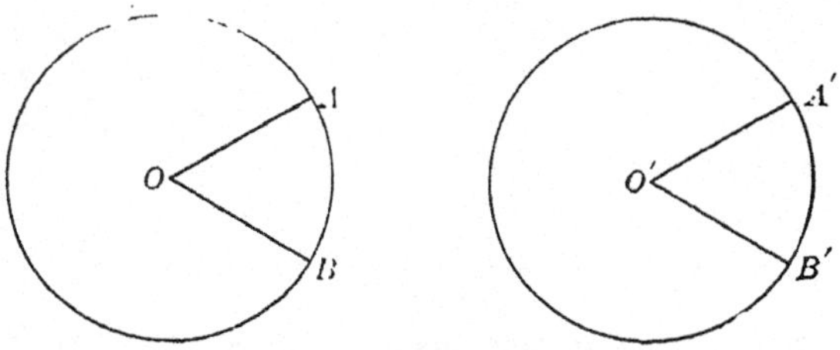

Hyp. In equal ⓢ, O and O',

$$\angle O = \angle O'.$$

To prove arc AB = arc $A'B'$.

HINT. — Prove by superposition. Compare (69).

CONVERSELY. — **Hyp.** In equal ⓢ, O and O',

$$\text{arc } AB = \text{arc } A'B'.$$

To prove $\angle O = \angle O'$.

HINT. — By superposition.

175. COR. In the same or in equal circles, the greater of two unequal central angles intercepts the greater arc, and conversely.

Ex. 295. Divide a circumference into four equal parts.

Ex. 296. Divide a circumference into eight equal parts.

Ex. 297. Divide a circumference into six equal parts.

Ex. 298. What is the means for proving the equality of arcs?

Ex. 299. If from a point A in a circumference a chord AB and a diameter AC are drawn, a radius parallel to AB bisects arc BC.

Ex. 300. If through a point equidistant from two points in the circumference a radius is drawn, the arc between the two points is bisected.

Ex. 301. If a secant is parallel to a diameter, the lines intercept equal arcs on the circumference.

Ex. 302. Any two parallel secants intercept equal arcs on a circumference. (Ex. 301.)

Ex. 303. If the perpendiculars drawn from a point in the circumference upon two radii are equal, the point bisects the arc intercepted by the two radii.

Ex. 304. If the line joining the midpoints of two radii is equal to the line joining the midpoints of two other radii, the radii intercept equal arcs respectively.

Ex. 305. If the perpendiculars from the center upon two chords are equal, the arcs subtended by the chords are equal.

176. Def. A polygon is **inscribed in a circle,** if all its vertices are in the circumference. The circle is then said to be **circumscribed about the polygon.**

Proposition II. Theorem

177. *In the same circle, or equal circles, equal arcs are subtended by equal chords; and, conversely, equal chords subtend equal arcs.*

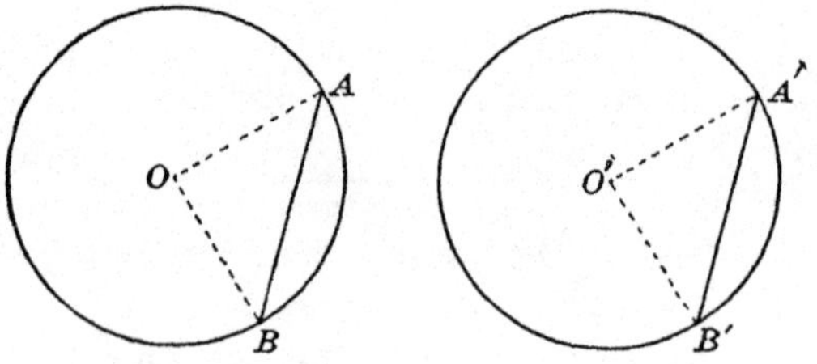

Hyp. In equal ⓢ, O and O',

$$\text{arc } AB = \text{arc } A'B'.$$

To prove $AB = A'B'$.

Proof. Draw radii, OA, OB, $O'A'$, $O'B'$.

$$\angle AOB = \angle A'O'B',$$

(*equal arcs subtend equal central* ∡ *in equal* ⓢ).

$$AO = A'O',$$

(*radii of equal* ⓢ).

$$BO = B'O',$$

(*radii of equal* ⓢ).

$$\therefore \triangle ABO = \triangle A'B'O',$$

(*s. a. s.* = *s. a. s.*).

$$\therefore \text{chord } AB = \text{chord } A'B'.$$

Q.E.D.

CONVERSELY. — **Hyp.** In equal ⓢ, O and O',

$$AB = A'B'.$$

To prove arc AB = arc $A'B'$.

HINT. — Draw radii OA, OB, $O'A'$, $O'B'$, and prove the equality of ▲ OAB and $O'A'B'$.

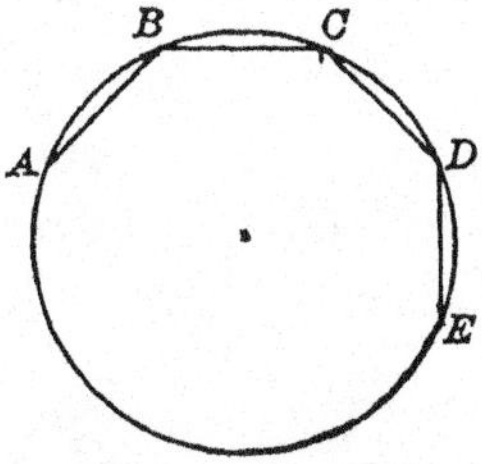

Ex. 306. If chords AB, BC, CD, DE are equal, then chords AC, BD, CE are equal.

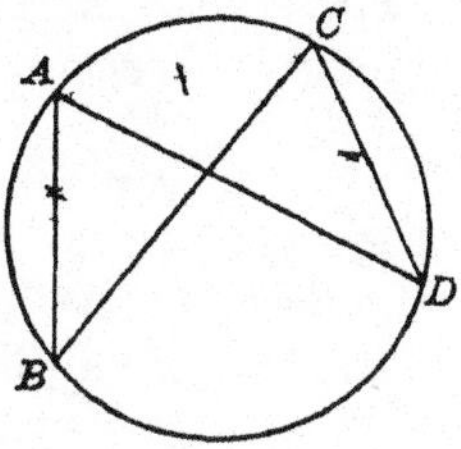

Ex. 307. If in the annexed diagram $AB = CD$, then $BC = AD$.

Ex. 308. If in the same diagram two intersecting chords, AD and BC, are equal, then $AB = CD$.

Ex. 309. The diagonals of an equilateral pentagon inscribed in a circle are equal.

Ex. 310. The radii drawn to the vertices of an inscribed equilateral hexagon divide the figure into six equilateral triangles.

Ex. 311. If two chords bisect each other, the arcs intercepted by the sides of a pair of vertical angles are equal.

Ex. 312. If two chords bisect each other, they are diameters.

PROPOSITION III. THEOREM

178. *A radius perpendicular to a chord bisects the chord and the subtended arc.*

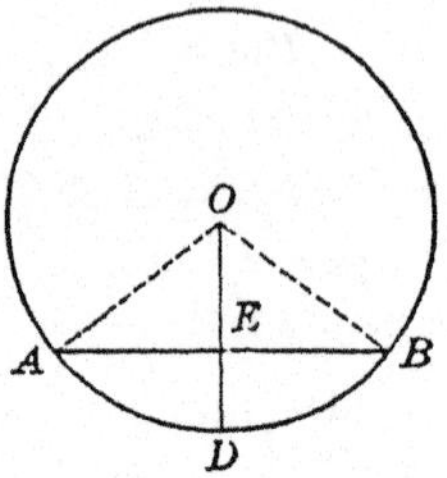

Hyp. In ⊙ O,

the radius $OD \perp AB$.

To prove $AE = EB$.

arc AD = arc DB.

HINT. — Draw radii AO and OB.

What is the usual means of proving:

(1) The equality of lines?

(2) The equality of arcs?

179. COR. A perpendicular bisector of a chord passes through the center of the circle.

Ex. 313. If two chords are equal, the perpendiculars from the center upon the chords are equal.

Ex. 314. If the perpendiculars from the center upon the sides of an inscribed polygon are equal, the polygon is equilateral.

Ex. 315. If from a point without a circle two equal lines are drawn to a circumference, the bisector of the angle they form passes through the center of the circle.

Ex. 316. Two points, each equidistant from the ends of a chord, determine a line passing through the center of the circle.

PROPOSITION IV. PROBLEM

180. *To circumscribe a circle about a given triangle.*

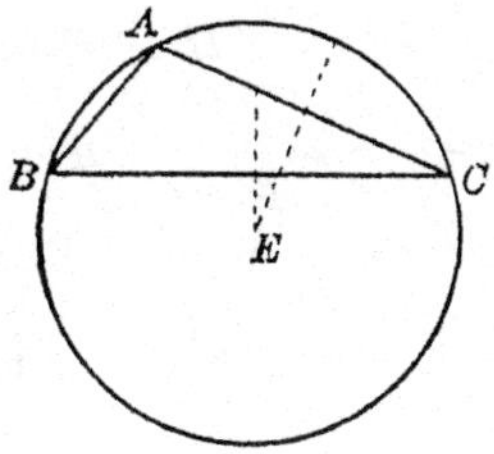

Given. $\triangle ABC$.

Required. To circumscribe a circle about $\triangle ABC$.

Construction. Erect perpendicular-bisectors upon BC and AC.

They will intersect at some point E. (Why?)

From E as a center, with a radius equal to EA, describe a ⊙. It will pass through A, B, and C. (Why?)

∴ ABC is the required ⊙. (Why?) Q.E.F.

181. COR. 1. **Three points not in a straight line determine a circumference.**

182. COR. 2. **A circumference cannot be drawn through three points which lie in the same straight line.**

183. COR. 3. **A straight line cannot intersect a circumference in more than two points.**

184. COR. 4. **Two circumferences cannot meet in more than two points.**

Ex. 317. To find the center of a given circle.

Ex. 318. To find the midpoint of a given arc.

PROPOSITION V THEOREM

185. *In the same circle, or in equal circles, equal chords are equally distant from the center; and, conversely, chords equally distant from the center are equal.*

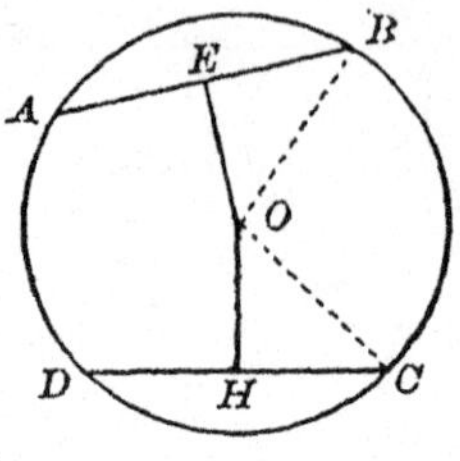

Hyp. — In $\odot ABCD$:

$$\text{chord } AB = \text{chord } CD, \quad OE \perp AB, \quad OH \perp CD.$$

To prove $OE = OH$.

HINT. — What is the means of proving the equality of lines?

CONVERSELY. **Hyp.** In $\odot ABCD$:

$$OE = OH, \quad OE \perp AB, \quad OH \perp CD.$$

To prove $AB = CD$.

The proof is similar to the above.

186. REMARK. — **The equality of two chords is usually established by means of equal distances from the center or equal subtended arcs.**

Ex. 319. If from any point in the circumference two chords are drawn making equal angles with the radius to the point, these chords are equal.

Ex. 320. If through any point in a radius two chords are drawn making equal angles with the radius, these chords are equal.

Ex. 321. A line joining the point of intersection of two equal chords to the center bisects the angle formed by the chords.

Ex. 322. In a given circle, to draw a chord equal and parallel to a given chord.

PROPOSITION VI. THEOREM

187. *In the same circle, or in equal circles, the greater of two minor arcs is subtended by a greater chord; and, conversely, the greater chord subtends the greater arc.*

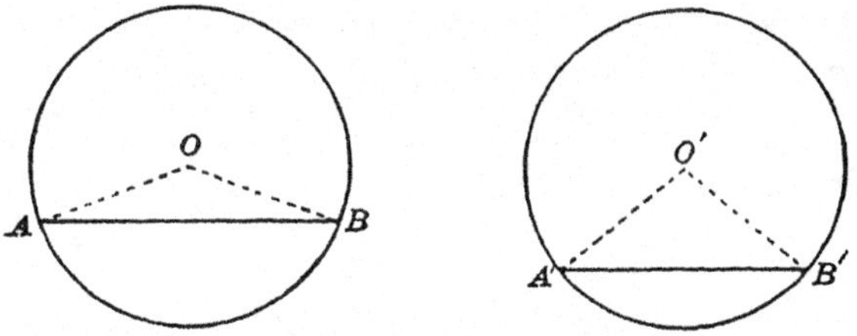

Hyp. In equal ⑤, O and O',

arc $AB >$ arc $A'B'$.

To prove $AB > A'B'$.

Proof. Draw radii OA, OB, $O'A'$, $O'B'$.

In ⚠ AOB and $A'O'B'$,

$\angle O > \angle O'$, (175)

but $AO = A'O'$ and $BO = B'O'$,

(Radii of equal ⑤).

$\therefore AB > A'B'$. (128) Q.E.D.

CONVERSELY.—**Hyp.** In equal ⑤, O and O',

$AB > A'B'$.

To prove arc $AB >$ arc $A'B'$.

Proof. Draw radii OA, OB, $O'A'$, $O'B'$.

In ⚠ AOB and $A'O'B'$,

$AO = A'O'$, $BO = B'O'$,

$AB > A'B'$. (Hyp.)

$\therefore \angle AOB > \angle A'O'B'$. (129)

$\therefore$ arc $AB >$ arc $A'B'$. (175) Q.E.D.

Ex. 323. In a given circle, to draw a chord equal and perpendicular to a given chord.

Ex. 324. In a given circle, to draw a chord equal to a given chord, and parallel to a given line.

Ex. 325. In a given circle, to draw a chord equal to one-half a given chord, and perpendicular to a given line.

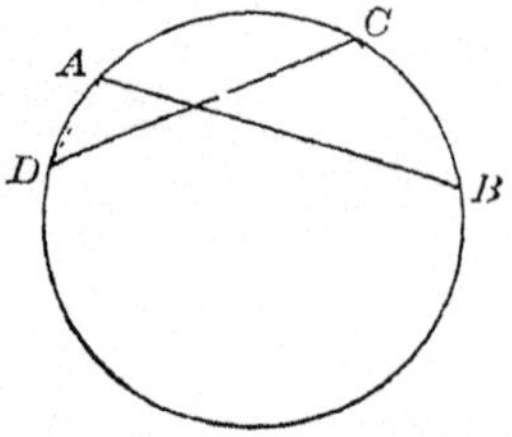

Ex. 326. In the annexed diagram, if $AB > CD$, and arcs ACB and DAC are minor arcs, prove that $CB > AD$.

Ex. 327. In the same diagram, if $CB > AD$, and arcs DAC and ACB are minor arcs, prove that $AB > CD$.

PROPOSITION VII. THEOREM

188. *In the same circle, or in equal circles, chords unequally distant from the center are unequal, the nearer one being the greater; and, conversely.*

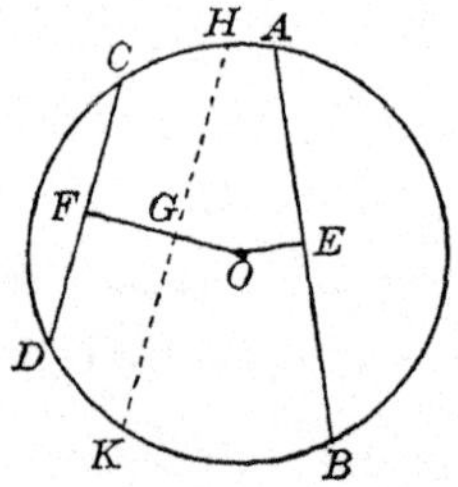

Hyp. In ⊙ $ABDC$, $OE \perp$ chord AB, $OF \perp$ chord CD, and

$$OF > OE.$$

To prove $AB > CD$.

Proof. On OF lay off $OG = OE$.

Through G, draw chord $HK \perp OF$.

Then $HK = AB$. (Why?)

Since $OG < OF$, G lies between O and F.

$\therefore$ KH lies between O and CD.

$\therefore$ arc $HK >$ arc CD. (Ax. 9.)

$\therefore$ $HK > CD$. (187)

$\therefore$ $AB > CD$. Q.E.D.

CONVERSELY. — **Hyp.** $AB > CD$.

To prove $OF > OE$.

Prove by the indirect method.

189. COR. The diameter is greater than any other chord.

190. REMARK. — **The inequality of chords is usually established by means of unequal distances from the center or by means of unequal arcs.**

NOTE. — The following table will be found convenient.

Prop. I, II, V, VI, VII, and several others, may be represented by the following schedule:

In the same circle or equal circles:

When **central angles** are	equal,	greater,	smaller,
minor arcs are	equal,	greater,	smaller,
chords are	equal,	greater,	smaller,
distances of chords from center are	equal,	smaller,	greater,
segments are	equal,	greater,	smaller,
sectors are	equal,	greater,	smaller.

If, *e.g.* chords are equal, then central angles, minor arcs, etc., are equal; similarly for unequal parts.

[DEF. **A segment of a circle** is a portion of a circle bounded by a chord and its arc.

A sector of a circle is a portion of a circle bounded by two radii and their intercepted arc.]

Ex. 328. The side of an equilateral hexagon inscribed in a circle is nearer to the center than the side of an equilateral heptagon inscribed in the same circle, and more remote from the center than the side of an equilateral pentagon inscribed in the same circle.

Ex. 329. The shortest chord which can be drawn through a point within a circle is perpendicular to the radius drawn through the point. Which is the longest?

Ex. 330. Two chords drawn from a point in the circumference are unequal if they make unequal angles with the radius drawn from that point. Which of the chords is the greater?

*** Ex. 331.** Two chords drawn through an interior point are unequal if they make unequal angles with the radius drawn through that point. Which is the greater one?

Proposition VIII. Theorem

191. *A straight line perpendicular to a radius at its extremity is a tangent to the circle.*

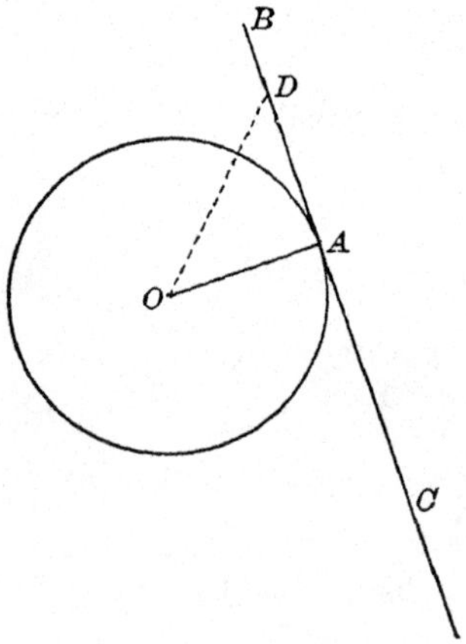

Hyp. In ⊙ O,

radius $OA \perp BC$ at A.

To prove BC is a tangent.

Proof. From O draw any line OD to BC.

$OA \perp BC$. (Hyp.)

$\therefore$ OD is oblique to BC. (98)

$\therefore$ $OD > OA$. (126)

$\therefore$ D lies without the circumference. (170)

$\therefore$ BC is a tangent to $\odot$ O. (165) Q.E.D.

192. Cor. 1. A tangent is perpendicular to the radius drawn to the point of contact.

193. Cor. 2. A perpendicular to a tangent at the point of contact passes through the center of the circle.

194. Cor. 3. A perpendicular from the center to a tangent meets it at the point of contact.

195. Cor. 4. At a given point of contact there can be one tangent only.

Proposition IX. Theorem

196. *The tangents drawn to a circle from a point without are equal.*

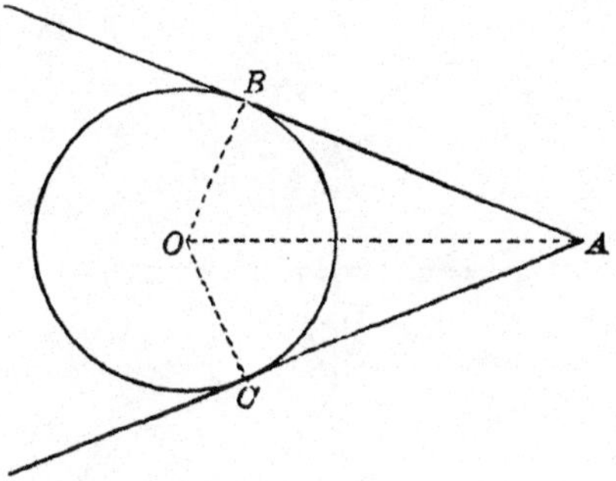

Hyp. In $\odot$ O,

AB and AC are tangent.

To prove $AB = AC$.

Hint. — What is the means of proving the equality of lines?

197. Def. A common tangent of two circles is called an **interior tangent** when it lies between the two circles; otherwise, it is called an **exterior tangent.** The **length** of a common tangent is the length of the segment between the **points of contact.**

198. Def. A polygon is **circumscribed about a circle** if all its sides are tangent to the circle. The circle is then said to be inscribed in the polygon.

Ex. 332. The common internal tangents of two circles are equal.

Ex. 333. The common external tangents of two circles are equal.

Ex. 334. A chord forms equal angles with the tangents drawn at its ends.

Ex. 335. The sum of two opposite sides of a circumscribed quadrilateral is equal to the sum of the other two opposite sides.

Ex. 336. Find a similar proposition for the circumscribed hexagon.

Ex. 337. If two tangents make an angle of 60°, the chord joining the points of contact equals the tangents.

Proposition X. Theorem

199. *If two circumferences intersect, a straight line joining their centers bisects their common chord at right angles.*

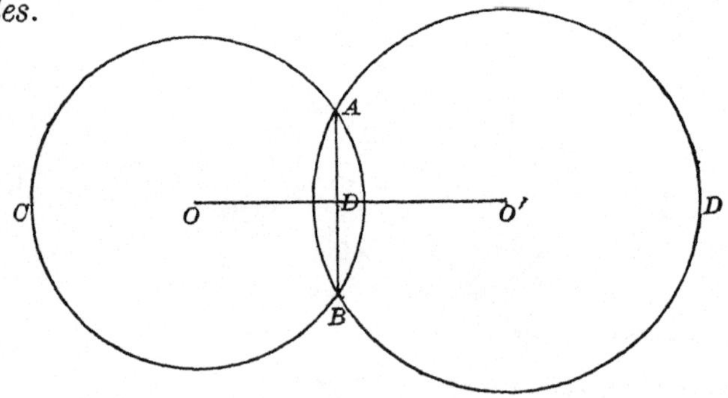

Hyp. Circumferences ACB and ADB intersect at A and B.

To prove OO', joining their centers, is the perpendicular-bisector of AB.

Proof. O and O' are each equally distant from A and B. (Why?)

∴ OO' is the perpendicular-bisector of AB. (111) Q.E.D.

200. Def. The **line of centers** is the line joining the centers of two circles.

201. Def. Two circles are **tangent to each other** if both are tangent to a straight line at the same point. They are tangent internally or externally, according as one circle lies within or without the other.

Proposition XI. Theorem

202. *If two circles are tangent to each other, their line of centers passes through the point of contact.*

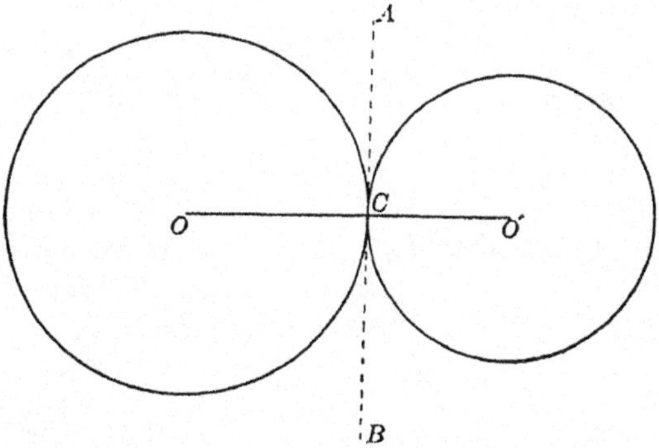

Hyp. OO' is the line of centers of circles, O and O', tangent at C.

To prove OO' passes through C.

Proof. At the point of contact, C, draw a perpendicular to the common tangent.

This perpendicular passes through O and O'. (193)

∴ OO' passes through C. Q.E.D.

Ex. 338. What are the relative positions of two circles, if the line of centers is

(*a*) Greater than the sum of the radii?
(*b*) Equal to the sum of the radii?
(*c*) Smaller than the sum but greater than the difference of the radii?
(*d*) Equal to the difference of the radii?
(*e*) Smaller than the difference of the radii?
(*f*) Equal to zero?

EXERCISES

Ex. 339. If a secant intersects two concentric circles, its segments, intercepted by the two circumferences, are equal.

Ex. 340. Two parallel chords, drawn through the extremities of a diameter, are equal.

Ex. 341. In the annexed diagram, if the radius OB is equal to AB, prove $\angle COD = 3\angle A$.

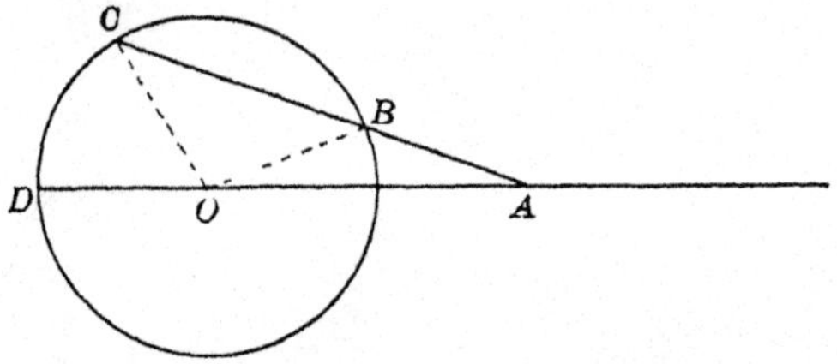

Ex. 342. If from A the tangents AB and AC are drawn to a given circle O, and a third tangent intersects AB and AC in D and F,

Prove (1) $AD + DF + AF = 2\,AB.$

(2) $\angle O = \text{rt.}\angle - \frac{A}{2}.$

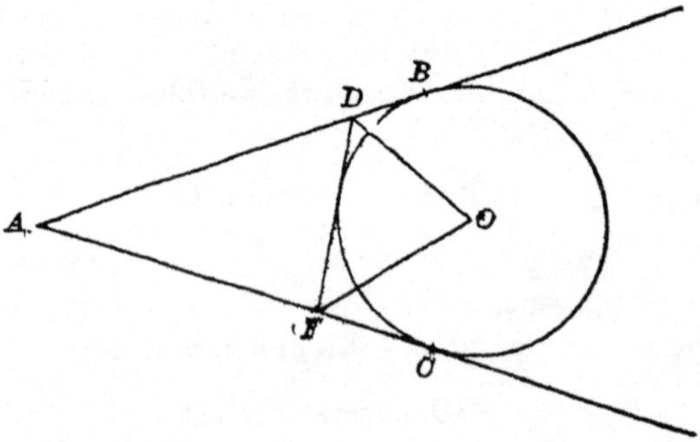

***Ex.** 343. The two common exterior tangents of two circles intercept on a common interior tangent a segment, *CD*, equal to the exterior tangent, *AB*.

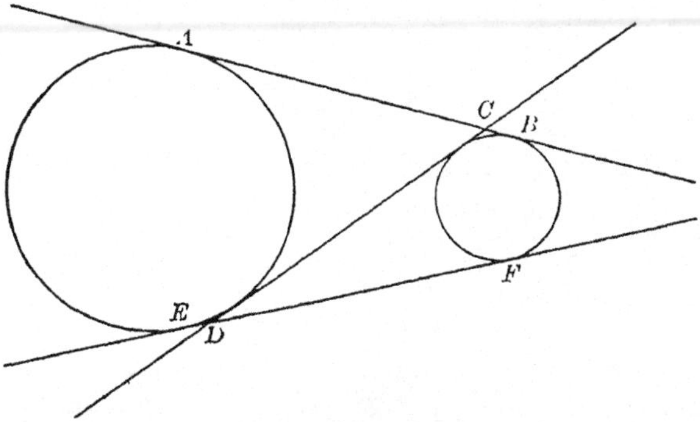

MEASUREMENT

203. A **ratio** of two quantities of the same kind is the quotient obtained by dividing the first quantity by the second. Thus, the ratio of two quantities, a and b, is $\frac{a}{b}$ or $a \div b$; the ratio of four yards and two yards is $\frac{4}{2}$ or 2. A ratio is used to compare the magnitude of two quantities.

204. To **measure** a quantity is to find how many times it contains another quantity of the same kind, called a **unit of measurement**. Thus a line is measured by finding that it contains a certain number of yards or inches.

205. A number which expresses how many times a quantity contains a unit is called the **numerical measure** of the quantity.

206. Two quantities are **commensurable** when there is a third quantity, called the *common measure*, which is contained an integral number of times in each.

207. Two quantities are **incommensurable** when there is no common measure. The ratio of such quantities is called **an incommensurable ratio.**

208. Such a ratio cannot be found exactly in figures, but an approximate value can be found that differs from the true value by less than any assigned value, however small. If AB and CD are two lines whose ratio is the square root of two,

then $$\frac{AB}{CD} = \sqrt{2} = 1.41421 + \cdot$$

Thus the true value lies between 1.41421 and 1.41422, and differs from either of the approximate values by less than 0.00001. It is evident that by continuing the decimal this difference may be diminished to one millionth or one billionth or any assigned value, however small.

If a quantity, A, be divided into m equal parts, and another quantity, B, is found to contain n of these parts, with a remainder less than one of the parts,

then $$\frac{B}{A} > \frac{n}{m} \text{ but } < \frac{n+1}{m}.$$

The error in taking one of these approximate values as a ratio is less than $\frac{1}{m}$.

As m may be made indefinitely large, the fraction may be made indefinitely small, or the value of $\frac{B}{A}$ may be found with any assigned degree of accuracy.

LIMITS

209. A **constant** is a quantity that maintains the same value throughout the same discussion.

210. A **variable** is a quantity whose value changes during the same discussion.

211. A **limit** of a variable is a constant which the variable does approach indefinitely near, but which it can never reach.

212. For example, suppose a point P to move from A towards B in such a way as to move in the first second over half of AB

to C, in the second second, over half of the remainder, CB, to D, in the third second over half of the new remainder, DB, to E, and so on indefinitely.

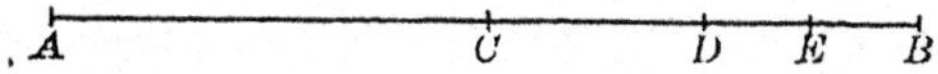

It is evident that the moving point P can never reach B, while it will approach nearer to B than any quantity which we can assign.

The distance from A to the moving point P is a variable whose value can be made to differ from AB by less than any assigned quantity, while it never can be made equal to AB.

AB is therefore the **limit of the variable.**

213. Theorem. *If two variables are always equal and each approaches a limit, the two limits are equal.*

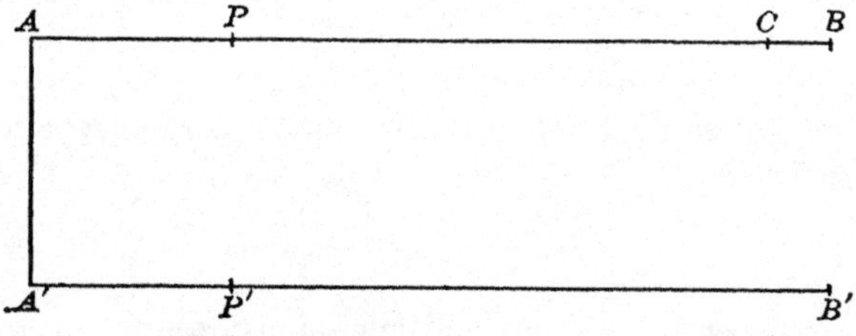

Hyp. The equal variables AP and $A'P'$, having the limits AB and $A'B'$ respectively.

To prove $AB = A'B'$.

Proof. Suppose AB is greater than $A'B'$, and on AB lay off $AC = A'B'$.

Then the variable AP attains values greater than AC, while the variable $A'P'$ is always less than AC, or the two variables are unequal, which is contrary to the hypothesis.

Whence AB cannot be greater than $A'B'$, and in like manner it can be shown that $A'B'$ cannot be greater than AB.

Therefore $$AB = A'B'.$$

Ex. 344. What is the limit of the circulating decimal .999... ?

PROPOSITION XII. THEOREM

214. *In the same circle, or in equal circles, two central angles have the same ratio as their intercepted arcs.*

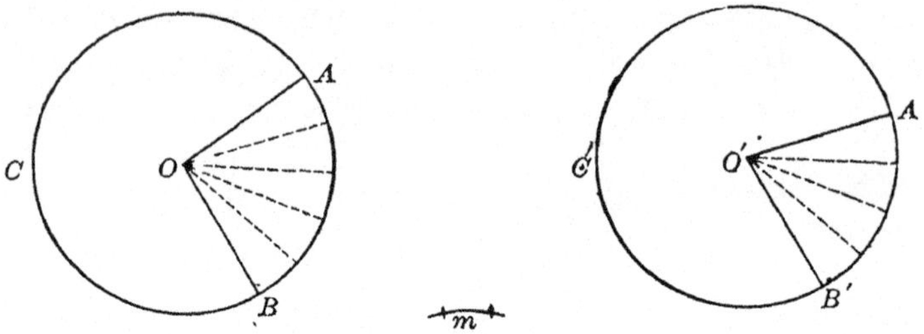

Hyp. In the equal circles ABC and $A'B'C'$, two central angles AOB and $A'O'B'$ intercept the arcs AB and $A'B'$ respectively.

To prove $$\frac{\angle AOB}{\angle A'O'B'} = \frac{AB}{A'B'}.$$

Proof. CASE I. The arcs are commensurable.

Let m be a common measure contained in AB five times and $A'B'$ four times.

Then $$\frac{AB}{A'B'} = \frac{5}{4}.$$

Connect the points of division with the center. Then $\angle AOB$ will have been divided into five parts and $\angle A'O'B'$ into four, all being equal. (174)

Whence $$\frac{\angle AOB}{\angle A'O'B'} = \frac{5}{4}.$$

$$\therefore \frac{\angle AOB}{\angle A'O'B'} = \frac{AB}{A'B'}.$$ (Ax. 1.) Q.E.D.

CASE II. The arcs are incommensurable.

Divide AB into any number of equal parts, and apply one of those parts to $A'B'$ as often as possible.

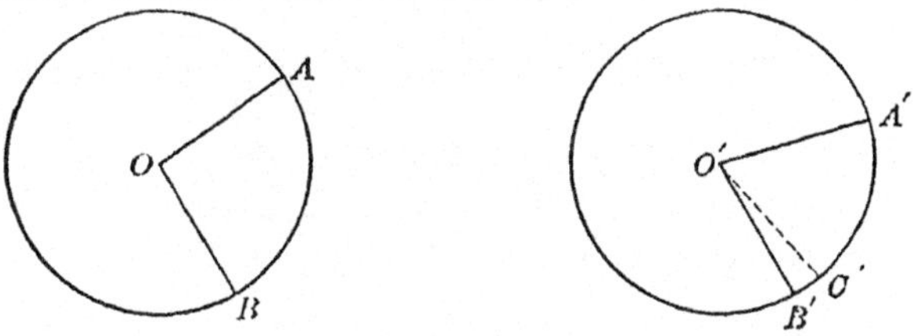

Since AB and $A'B'$ are incommensurable, there must be a remainder $C''B'$ less than one of the equal parts. Draw $O'C''$.

Since the arcs AB and $A'C''$ are commensurable,

$$\frac{A'C''}{AB} = \frac{\angle A'O'C''}{\angle AOB}.$$

By increasing the number of parts into which AB is divided, we can diminish the length of each part, and, therefore, the length of $C''B'$ indefinitely.

Hence the arc $A'C''$ approaches $A'B'$ as a limit, and $\angle A'O'C''$ approaches $\angle A'O'B'$ as a limit.

Therefore $\dfrac{A'C''}{AB}$ approaches $\dfrac{A'B'}{AB}$ as a limit, and $\dfrac{\angle A'O'C''}{\angle AOB}$ approaches $\dfrac{\angle A'O'B'}{\angle AOB}$ as a limit.

The variables $\dfrac{A'C''}{AB}$ and $\dfrac{\angle A'O'C''}{\angle AOB}$ being always equal, must have equal limits.

Whence $$\frac{A'B'}{AB} = \frac{\angle A'O'B'}{\angle AOB}.$$ Q.E.D.

215. SCHOLIUM. The circumference is divided into 360 equal parts called **degrees,** and therefore a central angle of 1° will intercept an arc of 1°.

The numerical measure of any central angle is equal to the numerical measure of the intercepted arc, or more briefly:

216. A central angle is measured by the intercepted arc.

DEFINITIONS

217. An **inscribed angle** of a circle is one whose vertex lies in the circumference and whose sides are chords.

218. A **segment** of a circle is a portion of a circle bounded by an arc and its chord.

219. An *angle* is said to be **inscribed in a segment** if its vertex lies in the arc and its sides pass through the extremities of that arc.

Ex. 345. How many degrees are there in a central angle which intercepts $\frac{1}{4}$ of the circumference? $\frac{1}{3}$ of a semicircumference?

Proposition XIII. Theorem

220. *An inscribed angle is measured by one-half the intercepted arc.*

Hyp. Angle ACB is inscribed in circle O.

To prove $\angle ACB$ is measured by $\frac{1}{2}$ arc AB.

Proof. Case I. One side of the angle is a diameter of the circle.

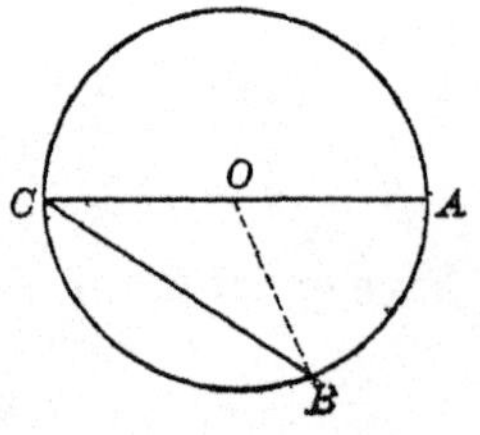

Draw OB.

Then $BO = OC$, (Why?)

and $\angle C = \angle B$, (Why?)

$\angle AOB = \angle C + \angle B.$

$\therefore \angle AOB = 2 \angle C.$

$\therefore \angle ACB = \frac{1}{2} \angle AOB.$

But $\angle AOB$ is measured by arc AB.

$\therefore \angle ACB$ is measured by $\frac{1}{2}$ arc AB. Q.E.D.

CASE II. The center lies within the angle.

Draw the diameter CD.

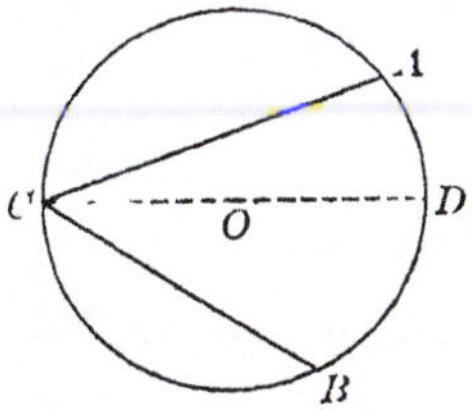

Then $\angle ACD$ is measured by $\frac{1}{2}$ arc AD,
and $\angle BCD$ is measured by $\frac{1}{2}$ arc BD.
(Case I.)

$\therefore \angle ACD + \angle BCD$ is measured by $\frac{1}{2}$ arc $(AD + BD)$,

or $\angle ACB$ is measured by $\frac{1}{2}$ arc AB. Q.E.D.

CASE III. The center lies without the angle.

Draw the diameter CD.

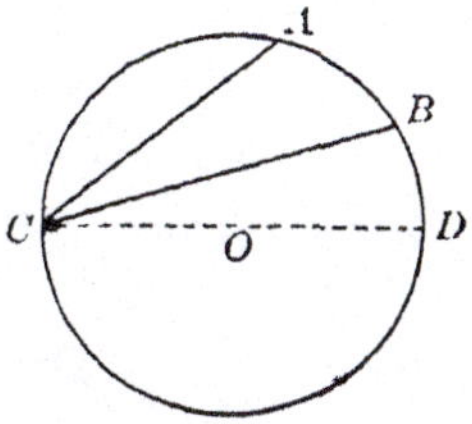

Then $\angle ACD$ is measured by $\frac{1}{2}$ arc AD,
and $\angle BCD$ is measured by $\frac{1}{2}$ arc BD.
(Case I.)

$\therefore \angle ACD - \angle BCD$ is measured by $\frac{1}{2}$ arc $(AD - BD)$,

or $\angle ACB$ is measured by $\frac{1}{2}$ arc AB. Q.E.D.

221. COR. I. Angles inscribed in the same segment, or in equal segments, are equal.

222. COR. 2. An angle inscribed in a semicircle is a right angle.

Ex. 346. If in the diagram for Case I, $\angle C = 30°$, how many degrees are in arc CB?

Ex. 347. If in the same diagram arc $BC = 3$ arc AB, find $\angle C$.

Ex. 348. If in the diagram for Case II arc AC is $\frac{1}{3}$, and arc BC is $\frac{1}{4}$ of the circumference, find $\angle ACB$, $\angle ACD$.

Ex. 349. If in the diagram for Case III, A is the midpoint of arc CD, and B is the midpoint of arc AD, how many degrees are there in $\angle ACB$?

Ex. 350. If a quadrilateral $ABCD$ be inscribed in a circle, and the two diagonals be drawn, find all the angles in the figure, if arc $AB = 80°$, arc $BC = 110°$, and arc $CD = 90°$.

Ex. 351. In the diagram of the preceding exercise, find four pairs of equal angles.

Ex. 352. The opposite angles of an inscribed quadrilateral are supplementary.

Ex. 353. If through one of the points of intersection of two equal circles a line be drawn to meet the circumferences, the extremities of that line are equidistant from the other point of intersection.

PROPOSITION XIV. THEOREM

223. *An angle formed by two chords intersecting within the circle is measured by one-half the sum of the intercepted arcs.*

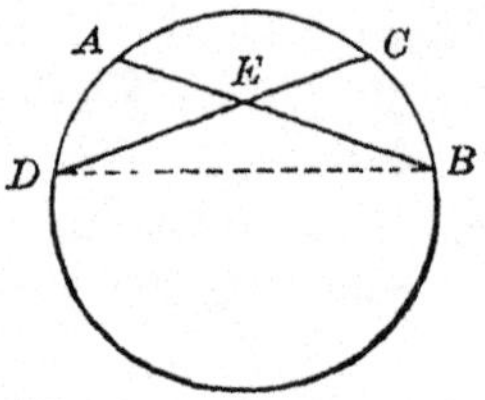

Hyp. Two chords AB and CD intersect in E.

To prove $\angle AED$ is measured by

$\frac{1}{2}$ (arc AD + arc CB).

Proof. Draw DB, and apply Prop. XIII.

PROPOSITION XV. THEOREM

224. *An angle formed by a tangent and a chord drawn from the point of contact is measured by half the intercepted arc.*

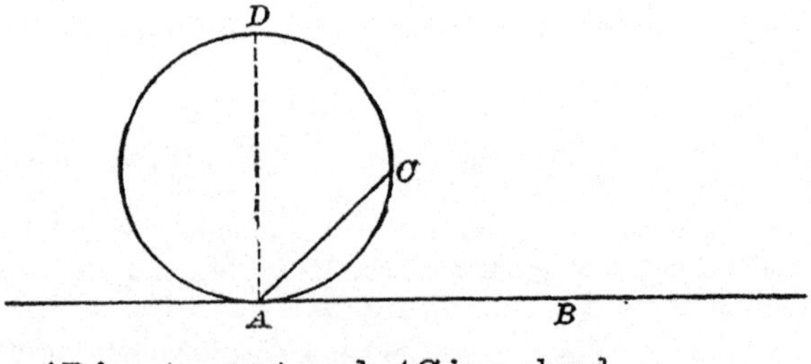

Hyp. AB is a tangent, and AC is a chord.

To prove $\angle CAB$ is measured by $\frac{1}{2}$ arc AC.

Proof. Draw the diameter AD.

The rt. $\angle DAB$ is measured by $\frac{1}{2}$ arc DCA.

$\angle DAC$ is measured by $\frac{1}{2}$ arc DC. (Prop. XIII.)

$\therefore \angle CAB$ is measured by $\frac{1}{2}$ arc CA. (Ax. 3.) Q.E.D.

Ex. 354. If in diagram for Prop. XIV arc $AD = 60°$, arc $BA = 140°$, and arc $CB = 20°$, find $\angle AEC$ and $\angle DAE$.

Ex. 355. In the same diagram, prove that $\angle ADE$ equals $\angle EBC$.

Ex. 356. If two perpendicular chords intersect within the circle, the sum of a pair of opposite intercepted arcs is equal to a semicircumference.

Ex. 357. Prove Prop. XV by demonstrating the equality of $\angle ADC$ and $\angle CAB$.

Ex. 358. If at the vertex of an inscribed square a tangent be drawn, what angle is formed by the tangent and adjacent side?

Ex. 359. If, in Prop. XV, arc $AC = 2$ arc CD, find $\angle CAB$.

Ex. 360. A chord is parallel to a tangent drawn through the midpoint of the subtended arc.

Proposition XVI. Theorem

225. *An angle formed by two secants, or two tangents, or a tangent and a secant, intersecting without a circle, is measured by half the difference between the intercepted arcs.*

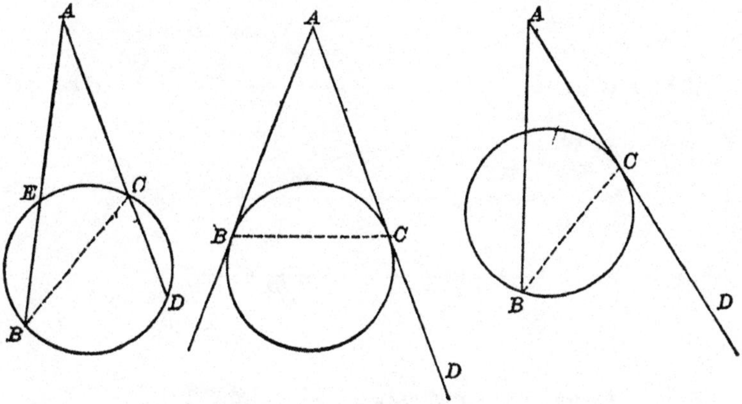

Hyp. AB and AD are secants drawn from an external point A.

To prove $\angle A$ is measured by $\frac{1}{2}$ (arc BD − arc EC).

Hint. $\angle A = \angle BCD - \angle CBA$.

Ex. 361. If the angle formed by two tangents is 60°, how many degrees are in each of the intercepted arcs?

Ex. 362. If in the diagram for Case I, arc $BD = 100°$ and $\angle A = 20°$, find arc EC.

Ex. 363. If in the same diagram arc $EC = 60°$ and arc $EB =$ arc $BD =$ arc CD, find $\angle A$.

Ex. 364. If an angle formed by a secant and a tangent is 20° and the greater of the intercepted arcs is 90°, how many degrees are in the other intercepted arc?

226. Scholium. If we consider an arc which intersects the sides of an angle as negative when it turns its convex side toward the vertex, and positive when it turns the concave side toward the vertex, or passes through the vertex, Props. XIII, XIV, XV, and XVI may be stated as follows:

227. Theorem. *If the sides of an angle (indefinitely produced) intersect or touch a circumference, the angle is measured by one-half the algebraic sum of the intercepted arcs.*

Proposition XVII. Theorem

228. *Parallels intercept equal arcs on a circumference.*

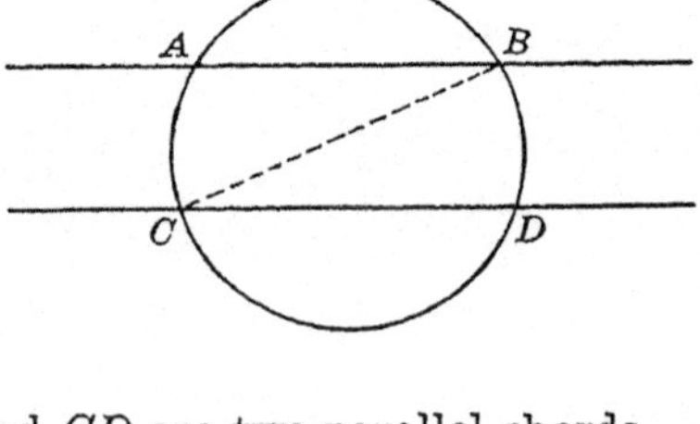

Hyp. AB and CD are two parallel chords.

To prove arc $AC =$ arc BD.

Draw BC and prove by means of Prop. XIII.

Ex. 365. Prove the above theorem if both lines are tangents.

Ex. 366. Prove the above theorem if one line is a tangent and the other a chord.

Ex. 367. State and prove the converse of the same proposition.

Ex. 368. In the diagram for Prop. XVII, find $\angle ABC$ if arc $AB = 80°$ and arc $CD = 120°$.

Ex. 369. A line bisecting an angle formed by a tangent and a chord bisects the intercepted arc.

Ex. 370. Prove that Prop. XVII is a limiting case of (225).

PROPOSITION XVIII THEOREM

229. *The opposite angles of an inscribed quadrilateral are supplementary.*

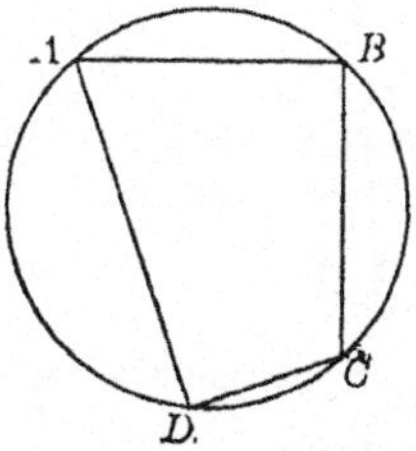

Hyp. $ABCD$ is an inscribed quadrilateral.

To prove $\angle A + \angle C = 2$ rt. $\angle$s.

HINT: — Find the arcs by which $\angle A$ and $\angle C$ are measured

---

Ex. 371. In the diagram for Prop. XVIII, prove that an exterior angle at C is equal to $\angle A$.

* **Ex. 372.** If the opposite angles of a quadrilateral are supplementary, its vertices are concyclic, *i.e.* a circumference can be described through them. HINT. — Apply the indirect method.

Ex. 373. Find the sum of three alternate angles of an inscribed hexagon.

Ex. 374. The corresponding segments of two equal intersecting chords are equal.

***Ex. 375.** If, through the points of intersection of two circumferences, parallels be drawn terminated by the circumferences, these parallels are equal.

Ex. 376. If the bisector of an inscribed angle be produced until it meets the circumference, and through this point of intersection a chord be drawn parallel to one side of the angle, it is equal to the other side.

Ex. 377. If in the greater of two concentric circles, chords be drawn touching the smaller circle, the chords are equal.

Ex. 378. If two equal chords intersect, the lines joining their ends form an isosceles trapezoid.

***Ex. 379.** If from the extremities of a diameter perpendiculars be drawn upon any chord (produced, if necessary), the feet of the perpendicular are equidistant from the center.

***Ex. 380.** If two unequal chords be produced to meet, the secants thus formed are unequal.

Ex. 381. Which is the shortest line that can be drawn from a point within to a given circumference? Which is the longest?

***Ex. 382.** Each angle formed by joining the feet of the three altitudes of a triangle is bisected by the corresponding altitude. (Ex. 372.)

***Ex. 383.** If from any point in the circumference of a circle perpendiculars be dropped upon the sides of an inscribed triangle (produced, if necessary), the feet of the perpendiculars are in a straight line. (Ex. 372.)

***Ex. 384.** The tangents drawn at the vertices of an inscribed rectangle enclose a rhombus.

***Ex. 385.** If from any point in the circumference of a circle, chords be drawn to the vertices of an inscribed equilateral triangle, the longest chord equals the sum of the smaller chords. (162)

***Ex. 386.** If the opposite sides of an inscribed quadrilateral be produced to meet in A and F, the bisectors of the angles A and F meet at right angles.

HINT. — Prove

$\angle BGK = \angle CHK$. (100)

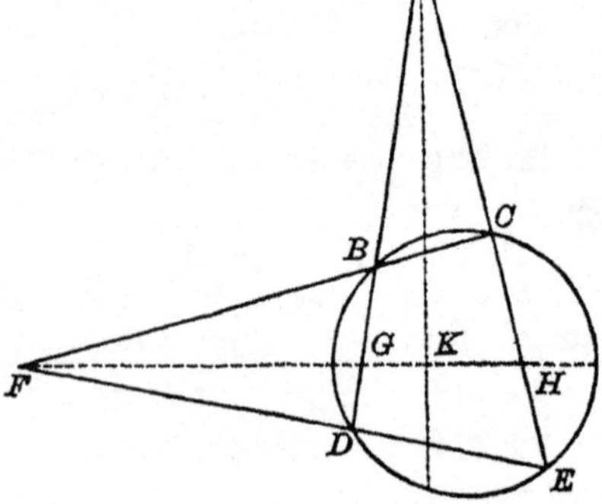

CONSTRUCTIONS

230. Note. — In the following examples, we shall denote the given parts of a triangle always in the same manner, the sides by a, b, c, the opposite angles by A, B, and C, the altitudes by h_a, h_b, and h_c, the medians by m_a, m_b, and m_c, and the bisectors by t_a, t_b, and t_c.

Proposition XIX. Problem

231. *To construct a triangle, when three sides are given.*

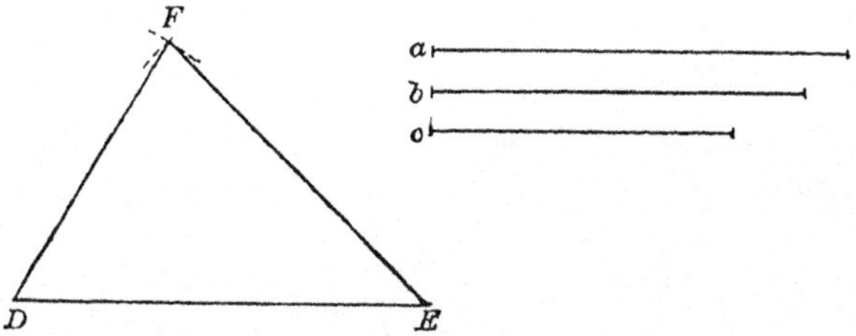

Given. Lines a, b, c.

Required. A triangle having sides equal to a, b, c.

Construction. Draw $DE = a$.

From E as a center, with a radius equal to b, draw an arc.
From D as a center, with a radius equal to c, draw an arc.

The arcs intersect at F. Join FE and FD.

$\triangle DEF$ is the required triangle. Q.E.F.

Discussion. The construction is impossible if one side is greater than the sum of the other two.

Ex. 387. Construct an equilateral triangle, having one side given.

Ex. 388. Construct an equilateral triangle, having given the perimeter.

Ex. 389. To find the third angle of a triangle, when two angles are given.

PROPOSITION XX. PROBLEM

232. *To construct a triangle, when one side and two adjacent angles are given.*

The solution is left to the student.

Ex. 390. Upon a given base, to construct an isosceles right triangle.

Ex. 391. Construct an isosceles triangle, having given the base and a base angle.

Ex. 392. To construct a quadrilateral, having given the four sides and one angle.

Ex. 393. Is it possible to solve the preceding exercise by constructing first a side not adjacent to the given angle?

233. REMARK. — **The possibility of a solution of a problem depends often upon the proper choice of the part which is drawn first.**

PROPOSITION XXI. PROBLEM

234. *To construct a triangle, when two sides and an included angle are given.*

The solution is left to the student.

Ex. 394. Construct an isosceles triangle, having given an arm and the vertical angle.

Ex. 395. Construct a right triangle, having given the two arms.

PROPOSITION XXII. PROBLEM

235. *To construct a triangle, when one side, one adjacent, and one opposite angle are given.*

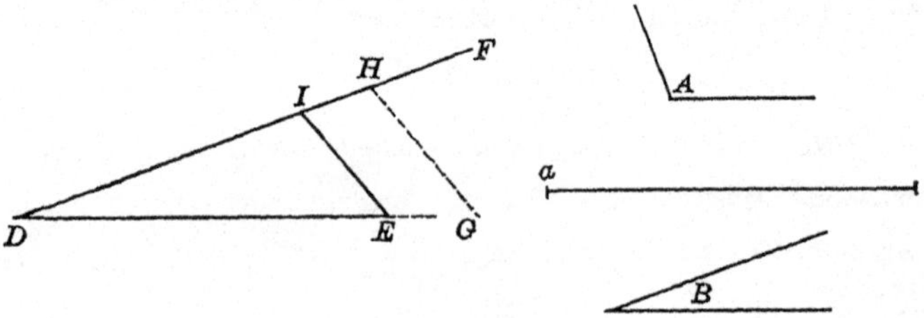

Given. Line a, angles A and B.

Required. A $\triangle$ having $\angle B$ adjacent to, and $\angle A$ opposite to, a side $= a$.

Construction. Draw DE equal to a.

At D draw $\angle EDF$ equal to $\angle B$.

At any point, H, in DF, construct $\angle DHG = \angle A$.

Through E, draw a line parallel to HG, meeting DH in I.

$\triangle DIE$ is the required $\triangle$.

Proof. $DE = a$. (Con.)

$\angle D = \angle B$. (Con.)

$\angle DIE = \angle DHG$,

(*corr.* $\angle$s *of* $\parallel$ *lines*).

$\angle A = \angle DHG$. (Con.)

$\therefore \angle DIE = \angle A$. (Ax. 1.)

Discussion. The construction is impossible if the sum of the given angles is greater than or equal to a straight angle.

Ex. 396. Construct by means of Prop. XXII a right triangle, having given the hypotenuse and an acute angle.

Ex. 397. Find a construction of the same problem which does not depend upon Prop. XXII.

Ex. 398. Construct a right triangle, having given an arm and the opposite angle.

Ex. 399. Construct an equilateral triangle, having given

(*a*) the altitude.

(*b*) the radius of the inscribed circle.

(*c*) the radius of the circumscribed circle.

PROPOSITION XXIII. PROBLEM

236. *To construct a triangle, when two sides and an angle opposite one of them are given.*

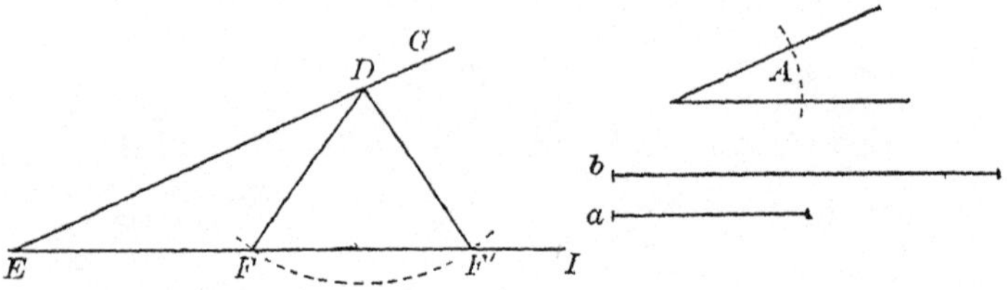

Given. Lines a and b, and $\angle A$.

Required. A triangle, two sides of which are equal to a and b, a being opposite an angle equal to A.

Construction. Draw $\angle GEI = \angle A$.

On EG, lay off $ED = b$.

From D as a center with a radius equal to a, draw an arc intersecting EI in F and F'.

Both $\triangle EDF$ and EDF' fulfil the required conditions.

DISCUSSION. If the arc intersects the base twice, there will be two solutions, and if it touches the line, but one. If it does not touch the line, a solution is impossible.

Ex. 400. In the Prop. XXIII, how many solutions are possible, when angle A is obtuse? right? acute?

Ex. 401. Construct a right triangle, having given the hypotenuse and one arm.

Ex. 402. Is it possible to solve the preceding problem by drawing the hypotenuse first?

237. A triangle may be constructed if the following parts are given:

(1) Three sides.
(2) Two sides and the included angle.
(3) Two angles and the included side.

(4) Two angles and a non-included side.

(5) Two sides and the angle opposite one of them.

To construct a triangle, three independent parts must be given.

Ex. 403. Are the three angles of a triangle three independent parts, and can a triangle be constructed when the three angles are given?

Proposition XXIV. Problem

238. *From a given point, to draw a tangent to a given circle.*

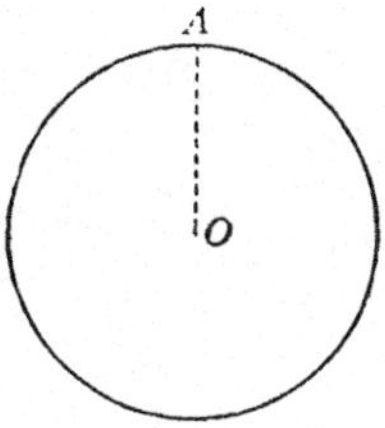

I. When the given point, *A*, is in the circumference.

Hint. — What is the angle formed by a radius and a tangent at its extremity?

II. When the given point, *A*, is without the circle.

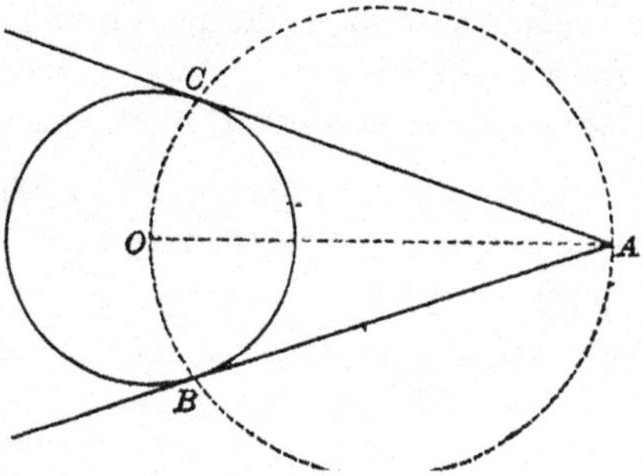

Construction. Join *A*, and *O* the center of the given circle.

On *OA* as a diameter, construct a circumference, intersecting the given circumference in *B* and *C*.

Then *AC* and *AB* are the required tangents.

HINT. — Show that ∡ *ACO* and *OBA* are right angles.

Ex. 404. Construct a line tangent to a given circle and parallel to a given line.

Ex. 405. Construct a line tangent to a given circle and perpendicular to a given line.

PROPOSITION XXV. PROBLEM

239. *To inscribe a circle in a given triangle.*

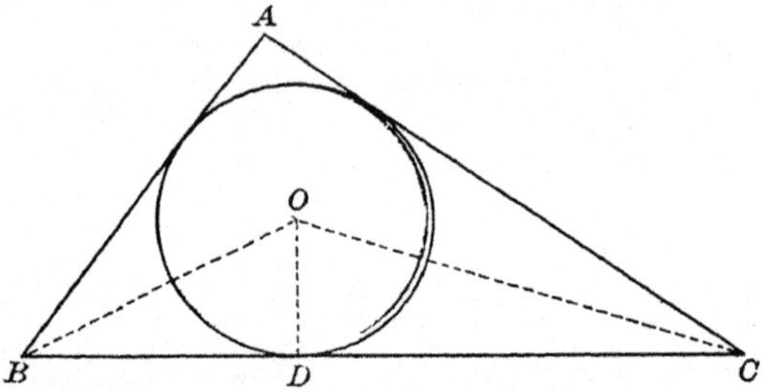

Given. △ *ABC*.

Required. To inscribe a circle in △ *ABC*.

Construction. Bisect the ∡ *B* and *C*.

From *O*, the intersection of the bisectors, draw *OD* ⊥ *BC*.

From *O* as a center, with a radius equal to *OD*, draw a circle, which is the required one.

[The proof is left to the student.]

240. DEF. A circle touching one side of a triangle and the prolongations of the other two sides is an **escribed circle**.

Ex. 406. Construct the three escribed circles of a triangle.

Ex. 407. The bisector of an angle of a triangle meets the circumference of the circumscribed circle in a point which is equidistant from the other two vertices of the triangle and the center of the inscribed circle.

PROPOSITION XXVI. PROBLEM

241. *Upon a given straight line, to construct a segment of a circle which shall contain a given angle.*

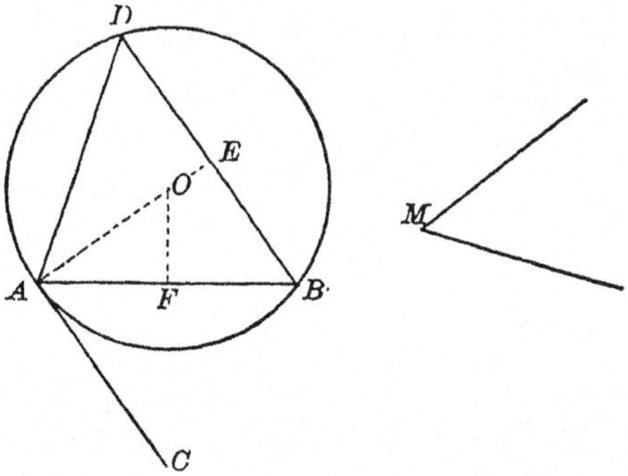

Given. Line AB and $\angle M$.

Required. To construct a segment of a circle on AB which shall contain $\angle M$.

Construction. At A, draw $\angle BAC = \angle M$.

At A, draw $AE \perp$ to AC.

Draw the perpendicular-bisector of AB, intersecting AE at O.

From O as a center, with a radius equal to OA, draw a circle ADB.

ADB is the required segment.

Proof. AC is a tangent to the circumference. (191)

$\therefore \angle CAB$ is measured by $\frac{1}{2}$ arc AB. (224)

But $\angle D$ is measured by $\frac{1}{2}$ arc AB; (220)

$\therefore \angle D = \angle CAB = \angle M$. Q.E.F.

ANALYSIS OF PROBLEMS

242. An **analysis** of a problem is a course of reasoning by which a construction is discovered. Although no rules can be given which apply to all constructions, the method explained in the following exercises may be used in many problems.

Ex. 408. *To construct a triangle having given one side, the corresponding median, and the altitude to another side.*

Given. Three lines, b, h_a, m_b.

Required. A triangle having one side equal to b, the corresponding median equal to m_b, and the altitude upon another side equal to h_a.

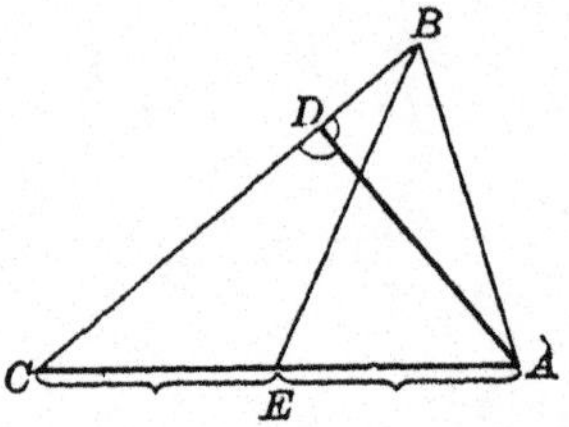

Analysis. (1) Suppose ABC is the required triangle.

(2) Then we know $CA (= b)$, CE and $EA \left(= \frac{b}{2}\right)$, $AD (= h_a)$, $BE (= m_b)$, and ∠s ADC and ADB (= rt. ∠s).

(The student is advised to mark the known parts as indicated in the diagram, or to draw them in a different color from the other lines.)

(3) Examine all triangles in the figure, and see if one can be constructed. The rt. △ ADC can be constructed, having given two sides.

(4) Make this triangle the basis of the construction.

Hence,

Construction. Draw $DA = h_a$.

At D, draw $FH \perp$ to AD.

From A as a center, with a radius equal to b, draw an arc intersecting DH in C. Draw AC.

Bisect AC in E, and from E as a center, with a radius equal to m_b, describe an arc, meeting FH in B.

ABC is the required triangle.

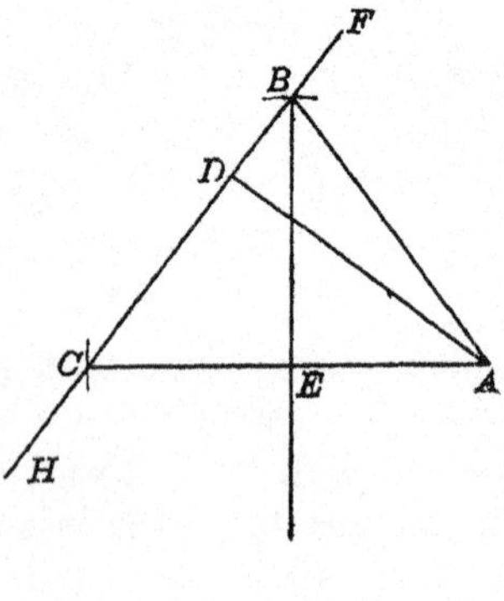

Proof. $AD = h_a$.

$BE = m_b$.

$AC = b$.

AD is an altitude as $\angle ADC =$ rt. $\angle$.

$CE = EA$.

$\therefore$ EB is a median.

243. The following rules express the procedure in a general form:

(1) Construct a diagram as if the construction were completed.

(2) Find all known lines and angles of the figure (and mark them).

(3) Try to construct some part of the figure, **usually a triangle.**

(4) Make this triangle (or other figure) the basis of the construction.

(4 *a*) If it is impossible to find such a triangle, draw additional lines and proceed as before.

Ex. 409. *To construct a triangle, having given the median and altitude to one side and another median.*

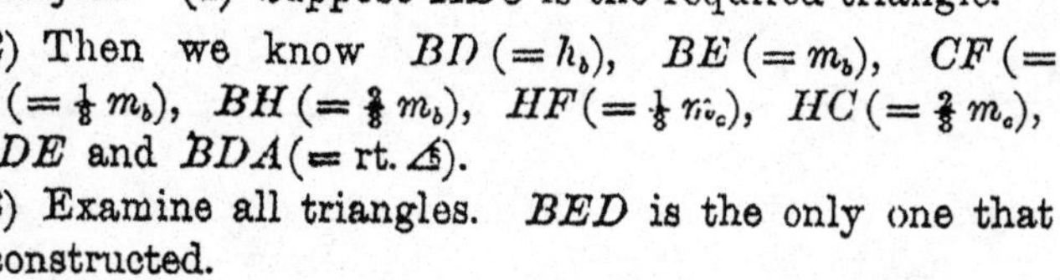

Analysis. (1) Suppose ABC is the required triangle.

(2) Then we know $BD\ (= h_b)$, $BE\ (= m_b)$, $CF\ (= m_c)$, $EH\ (= \frac{1}{3} m_b)$, $BH\ (= \frac{2}{3} m_b)$, $HF\ (= \frac{1}{3} m_c)$, $HC\ (= \frac{2}{3} m_c)$, and $\triangle BDE$ and BDA (= rt. $\triangle$).

(3) Examine all triangles. BED is the only one that can be constructed.

(4) Make the $\triangle BED$ the basis of the construction.

Some problems require the drawing of additional lines (as stated above under 4 *a*).

244. Remark. — **When a sum or a difference is given, construct such sum or difference in the analysis.**

Ex. 410. *To construct a triangle, having given the base, the sum of the other two sides, and the angle included by the two.*

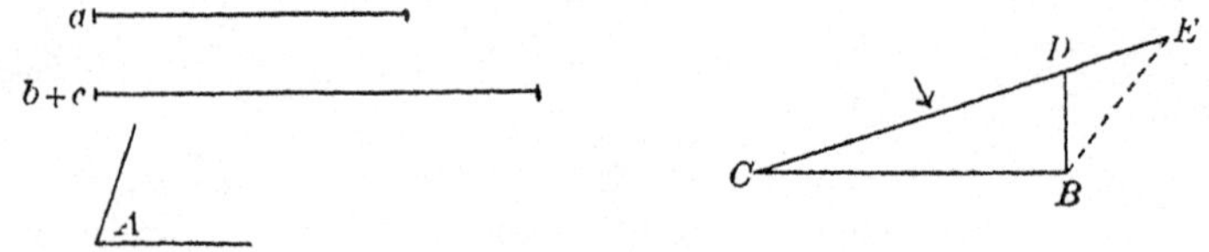

Analysis. (1) Suppose BCD is the required triangle.

(2) Then we know two parts only.

$\therefore$ produce CD to E so that $DE = DB$.

Draw EB. We now know $CB\ (= a)$, $CE\ (= b + c)$, $\angle CDB\ (= A)$, $\angle E\left(= \frac{A}{2}\right)$, and $\angle DBE\left(= \frac{A}{2}\right)$.

(3) Examine all triangles. As BCE can be constructed, make this the basis of the construction.

Ex. 411. *To construct a trapezoid, having given the sum of the bases (s), a diagonal (d), a base angle (A), and the angle formed by the diagonals (O).*

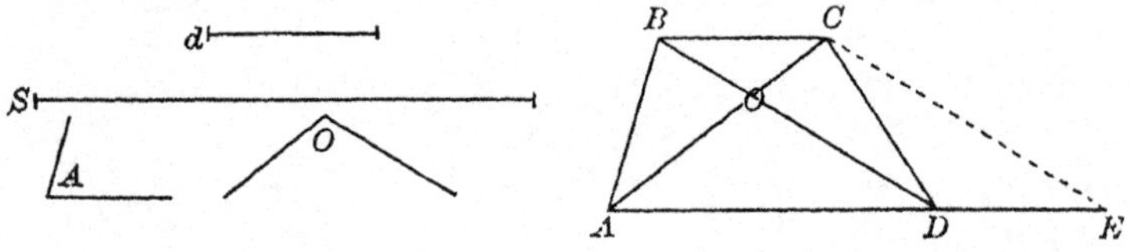

Analysis. (1) Suppose $ABCD$ is the required trapezoid. Produce AD to E so that $DE = BC$. Draw CE.

(2) We know $AE\,(=s)$, $AC\,(=d)$, $\angle AOD\,(=O)$, $\angle BAD\,(=A)$, $\angle ACE\ (= AOD = O)$.

(3) $\triangle ACE$ can be constructed.

Ex. 412. Construct an isosceles triangle, having given the base and the vertical angle.

Ex. 413. Construct an isosceles triangle, having given the sum of base and an arm, and a base angle.

Ex. 414. Construct a triangle, having given an angle, an adjacent side, and the difference of the other two sides.

Ex. 415. Construct a triangle, having given the base, the difference of the other two sides, and the angle included by the two.

Construct a right triangle, having given:

Ex. 416. One arm and the altitude upon the hypotenuse.

Ex. 417. The hypotenuse and the difference between the arms.

Ex. 418. The hypotenuse and the sum of the arms.

Construct a triangle, having given: (See note 230.)

Ex. 419. a, b, m_b.

Ex. 420. a, b, h_b.

Ex. 421. a, h_a, m_a.

Ex. 422. a, m_a, $\angle B$.

Ex. 423. m_a, h_a, $\angle B$.

Ex. 424. b, c, h_a.

Ex. 425. a, h_a, h_c.

Ex. 426. h_a, $\angle B$, $\angle C$.

Ex. 427. h_a, h_c, $\angle C$.

Ex. 428. a, b, $\angle A + \angle B$.

LOCI

245. A **locus of a point** in a plane is a line or a group of lines, all points of which fulfil a certain condition, fulfilled by no other points.

246. All points equidistant from two given points lie in the perpendicular-bisector of the line joining the points, and there is no point without the perpendicular-bisector which is equidistant from these points.

∴ The perpendicular-bisector is a *locus*.

247. To prove therefore that a certain line is a locus, we must establish that:

(1) **Every point in the line satisfies the given condition.**
(2) **No point without the line satisfies the given condition.**

THEOREMS

248. The locus of a point equidistant from the ends of a given line is the perpendicular-bisector of that line.

249. The locus of a point at a given distance from a given point is the circumference described from the point with the given distance as radius.

250. The locus of a point that is at a given distance from a given straight line consists of two lines parallel to the given line at the given distance.

251. The locus of a point equidistant from two given parallel lines is a third parallel, bisecting any line ending in the given parallels.

252. The locus of a point equidistant from two intersecting straight lines, consists of the bisectors of the included angles.

Ex. 429. Find the locus of the vertex of all right angles whose sides pass through two given points.

Ex. 430. Find the locus of the midpoints of the radii of a given circle.

EXERCISES

In the following exercises, state under what conditions no point, one point, or several points may be found.

Ex. 431. In a given line, *AB*, find a point at a given distance, *d*, from a given point, *C*.

Ex. 432. In a given line, *AB*, find a point at a given distance, *d*, from a given line, *CD*.

Ex. 433. In a given line, *AB*, find a point equidistant from two given points, *P* and *Q*.

Ex. 434. In a given circumference, find a point at a given distance, *d*, from a given point, *C*.

Ex. 435. In a given circumference, find a point equidistant from two given parallel lines, *CD* and *EF*.

Ex. 436. In a given circumference, find a point equidistant from two given intersecting lines, *CD* and *EF*.

Ex. 437. Find a point equidistant from two given intersecting lines, *AB* and *CD*, and at a given distance from a given point, *E*.

Ex. 438. Find a point equidistant from two given intersecting lines, *AB* and *CD*, and at a given distance from a given line, *EF*.

Ex. 439. Find a point equidistant from two given intersecting lines, *AB* and *CD*, and equidistant from two given points, *E* and *F*.

Ex. 440. Find a point equidistant from two given points, and having a given distance from a given point, *E*.

Ex. 441. Find a point equidistant from two given points and equidistant from two given parallel lines, *EF* and *GH*.

Ex. 442. Find a point equidistant from two given parallel lines and equidistant from two given intersecting lines, *EF* and *GH*.

Ex. 443. Find a point at a given distance, *d*, from a given line, *AB*, and equidistant from two given points, *E* and *F*.

Ex. 444. Find a point having a given distance, *d*, from a given line, *AB*, and equidistant from two given parallel lines, *EF* and *GH*.

Ex. 445. Find a point having a given distance, *d*, from a given line, *AB*, and equidistant from two given intersecting lines, *EF* and *GH*.

Ex. 446. Find the locus of the center of a circle that passes through two given points.

Ex. 447. Find the locus of the center of a circle that touches two given lines.

Ex. 448. Find the locus of the center of a circle which has a given radius and touches a given line.

Ex. 449. Find the locus of the center of a circle which has a given radius and touches a given circle.

Ex. 450. Find the locus of the center of a circle touching a given line at a given point.

Ex. 451. Find the locus of the center of a circle that touches a given circle in a given point.

To construct a circle having a given radius:

Ex. 452. Touching a given line and passing through a given point.

Ex. 453. Touching two given lines.

T

Ex. 454. Passing through a point and touching a given circle.

Ex. 455. Touching two given circles.

Ex. 456. Touching a given circle and a given line.

253. No general method can be given for the solution of exercises; a great many, however, can be solved

(1) **By a gradual putting together of the given parts.** (Remark 233.)

(2) **By means of an analysis.**

(3) **By means of loci.**

MISCELLANEOUS EXERCISES

Ex. 457. Through a given point, to draw a line cutting off equal parts on the sides of a given angle.

Ex. 458. Through a given point, to draw a line making a given angle with a given line.

Construct an isosceles triangle, having given:

Ex. 459. The base and the altitude upon an arm.

Ex. 460. The altitude upon the base and the vertical angle.

Ex. 461. The vertical angle and the sum of one arm and the base.

Ex. 462. The perimeter and the base angles.

Construct a right triangle, having given:

Ex. 463. One acute angle and the altitude upon the hypotenuse.

Ex. 464. The altitude upon the hypotenuse and one of the segments of the hypotenuse.

Ex. 465. The sum of the arms and one acute angle.

Ex. 466. To find a point in one side of a triangle which is equidistant from the other two sides.

Ex. 467. Find the locus of the vertex of a right triangle, having a given hypotenuse.

Ex. 468. From a point P, in the circumference of a circle, to draw a chord, having a given distance from the center.

Ex. 469. In a given circle, to draw a diameter having a given distance from a given point.

Ex. 470. Through a given point without a circle, to draw a secant having a given distance from the center.

Ex. 471. Through two given points in a circumference, to draw two equal parallel chords.

Ex. 472. Trisect a given straight angle.

Ex. 473. Trisect a given right angle.

Ex. 474. Through a given point, to draw a line of given length terminating in two given parallel lines.

Ex. 475. Through a given point, to draw a line making equal angles with two given lines.

* **Ex. 476.** To bisect an angle formed by two lines, without producing them to their intersection.

To construct a triangle, having given: (Note 230.)

Ex. 477. $a, \angle B, h_a$.

Ex. 478. $a, \angle A, h_a$.

Ex. 479. $a + b + c, \angle B, \angle C$.

Ex. 480. $a, \angle A, m_a$.

* **Ex. 481.** a, b, m_c.

Ex. 482. $a, \angle B, m_c$.

* **Ex. 483.** m_a, m_b, h_c.

* **Ex. 484.** m_a, h_b, h_c.

* **Ex. 485.** m_a, m_b, m_c.

To construct a square, having given:

Ex. 486. The diagonal.

Ex. 487. The difference between the diagonal and the side.

Ex. 488. The sum of the diagonal and the side.

To construct a rectangle, having given:

Ex. 489. One side and the diagonal.

Ex. 490. The angle formed by the diagonals and one side.

Ex. 491. The perimeter and the diagonal.

To construct a rhombus, having given:

Ex. 492. The two diagonals.

Ex. 493. The perimeter and one diagonal.

Ex. 494. One angle and a diagonal.

Ex. 495. The altitude and the base.

Ex. 496. The altitude and one angle.

To construct a parallelogram, having given:

Ex. 497. Two sides and one altitude.

Ex. 498. Two sides and an angle.

Ex. 499. One side and the two diagonals.

Ex. 500. One side, one angle, and one diagonal.

Ex. 501. The diagonals and the angle formed by the diagonals.

254. In the analysis of a problem relating to a trapezoid, draw a line through one vertex, *A*, either parallel to the opposite arm, *DC*, or parallel to a diagonal, *DB*.

To construct a trapezoid, having given:

Ex. 502. The four sides.

Ex. 503. The bases and the base angles.

Ex. 504. The bases, another side, and one base angle.

Ex. 505. The bases and the diagonals.

Ex. 506. One base, the diagonals, and the angle formed by the diagonals.

Ex. 507. To draw a common exterior tangent to two given circles.

Ex. 508. To draw a common interior tangent to two given circles.

Ex. 509. About a given circle, to circumscribe a triangle, having given the angles.

Ex. 510. Find the locus of the midpoints of the chords that pass through a given point in the circumference.

Ex. 511. Find the locus of the midpoints of the secants that pass through a given point without a circle.

Ex. 512. In a given circle, to inscribe a triangle, having given the angles.

***Ex. 513.** From a given point in a circumference, to draw a chord that is bisected by a given chord.

Ex. 514. Given a point, A, between a circumference and a straight line. Through A, to draw a line terminated by the circumference and the given line, and bisected in A.

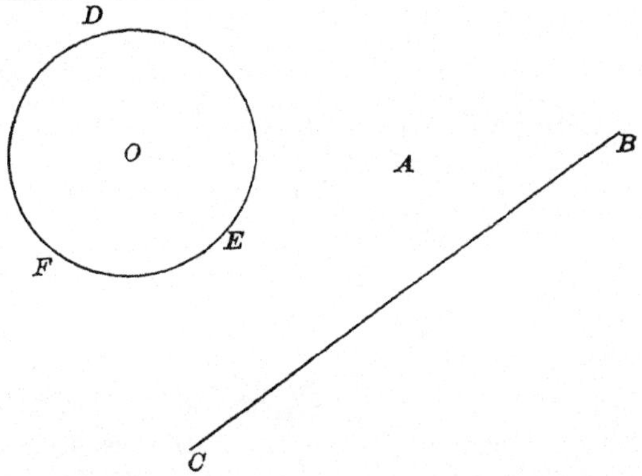

Ex. 515. Given two points, A and B, on the same side of a line, CD. To find a point, X, in CD, such that $\angle AXC = \angle BXD$.

Ex. 516. The bisectors of the angles of a circumscribed quadrilateral meet in a point.

BOOK III

PROPORTION. SIMILAR POLYGONS

255. A **proportion** is a statement expressing the equality of two ratios, as $\frac{a}{b} = \frac{c}{d}$ or $a : b = c : d$.

256. The first and the fourth terms of a proportion are called the **extremes**, the second and the third, the **means**.

257. The first and the third terms are called the **antecedents**, the second and the fourth the **consequents**.

Thus, in the proportion, $a : b = c : d$, a and d are the extremes, b and c the means, a and c the antecedents, and b and d the consequents.

258. When the means of a proportion are equal, *either* mean is said to be the **mean proportional** between the first and the last terms, and the *last* term is said to be the **third proportional** to the first and the second terms.

Thus, in the proportion, $a : b = b : c$, b is the mean proportional between a and c, and c is the third proportional to a and b.

259. The last term is the **fourth proportional** to the first three. Thus, in the proportion, $a : b = c : d$, d is the fourth proportional to a, b, and c.

260. A series of equal ratios is called a **continued proportion**.

261. The two terms of a ratio must be either quantities of the same kind, or the quantities must be represented by their numerical measures only.

PROPOSITION I. THEOREM

262. *In any proportion, the product of the means is equal to the product of the extremes.*

Hyp. $a : b = c : d.$

To prove $ad = bc.$

Proof. $\frac{a}{b} = \frac{c}{d}.$

Multiplying both members by bd,

$$ad = bc.$$

Q.E.D.

263. COR. If three terms of a proportion are respectively equal to the three corresponding terms of another proportion, the fourth terms are also equal.

264. NOTE. — The product of two quantities, in Geometry, means the product of the numerical measures of the quantities.

Ex. 517. Find the value of x if $3 : x = 4 : 8$.

Ex. 518. Find the value of x if $a : m = x : n$.

PROPOSITION II. THEOREM

265. *If the product of two numbers is equal to the product of two other numbers, either two may be made the means, and the other two the extremes of a proportion.*

Hyp. $mn = pq.$

To prove $m : p = q : n.$

Proof. $mn = pq.$

Dividing both members by np

$$\frac{m}{p} = \frac{q}{n}.$$

Q.E.D.

Ex. 519. If $ab = mn$, find all possible proportions consisting of a, b, m, and n.

PROPOSITION III. THEOREM

266. *A mean proportional between two quantities is equal to the square root of their product.*

Hyp. $a : b = b : c.$

To prove $b = \sqrt{ac}.$

Proof. $b^2 = ac.$ (262)

Extracting the square root of both members

$$b = \sqrt{ac}.$$ Q.E.D.

Ex. 520. Find the mean proportional between 2 and 50, between $a + m$ and $a - m$.

Ex. 521. Find the third proportional to m and n.

PROPOSITION IV. THEOREM

267. *If four quantities are in proportion, they are in proportion by alternation,* i.e. *the first term is to the third as the second is to the fourth.*

Hyp. $a : b = c : d.$

To prove $a : c = b : d.$

Proof. $\frac{a}{b} = \frac{c}{d}.$

Multiplying by $\frac{b}{c}$

$$\frac{a}{c} = \frac{b}{d}.$$ Q.E.D.

268. COR. If $a : b = c : d$, and $a = kc$, then $b = kd$.

HINT. — Since $a = kc$, $\frac{a}{c} = k$, etc.

Proposition V. Theorem

269. *If four quantities are in proportion, they are in proportion by inversion,* i.e. *the second term is to the first as the fourth is to the third.*

Hyp. $a : b = c : d.$

To prove $b : a = d : c.$

Proof. $ad = bc.$ (262)

$\therefore\ b : a = d : c.$ (265) Q.E.D.

Ex. 522. Transform the proposition, $m : x = p : q$, so that x becomes the fourth term.

Proposition VI. Theorem

270. *If four quantities are in proportion, they are in proportion by composition,* i.e. *the sum of the first two terms is to the second term as the sum of the last two terms is to the fourth term.*

Hyp. $a : b = c : d.$

To prove $a + b : b = c + d : d.$

Proof. $\frac{a}{b} = \frac{c}{d}.$

$\frac{a}{b} + 1 = \frac{c}{d} + 1,$ (Ax. 2.)

or $\frac{a+b}{b} = \frac{c+d}{d}.$ Q.E.D.

Proposition VII. Theorem

271. *If four quantities are in proportion, they are in proportion by division,* i.e. *the difference of the first two terms is to the second term as the difference between the last two terms is to the fourth term.*

Hyp. $a : b = c : d.$

To prove $a - b : b = c - d : d.$

Proof. $\frac{a}{b} = \frac{c}{d}.$

$$\frac{a}{b} - 1 = \frac{c}{d} - 1, \quad \text{(Ax. 3.)}$$

or $$\frac{a-b}{b} = \frac{c-d}{d}.$$ Q.E.D.

Ex. 523. If $x + y : y = 7 : 3$, find the ratio of x and y.

Ex. 524. If $x - y : y = 2 : 3$, find the ratio of x and y.

Proposition VIII. Theorem

272. *If four quantities are in proportion, they are in proportion by composition and division,* i.e. *the sum of the first two terms is to their difference as the sum of the last two terms is to their difference.*

Hyp. $a : b = c : d.$

To prove $a + b : a - b = c + d : c - d.$

Proof. $$\frac{a+b}{b} = \frac{c+d}{d}. \quad (270)$$

$$\frac{a-b}{b} = \frac{c-d}{d}. \quad (271)$$

Dividing member by member,

$$\frac{a+b}{a-b} = \frac{c+d}{c-d}.$$ Q.E.D.

Ex. 525. If $x + y : x - y = 12 : 5$, find the ratio of x and y.

Ex. 526. If $x + y : x - y = a : b$, find the ratio of x to y.

PROPOSITION IX. THEOREM

273. *In a continued proportion the sum of any number of antecedents is to the sum of the corresponding consequents as any antecedent is to its consequent.*

Hyp. $a:b=c:d=e:f.$

To prove $a+c+e:b+d+f=a:b.$

Proof. $ab=ab.$

$ad=bc.$ (262)

$af=be.$ (262)

By adding the equations

$$ab+ad+af=ab+bc+be,$$

or $a(b+d+f)=b(a+c+e).$

$\therefore\ a+c+e:b+d+f=a:b.$ (265) Q.E.D.

Ex. 527. If $a:b=c:d=e:f=5:7$, find $\dfrac{a+c+e}{b+d+f}$.

PROPOSITION X. THEOREM

274. *The products of the corresponding terms of two or more proportions are in proportion.*

Hyp. $a:b=c:d.$

$m:n=p:q.$

To prove $am:bn=cp:dq.$

Proof. $\dfrac{a}{b}=\dfrac{c}{d}.$ (Hyp.)

$\dfrac{m}{n}=\dfrac{p}{q}.$ (Hyp.)

Multiplying the corresponding members of the equations,

$$\frac{am}{bn} = \frac{cp}{dq}.$$ Q.E.D.

275. Cor. If $$\frac{a}{b} = \frac{c}{d},$$

then $$\frac{ma}{nb} = \frac{mc}{nd}.$$

Ex. 528. If $2 : 3 = \frac{1}{y} : x$ and $2 : 27 = 4 : y$, find x.

Proposition XI. Theorem

276. *If four quantities are in proportion, like powers or like roots of these quantities are in proportion.*

Hyp. $$a : b = c : d.$$

To prove $$a^n : b^n = c^n : d^n$$

and $$\sqrt[m]{a} : \sqrt[m]{b} = \sqrt[m]{c} : \sqrt[m]{d}.$$

Proof. $$\frac{a}{b} = \frac{c}{d}.$$

Raising both members to the nth power,

$$\frac{a^n}{b^n} = \frac{c^n}{d^n}.$$

Similarly, it can be shown that

$$\frac{\sqrt[m]{a}}{\sqrt[m]{b}} = \frac{\sqrt[m]{c}}{\sqrt[m]{d}}.$$ Q.E.D.

Ex. 529. If $x^3 : y^3 = 64 : 125$, find $\frac{x}{y}$.

Ex. 530. If $\sqrt[3]{x} : 1 = \sqrt[3]{y} : 2$, find $\frac{x}{y}$.

PROPOSITION XII. THEOREM

277. *Equimultiples of two quantities are in the same ratio as the quantities.*

Hyp. a and b are two quantities.

To prove $ma : mb = a : b.$

[The proof is left to the student.]

278. DEF. If in a line AB, or its prolongation, a point C be taken, AC and BC are called **segments of the line.**

279. The segments are **internal** or **external** ones, according as C lies in AB or in the prolongation of AB.

PROPORTIONAL LINES

PROPOSITION XIII. THEOREM

280. *A line parallel to one side of a triangle divides the other two sides proportionally.*

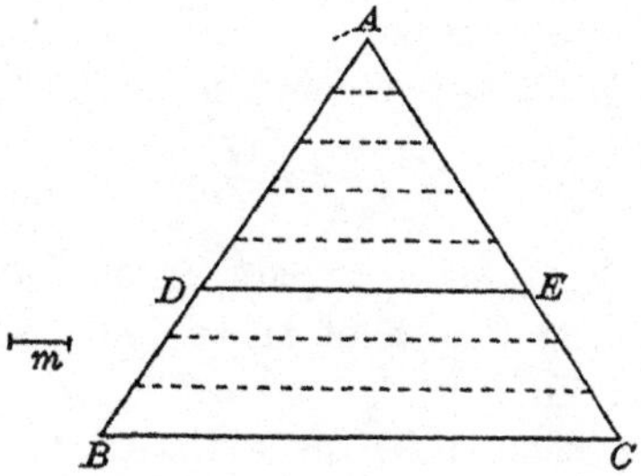

Hyp. In ▲ ABC, DE is parallel to BC.

To prove $AD : DB = AE : EC.$

Proof. CASE I. AD and DB are commensurable.

Let m be a common measure contained in AD five times and in DB three times.

Then $$\frac{AD}{DB}=\frac{5}{3}.$$

Through the points of division of AB draw parallels to BC. These lines divide AE into five parts and EC into three parts, all being equal. (144)

Whence $$\frac{AE}{EC}=\frac{5}{3}.$$

$$\therefore \frac{AD}{DB}=\frac{AE}{EC}.$$ (Ax. 1.) Q.E.D.

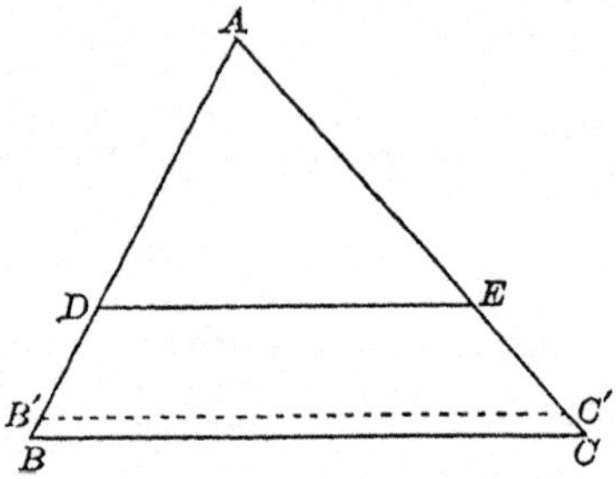

CASE II. AD and DB are incommensurable.

Divide AD into any number of equal parts, and apply one of those parts to DB as often as possible.

As the lines AD and DB are incommensurable, there must be a remainder, $B'B$, less than one of the equal parts.

Draw $B'C'$ parallel to BC.

The lines AD and DB' are commensurable.

$$\therefore \frac{AD}{DB'}=\frac{AE}{EC'}.$$

By increasing the number of parts into which AD is divided, we can diminish the length of these parts, and therefore the length of $B'B$ indefinitely.

Hence DB' approaches DB as a limit, and EC' approaches EC as a limit.

The variables $\frac{AD}{DB'}$ and $\frac{AE}{EC'}$, being always equals, must have equal limits.

Whence $$\frac{AD}{DB}=\frac{AE}{EC}.$$ Q.E.D.

281. Scholium. If the transversal intersects two sides of the triangle, these sides are divided internally, if it meets their prolongations, the sides are divided externally in the same ratio.

282. Cor. If a line parallel to one side of a triangle intersects the other two sides, either side is to one of its segments as the other side is to its corresponding segment.

For $$AD : DB = AE : EC.$$

By composition

$$AD + DB : DB = AE + EC : EC,$$

or $$AB : DB = AC : EC.$$

Ex. 531. In the diagram for Prop. XIII, if $AD = 4$, $DB = 8$, $AE = 3$, find EC.

Ex. 532. In the same diagram, find DB, if $AD = a$, $AE = b$, and $EC = c$.

Ex. 533. In the same diagram, find AE, if $AB = 12$, $AD = 8$, and $AC = 9$.

Ex. 534. In the same diagram, find EC, if $AB = m$, $AD = n$, and $AC = p$.

Ex. 535. In the same diagram, find AD, if $AD = EC$, $DB = 4$, and $AE = 9$.

Ex. 536. In the same diagram, find AE, if $AE = 2\,DB$, $AD = 10$, and $EC = 20$.

Ex. 537. Three or more parallels make proportional segments on the sides of an angle.

Ex. 538. In the diagram for Prop. XIII, Case II, find $B'B$ if $AD = 3$, $AE = 4$, and $C'C = 1$.

Ex. 539. In the same diagram find EC' if $AD = 4$, $DB = 3$, $AE = 8$, and $C'C = 1$.

Ex. 540. Three or more parallels make proportional segments on any two transversals.

Ex. 541. If in the diagram for Prop. XIII, $AD = 2$, $DB = 3$, $AE = 4$, and $EC = 4$, is DE parallel to BC?

PROPOSITION XIV. PROBLEM

283. *To find the fourth proportional to three given lines.*

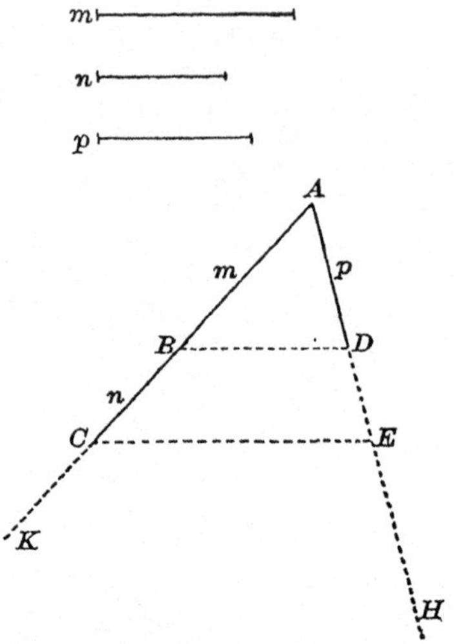

Given. Three lines m, n, and p.

Required. The fourth proportional to m, n, and p.

Construction. Draw any angle KAH.

On AK, make $AB = m$, $BC = n$; on AH, make $AD = p$.

Draw BD.

Through C, draw a line parallel to BD, meeting AH in E.

DE is the required fourth proportional.

[The proof is left to the student.]

Ex. 542. Find the third proportional to two given lines.

Ex. 543. If a, b, and c are given lines, construct a line x, so that

$$a : b = x : c.$$

Ex. 544. If a, b, and c are given lines, construct a line equal to $\frac{bc}{a}$.

PROPOSITION XV. PROBLEM

284. *To divide a given line into segments proportional to any number of given lines.*

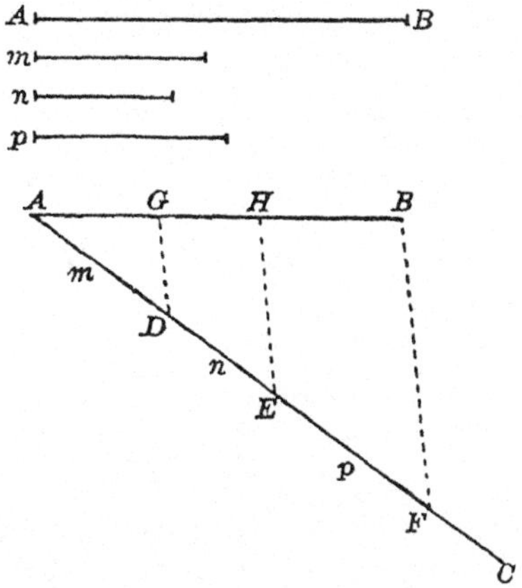

Given. Lines AB, m, n, and p.

Required. To divide AB into segments proportional to m, n, and p.

Construction. At any angle with AB draw AC.

On AC, cut off $AD = m$, $DE = n$, $EF = p$.

Draw FB.

Through E and D draw parallels to BF, intersecting AB in H and G, respectively.

Then AB is divided as required.

[The proof is left to the student.]

285. DEF. A line is said to be divided **harmonically** if it is divided internally and externally in the same ratio.

Ex. 545. To divide a given line externally into segments proportional to two other given lines.

Ex. 546. To divide a given line AB harmonically in the ratio of two given lines m and n.

K

Ex. 547. If a line AB is divided harmonically by two points, C and D, then CD is divided harmonically by A and B.

Ex. 548. Construct two lines when their sum and their ratio are given.

Ex. 549. Construct two lines when their difference and their ratio are given.

Ex. 550. In a given line, AB, to find a point, C, so that

$$AB : AC = m : n,$$

when m and n are two given lines.

Proposition XVI. Theorem

286. *If a line divides two sides of a triangle proportionally, it is parallel to the third side.*

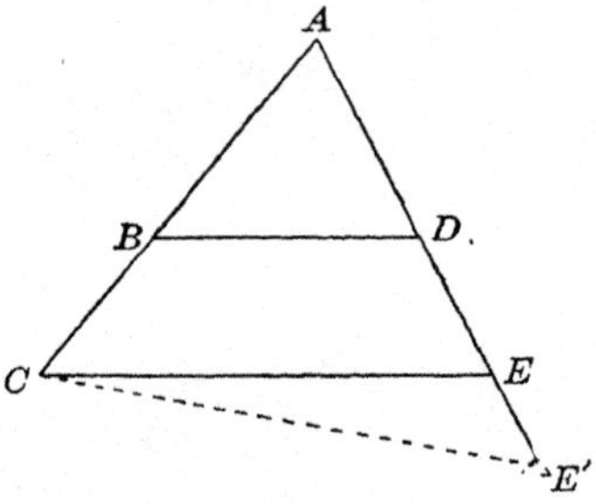

Hyp. In $\triangle AEC$, $AB : BC = AD : DE$.

To prove, DB parallel to EC.

Proof. Through C, draw CE' parallel to BD meeting AE in E'.

$AB : BC = AD : DE$. (Hyp.)

$AB : BC = AD : DE'$. (280)

$\therefore DE = DE'$. (263)

$\therefore CE$ and CE' coincide.

$\therefore BD \parallel CE$. Q.E.D.

Ex. 551. In the diagram for Prop. XVI, if $AB = 12$, $BC = 16$, $AD = 15$, $DE = 20$, is $BD \parallel CE$?

Ex. 552. Demonstrate that there is only one point that divides a given line internally in a given ratio.

PROPOSITION XVII. THEOREM

287. *The bisector of an angle of a triangle divides the opposite side into segments having the same ratio as the other two sides.*

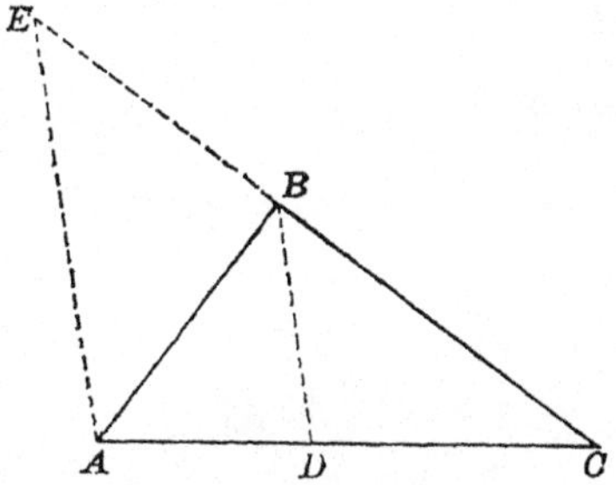

Hyp. In $\triangle ABC$, BD bisects $\angle ABC$.

To prove $AB : BC = AD : DC$.

Proof. Draw $AE \parallel DB$, to meet CB produced in E.

$\angle E = \angle CBD$. (89)

$\angle CBD = \angle ABD$. (Hyp.)

$\angle ABD = \angle EAB$. (87)

$\therefore \angle E = \angle EAB$. (Ax. 1.)

$\therefore AB = BE$. (103)

But $EB : BC = AD : DC$. (280)

$\therefore AB : BC = AD : DC$. Q.E.D.

Ex. 553. In the diagram for Prop. XVII, find DC, if $AB = 3$, $BC = 4$, and $AD = 2$.

Ex. 554. In the same diagram, find DC, if $AB = 4$, $BC = 5$, and $AC = 6$.

Ex. 555. In the same diagram, find AD, if $AB = m$, $BC = n$, and $DC = p$.

Ex. 556. In the same diagram, find AD and DC, if $BC = a$, $CA = b$, and $AB = c$.

Ex. 557. In the same diagram find AB, if $AB = DC$, $AD = 4$, and $BC = 16$.

Ex. 558. In the same diagram find AB, if $AB = DC$, $AD = m$, and $BC = n$.

Ex. 559. To divide one side of a triangle into segments proportional to the other two sides.

Ex. 560. State and prove the converse of Prop. XVII.

Proposition XVIII. Theorem

288. *The bisector of an exterior angle of a triangle divides the opposite side externally into segments which are proportional to the other two sides.*

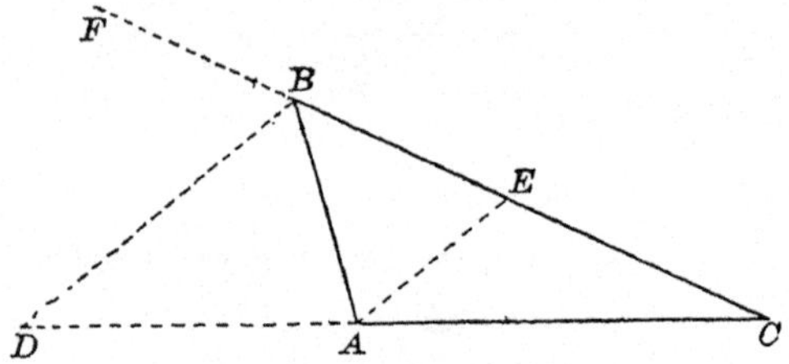

Hyp. In $\triangle ABC$, BD bisects the exterior $\angle ABF$.

To prove $AB : BC = AD : DC$.

Hint. — The proof is identical with the proof of Prop. XVII, excepting that $\angle s$ BEA and FBD must be substituted for $\angle s$ E and DBC.

Ex. 561. The bisectors of an interior and an adjacent exterior angle of a triangle divide the opposite side harmonically.

Ex. 562. If in the diagram for Prop. XVIII $BA = 3$, $BC = 4$, $AC = 5$, find DA.

Ex. 563. In the same diagram find DA, if $DA = BC$, $BA = 4$, and $DC = 9$.

Ex. 564. In the same diagram find DC, if $BC = a$, $AC = b$, and $AB = c$.

Ex. 565. State and prove the converse of Prop. XVIII.

SIMILAR POLYGONS

289. Def. Two polygons are **similar** if they are mutually equiangular and have their homologous sides proportional.

Thus, polygons $ABCDE$ and $A'B'C'D'E'$ are similar, if,

$$(1) \quad \angle A = \angle A', \ \angle B = \angle B', \text{ etc.}$$

and $$(2) \quad \frac{AB}{A'B'} = \frac{BC}{B'C'} = \frac{CD}{C'D'}, \text{ etc.}$$

Proposition XIX. Theorem

290. *Triangles which are mutually equiangular are similar.*

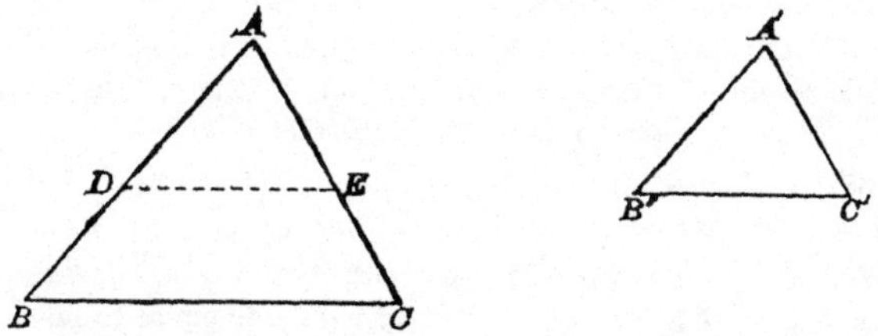

Hyp. In $\triangle ABC$ and $A'B'C'$,

$$\angle A = \angle A', \ \angle B = \angle B', \text{ and } \angle C = \angle C'.$$

To prove $$\triangle ABC \sim \triangle A'B'C'.$$

Proof. Place $\triangle A'B'C'$ upon $\triangle ABC$ so that $\angle A'$ coincides with $\angle A$. Then B' will fall on some point D and C' on a point E.

Since $\angle ADE = \angle B$, (Hyp.)

$DE \parallel BC$. (83)

$\therefore AB : AD = AC : AE$, (282)

or $AB : A'B' = AC : A'C'$.

In a similar manner it may be shown that

$$AB : A'B' = BC : B'C'.$$

$\therefore \triangle ABC$ and $A'B'C'$ have their homologous sides proportional, and are similar. Q.E.D.

291. COR. 1. Two triangles are similar if two angles of the one are respectively equal to two angles of the other.

292. COR. 2. Two right triangles are similar if an acute angle of the one is equal to an acute angle of the other.

Ex. 566. Two isosceles triangles are similar if an angle of the one is equal to the homologous angle of the other.

Ex. 567. If two chords AB and CD intersect in E, the triangle AEC is similar to the triangle BED.

Ex. 568. If from a point A, without a circle, two secants are drawn to meet the circumference in B and C, and D and E respectively, the triangle ABE is similar to the triangle ACD.

Ex. 569. The sides of a polygon are 3, 4, 5, 6, and 7. Find the sides of a similar polygon if the side corresponding with 3 equals 15.

Ex. 570. The sides of a triangle are a, b, and c. Find the sides of a similar triangle if the side corresponding with a is equal to m.

Ex. 571. If the altitudes AD and BE of the triangle ABC intersect in F, the triangle AFE is similar to the triangle BFD.

Ex. 572. If the bisector AD of the inscribed triangle ABC be produced to meet the circumference in E, the triangle ABD is similar to the triangle AEC.

293. REMARK. — **Similar triangles are the usual means of proving that lines are proportional.** To prove, therefore, that four lines are proportional.

(1) *Select two triangles so that each contains two of the given lines.* (It is advisable to mark the lines as indicated in Ex. 408.)

(2) *Prove the similarity of the two triangles.* (If triangles are not similar select another pair.)

(3) *Derive the proportion.*

(4) (*Apply alternation and inversion,* if necessary.)

Ex. 573. If in the triangle ABC the altitudes AD and BE be drawn, prove that $AC : BC = DC : EC$.

Ex. 574. In the same diagram, if AD and BE meet in F, prove $BF : FA = DF : FE$.

Ex. 575. If from the vertex A of an inscribed triangle ABC the altitude AD and the diameter AF be drawn, then $AB : AD = AF : AC$.

Ex. 576. If from a point without a circle a tangent and a secant be drawn, the tangent is the mean proportional between the secant and its exterior segment.

Ex. 577. If a diameter AB be produced to C, at C a perpendicular be erected, and through B a line be drawn to meet the circumference and the perpendicular in D and E, respectively, then $AB : BE = DB : BC$.

Ex. 578. In similar triangles homologous bisectors have the same ratio as any two homologous sides.

Ex. 579. In similar triangles homologous altitudes have the same ratio as any two homologous sides.

Ex. 580. If two circles are tangent internally, chords of the greater circle drawn from the point of contact are divided proportionally.

Ex. 581. The diagonals of a trapezoid divide each other proportionally.

Ex. 582. If in a right triangle ABC the altitude AD be drawn upon the hypotenuse, $AD : AB = AC : BC$.

Ex. 583. In the same diagram, $AD : AB = DC : AC$.

Ex. 584. In similar triangles the radii of the inscribed circles have the same ratio as any two homologous sides.

294. REMARK. — **To prove that the product of two lines equals the product of two other lines,** *use the method of* (293), *and take the products of the means and extremes of the resulting proportion.*

Ex. 585. If two chords intersect within a circumference, the product of the segments of one is equal to the product of the segments of the other.

Ex. 586. If from any point E in the chord AB the perpendicular EC be drawn upon the diameter AD, then

$$AC \times AD = AB \times AE.$$

Ex. 587. In the diagram for Prop. XIII, prove that

$$AD \times BC = AB \times DE.$$

Ex. 588. The product of two arms of a right triangle is equal to the product of the hypotenuse and the altitude upon the hypotenuse.

Ex. 589. The product of any altitude of a triangle and its corresponding side is equal to the product of any other altitude and its corresponding side.

Ex. 590. If in the triangle ABC the altitudes AD and BE meet in F, then
$$BD \times DC = DF \times AD.$$

Ex. 591. In the same diagram,
$$BD \times AC = BF \times AD.$$

Ex. 592. If AB is a diameter, BD the tangent at B, and DA meets the circumference in E, then
$$\overline{AB}^2 = AE \times AD.$$

Ex. 593. In the same diagram,
$$\overline{BE}^2 = AE \times ED.$$

Ex. 594. If in an arm AB of an isosceles triangle ABC a point D be taken so that CD equals the base, CB, then
$$\overline{CB}^2 = BD \times BA.$$

Proposition XX. Theorem

295. *Two triangles are similar if an angle of the one is equal to an angle of the other, and the sides including this angle are proportional.*

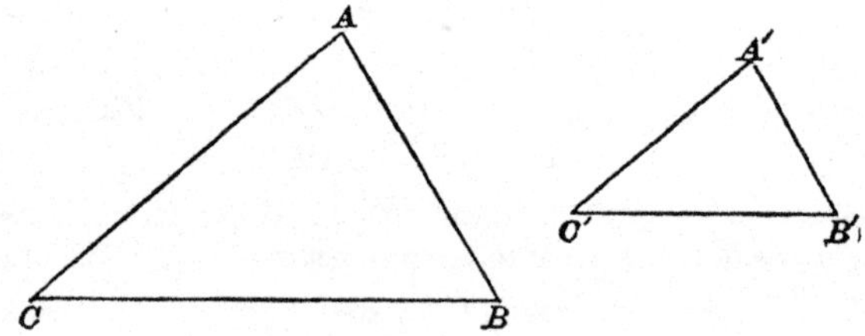

Hyp. In $\triangle$ ABC and $A'B'C'$,
$$\angle A = \angle A', \text{ and } AB : A'B' = AC : A'C'.$$

To prove $\triangle ABC \sim \triangle A'B'C'$.

Hint. — Place $\triangle A'B'C'$ upon $\triangle ABC$, so that A' coincides with A, and prove that $C'B'$ will become $\parallel$ CB.

Ex. 595. Two isosceles triangles are similar if their vertical angles are equal.

Ex. 596. In similar triangles, homologous medians have the same ratio as any two homologous sides.

Ex. 597. Two triangles, ABC and $A'B'C'$, are similar if altitude AD : altitude $A'D' = BC : B'C'$, and $\angle B = \angle B'$.

Ex. 598. In similar triangles, the radii of the circumscribed circles have the same ratio as any two homologous sides.

Proposition XXI. Theorem

296. *Two triangles are similar if their homologous sides are proportional.*

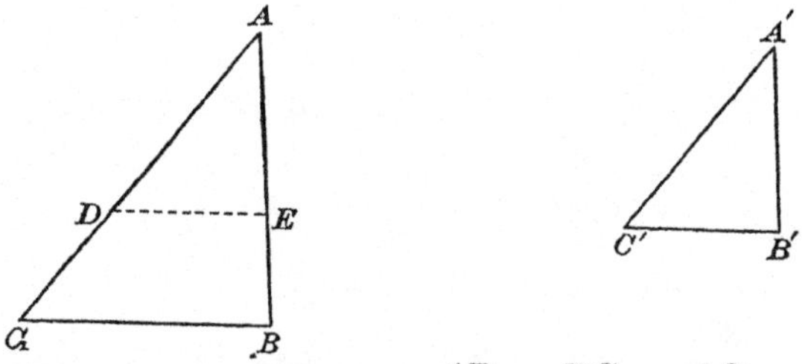

Hyp. In $\triangle ABC$ and $A'B'C'$, $\dfrac{AB}{A'B'} = \dfrac{BC}{B'C'} = \dfrac{AC}{A'C'}$.

To prove $\triangle ABC \sim \triangle A'B'C'$.

Proof. On AC and AB respectively, lay off $AD = A'C'$ and $AE = A'B'$, and draw DE.

Then $\triangle ADE \sim \triangle ACB$. (295)

$\therefore AC : AD = CB : ED$,

(*hom. sides of* $\sim \triangle$),

but $AC : A'C' = BC : B'C'$. (Hyp.)

$\therefore$ since $AD = A'C'$,

$DE = B'C'$. (263)

$\therefore \triangle A'B'C' = \triangle ADE$,

(*s. s. s.* = *s. s. s.*).

But $\triangle ADE$ has been proven similar to $\triangle ABC$.

$\therefore \triangle A'B'C' \sim \triangle ABC$. Q.E.D.

Ex. 599. Two triangles are similar if two sides and the median to one of these sides of one are proportional to the homologous parts of the other triangle.

***Ex. 600.** Two triangles are similar if two sides and the radius of the circumscribed circle of the one are proportional to the homologous parts of the other.

***Ex. 601.** Two right triangles are similar if the hypotenuse and an arm of one triangle are proportional to the hypotenuse and an arm of the other.

PROPOSITION XXII. THEOREM

297. *Two triangles are similar if the sides of the one are respectively parallel to the sides of the other.*

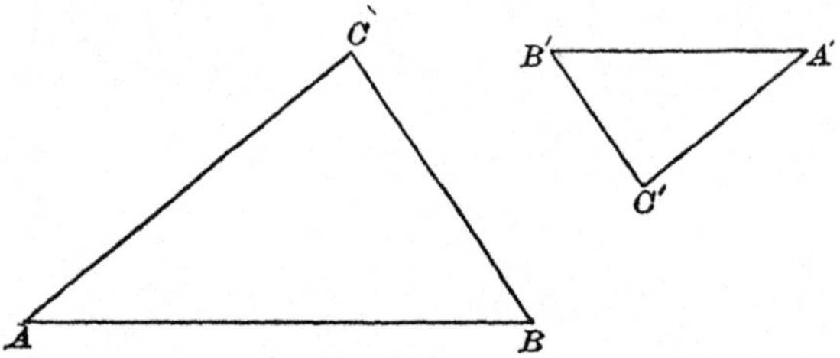

Hyp. In ⊿s ABC and $A'B'C'$, $AB \parallel A'B'$, $AC \parallel A'C'$, and $BC \parallel B'C'$.

To prove $\triangle ABC \sim \triangle A'B'C'$.

Proof. ∠s A and A' are either equal or supplementary. (91)

In like manner, ∠s B and B' and C and C' are either equal or supplementary.

Hence, one of the following possibilities must be true:

(1) The three homologous angles are supplementary,

i.e. $\angle A + \angle A' = 2$ rt. ∠s, $\angle B + \angle B' = 2$ rt. ∠s, $\angle C + \angle C' = 2$ rt. ∠s.

(2) Two angles are supplementary, one is equal to the homologous one,

e.g. $\angle A + \angle A' = 2$ rt. ∠s, $\angle B + \angle B' = 2$ rt. ∠s, $\angle C = \angle C'$.

(3) Two angles are equal, one is supplementary to the homologous ones,

$$e.g.\ \angle A + \angle A' = 2 \text{ rt. } \angle s,\ \angle B = \angle B',\ \angle C = \angle C'.$$

(4) The angles are respectively equal,

$$i.e.\ \angle A = \angle A',\ \angle B = \angle B',\ \angle C = \angle C'.$$

The first two statements cannot be true, for the sum of the angles of the two triangles would exceed four right angles.

Therefore, two angles of one triangle are equal to two angles of the other.

$$\therefore \triangle ABC \sim \triangle A'B'C'. \qquad \text{Q.E.D.}$$

Ex. 602. Two triangles are similar if the sides of one are respectively perpendicular to the sides of the other.

Ex. 603. If in the diagram for Prop. XXII BB' and CC' be drawn, BB' is divided in the ratio AB to $A'B'$.

Ex. 604. In the same diagram, the lines joining corresponding vertices meet in a point.

Proposition XXIII. Theorem

298. *If two triangles are similar, homologous altitudes have the same ratio as any two homologous sides.*

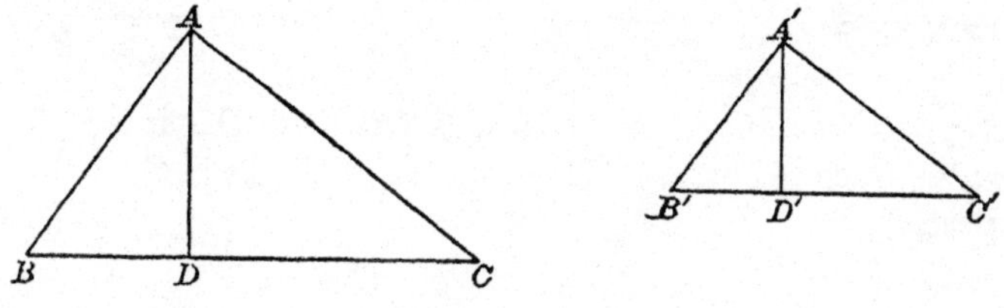

Hyp. AD and $A'D'$ are the homologous altitudes of the similar triangles ABC and $A'B'C'$.

To prove $AD : A'D' = AB : A'B'$.

Hint. — What is the means of proving that four lines are proportional?

299. Scholium. In general, it can be proved that any two homologous lines of two similar triangles have the same ratio as any two homologous sides.

Ex. 605. The base of a triangle is 2 ft. and the altitude 9 in. If the homologous base of a similar triangle is 6 in., find the homologous altitude.

300. Remark. — **To prove the proportionality of four lines which do not form similar triangles,** *find a third ratio equal to each of the given ones.*

Proposition XXIV. Theorem

301. *If two parallel lines are cut by three or more transversals passing through a common point, the corresponding segments are proportional.*

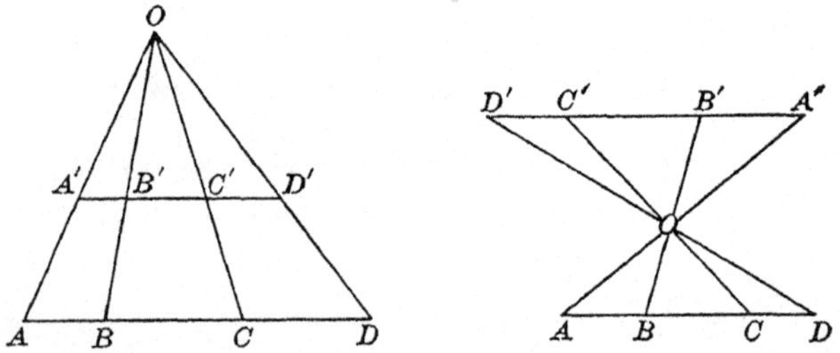

Hyp. The transversals OA, OB, OC, and OD intersect the parallel lines AD and $A'D'$ in A, B, C, D, and A', B', C', D', respectively.

To prove $$\frac{AB}{A'B'} = \frac{BC}{B'C'} = \frac{CD}{C'D'}.$$

Hint. — Which method for demonstrating the proportionality of the first four lines must be applied? Why?

$$\left(\frac{OB}{OB'} \text{ is the third ratio.}\right)$$

Ex. 606. In the diagram for Prop. XXIV, if $AB = B'C'$, $A'B' = 4$, and $BC = 9$, what is the value of AB?

***Ex. 607.** If in a parallelogram $ABCD$ from A a line be drawn intersecting the diagonal in E, and the sides (produced, if necessary) in F and G, AE is the mean proportional between FE and GE.

PROPOSITION XXV THEOREM

302. *If three or more non-parallel transversals intercept proportional segments on two parallel lines, they intersect in a common point.*

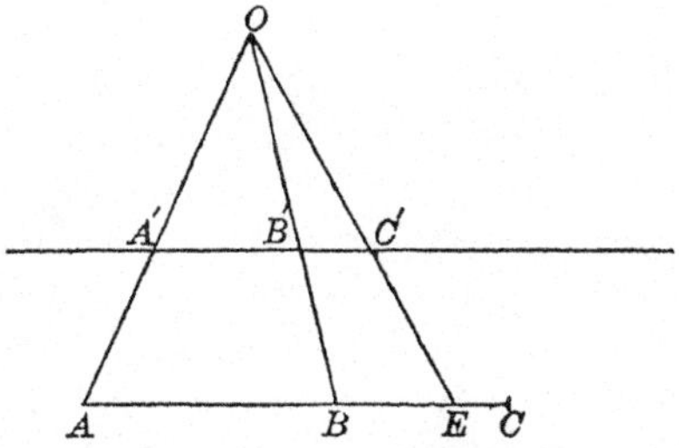

Hyp. AA', BB', and CC' intersect the parallels AC and $A'C'$, so that

$$AB : A'B' = BC : B'C'.$$

To prove AA', BB', and CC' intersect in a common point.

Proof. Let AA' and BB' intersect in O.

Draw OC', and suppose that its prolongation meets AC in E.

Then $AB : A'B' = BC : B'C'$. (Hyp.)

$AB : A'B' = BE : B'C'$. (301)

$\therefore BE = BC$, (263)

or E coincides with C.

$\therefore C$, C', and O lie in a straight line,

or CC', produced, passes through O. Q.E.D.

Ex. 608. Given two lines, AB and CD, and point E. Without producing AB and CD to their point of intersection, to draw a line XY through E, so that AB, CD, and XY would meet in a common point.

Ex. 609. If the sides of two triangles are respectively parallel, the lines joining their homologous vertices meet in a point.

Ex. 610. Two polygons are similar if their homologous sides are parallel and the lines joining their homologous vertices meet in a point.

Proposition XXVI. Theorem

303. *Two similar polygons may be divided into the same number of similar triangles similar each to each and similarly placed.*

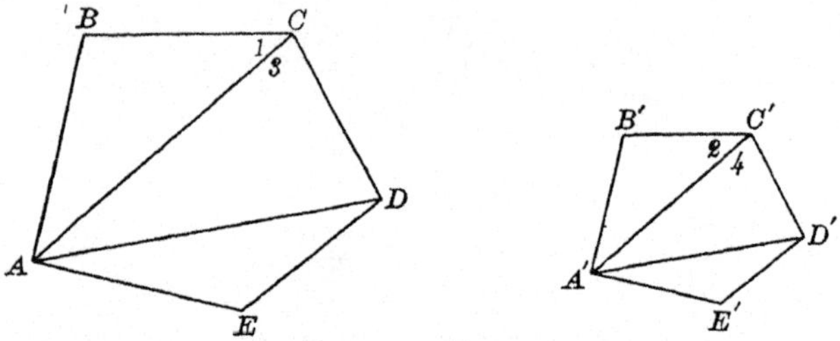

Hyp. Polygon $ABCDE \sim$ polygon $A'B'C'D'E'$.

To prove $\triangle ABC \sim \triangle A'B'C'$,

$\triangle ACD \sim \triangle A'C'D'$, etc.

Proof. Since the polygons are similar,

$$\frac{AB}{A'B'} = \frac{BC}{B'C'} \text{ and } \angle B = \angle B'.$$

Hence $\triangle ABC \sim \triangle A'B'C'$, (295)

$\angle BCD = \angle B'C'D'$, (Hyp.)

$\angle 1 = \angle 2$,

(*homol.* $\angle$*s of* $\sim \triangle$*s*).

$\therefore \angle 3 = \angle 4$ (Ax. 3.)

Since the polygons are similar, and $\triangle ABC \sim \triangle A'B'C'$,

$$\frac{BC}{B'C'} = \frac{CD}{C'D'},$$

and $$\frac{BC}{B'C'} = \frac{AC}{A'C'}.$$

$$\therefore \frac{CD}{C'D'} = \frac{AC}{A'C'}. \qquad \text{(Ax. 1.)}$$

$$\therefore \triangle ACD \sim \triangle A'C'D'. \qquad (295)$$

In like manner, $\triangle s$ ADE and $A'D'E'$ are similar. Q.E.D.

Proposition XXVII. Theorem

304. *Two polygons are similar if they are composed of the same number of triangles, similar each to each, and similarly placed.*

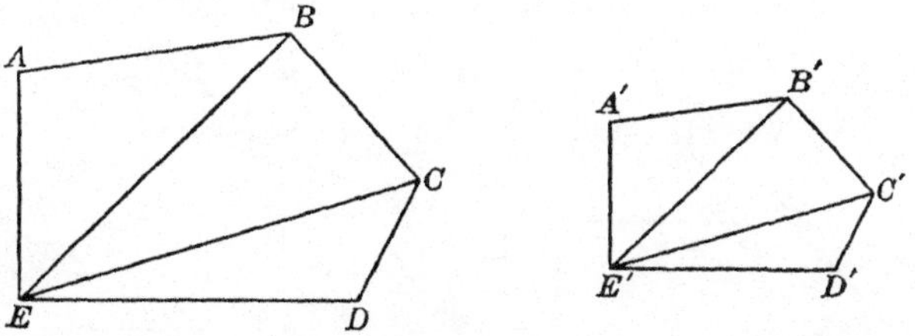

Hyp. In the polygons $ABCDE$ and $A'B'C'D'E'$,

$$\triangle ABE \sim \triangle A'B'E',$$
$$\triangle BCE \sim \triangle B'C'E',$$
$$\triangle CDE \sim \triangle C'D'E'.$$

To prove polygon $ABCDE \sim$ polygon $A'B'C'D'E'$.

Hint. — The polygons are mutually equiangular by Axiom 2.

The ratio of any pair of homologous sides is equal to the ratio of the next pair, for either ratio is equal to the ratio of the included homologous diagonals.

Proposition XXVIII. Problem

305. *To construct a polygon similar to a given polygon upon a line homologous to a side of the given polygon.*

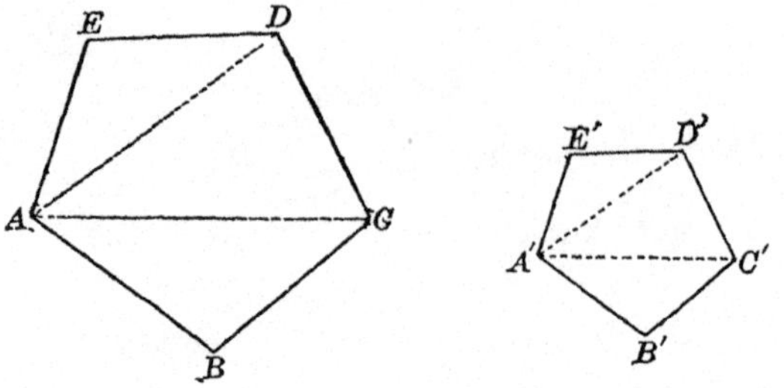

Given. Polygon, $ABCDE$, and line, $A'B'$.

Hint. — Draw AD and AC and make corresponding $\angle$s equal.

Proposition XXIX. Theorem

306. *The perimeters of two similar polygons are to each other as any two homologous sides.*

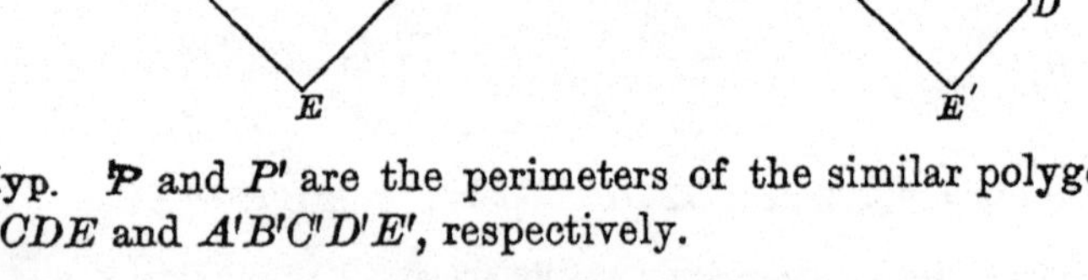

Hyp. P and P' are the perimeters of the similar polygons, $ABCDE$ and $A'B'C'D'E'$, respectively.

To prove $P : P' = AB : A'B'$.

Hint. — $AB : A'B' = BC : B'C' = CD : C'D' = DE : D'E' = EA : E'A'$. Apply (273).

Ex. 611. The sides of a polygon are 4, 5, 6, 7, and 8, respectively. Find the perimeter of a similar polygon, if the side corresponding to 5 is 7.

Ex. 612. The perimeters of two similar polygons are 20 and 25 in., respectively. If a side of the first polygon is four inches, find the homologous side of the second polygon.

PROPOSITION XXX. THEOREM

307. *In a right triangle, the altitude upon the hypotenuse is the mean proportional between the segments of the hypotenuse, and either arm is the mean proportional between the hypotenuse and the adjacent segment.*

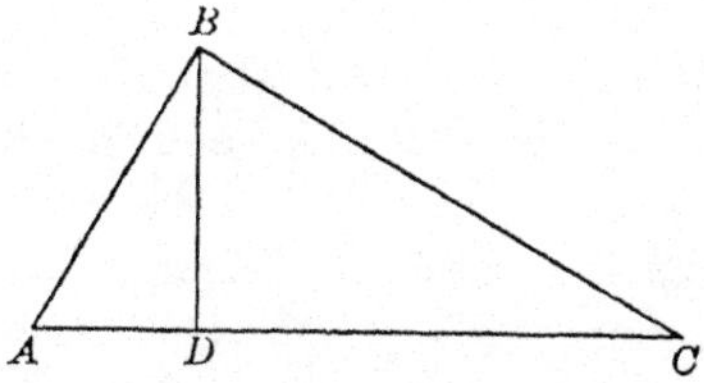

Hyp. In the rt. $\triangle ABC$, BD is the altitude upon the hypotenuse AC.

To prove (1) $AD : DB = DB : DC$.

(2) $AD : AB = AB : AC$.

HINT. $\triangle ABD \sim \triangle ABC$. (Why ?)

$\triangle CBD \sim \triangle ABC$. (Why ?)

$\therefore \triangle ABD \sim \triangle CBD$.

$\therefore AD : DB = DB : DC$,

and $AD : AB = AB : AC$. Q.E.D.

308. COR. The perpendicular from any point in the circumference upon the diameter is the mean proportional between the segments of the diameter; and the chord joining the point to either extremity is the mean proportional between the diameter and the adjacent segment.

Ex. 613. In the diagram for Prop. XXX, if $AD = 4$ and $AC = 9$, what is the value of AB?

Proposition XXXI. Problem

309. *To construct the mean proportional between two given lines.*

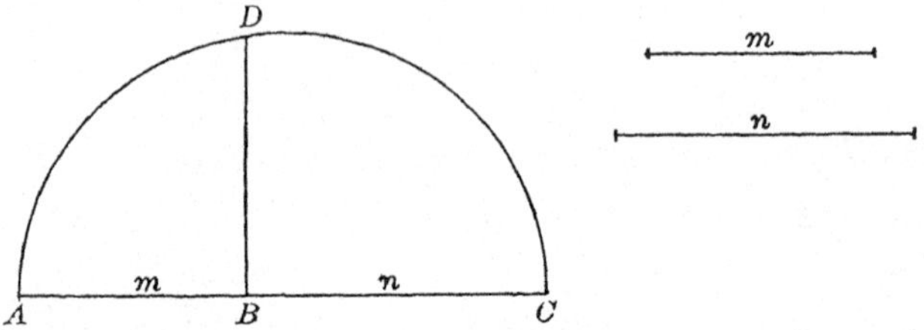

Given. Lines m and n.

Required. The mean proportional between m and n.

Construction I. Draw $AB = m$, and produce AB to C so that $BC = n$.

On AC as a diameter, describe a semicircle.

At B, erect a perpendicular upon AC, meeting the circumference in D.

BD is the required mean proportional.

[The proof is left to the student.]

Construction II. Draw $AB = n$.

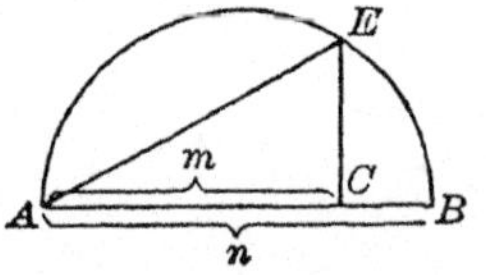

On AB as a diameter, describe a semicircle.

On AB, lay off $AC = m$.

Draw $CE \perp AB$. Join AE.

AE is the required mean proportional.

[The proof is left to the student.]

Ex. 614. If a and b are given lines, construct $\sqrt{ab}$.

Ex. 615. Construct $\sqrt{6\,ab}$, if a and b are two given lines.

Ex. 616. Construct a line equal to $a\sqrt{2}$, if a is a given line.

(Hint, $a\sqrt{2} = \sqrt{(2\,a)\cdot a}$.)

Ex. 617. Construct a line equal to $a\sqrt{5}$, if a is a given line.

Ex. 618. Construct a line equal to $a\sqrt{\frac{5}{7}}$, if a is a given line.

PROPOSITION XXXII. THEOREM

310. *The sum of the squares of the arms of a right triangle is equal to the square of the hypotenuse.*

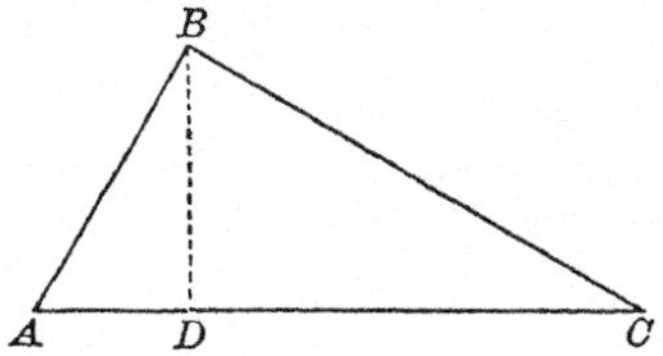

Hyp. ABC is a rt. $\triangle$, having its rt. $\angle$ at B.

To prove $\overline{AB}^2 + \overline{BC}^2 = \overline{AC}^2$.

Proof. Draw $BD \perp AC$.

$AD : AB = AB : AC.$ (307)

$\overline{AB}^2 = AD \times AC.$ (262)

Similarly $\overline{BC}^2 = DC \times AC.$

By adding $\overline{AB}^2 + \overline{BC}^2 = AC(AD + DC),$

or $\overline{AB}^2 + \overline{BC}^2 = \overline{AC}^2.$

Q.E.D.

311. COR. The square of either arm of a right triangle is equal to the square of the hypotenuse, diminished by the square of the other arm.

Ex. 619. Find the hypotenuse of a right triangle, whose arms are 1 ft. and 5 in. respectively.

Ex. 620. The hypotenuse of a right triangle is 25, one arm equals 20, find the other arm.

Ex. 621. Find the altitude of an equilateral triangle whose side is equal to 8 in.

Ex. 622. Find the altitude of an equilateral triangle whose side is equal to a.

Ex. 623. Find the altitude of an isosceles triangle whose base equals 8 and whose arm equals 5.

Ex. 624. If the hypotenuse of an isosceles right triangle equals 8 in., what is the length of an arm?

Ex. 625. The squares of the two arms of a right triangle have the same ratio as the adjacent segments of the hypotenuse.

Ex. 626. If AD is an altitude of a triangle ABC,

$$\overline{AB}^2 - \overline{AC}^2 = \overline{BD}^2 - \overline{CD}^2.$$

Ex. 627. If the diagonals of a quadrilateral are perpendicular to each other, then the sum of the squares of two opposite sides is equal to the sum of the squares of the other two.

Ex. 628. If the square of one side of a triangle is equal to the sum of the squares of the other two sides, the triangle is a right triangle.

* **Ex. 629.** If the square of one side of a triangle is greater than the sum of the squares of the other two, the triangle is obtuse.

* **Ex. 630.** If from a point A two equal tangents AB and AC, are drawn to two circles O and O', and AD is perpendicular to OO', then

$$\overline{OD}^2 - \overline{O'D}^2 = \overline{OB}^2 - \overline{O'C}^2.$$

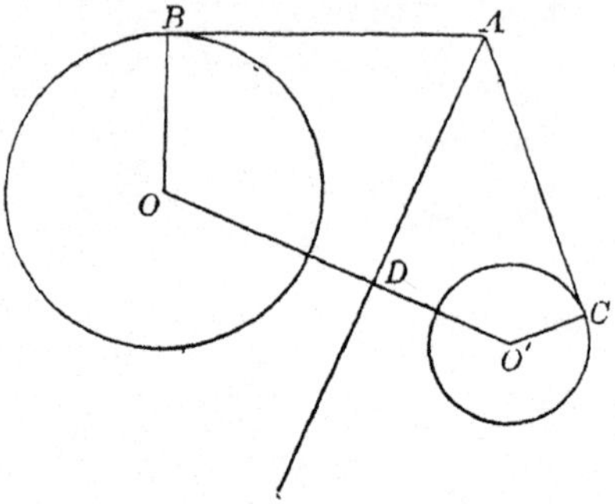

* **Ex. 631.** Conversely, if, in the same diagram, D be taken so that

$$\overline{OD}^2 - \overline{O'D}^2 = \overline{OB}^2 - \overline{O'C}^2,$$

then the tangents drawn from any point in the perpendicular, AD, to the circles are equal.

Proposition XXXIII. Theorem

312. *If two chords intersect within a circle, the product of the segments of one is equal to the product of the segments of the other.*

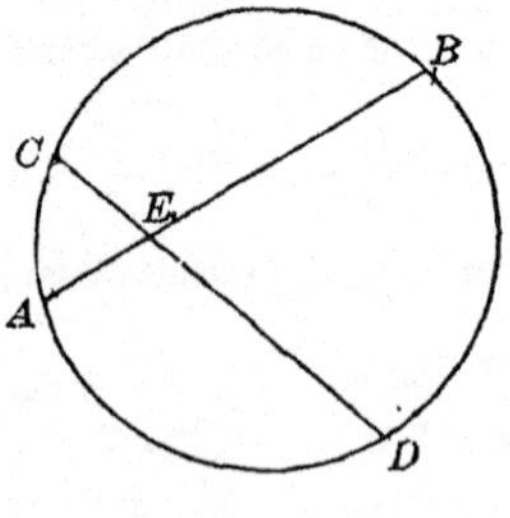

Hyp. The chords AB and CD meet in E.

To prove $AE \times EB = CE \times ED$.

HINT. — What is the means of proving that the product of two lines is equal to the product of two other lines?

Ex. 632. In the diagram for Prop. XXXIII, if $AE = 3$, $EB = 4$, $ED = 6$, find CE.

Ex. 633. In the same diagram, if $AE = a$, $EB = b$, and $ED = c$, find CE.

Ex. 634. In the same diagram, if $AE = 4$, $EB = 9$, and $CE = ED$, find ED.

Ex. 635. If the prolongations of two chords meet without a circle, is Prop. XXXIII correct for the external segments of the chords?

Ex. 636. If two lines AB and CD intersect in E so that

$$AE \times EB = CE \times ED,$$

then A, B, C, and D are concyclic. (See Ex. 372.)

PROPOSITION XXXIV. THEOREM

313. *If, from a point without a circle, two secants are drawn, the product of one secant and its external segment is equal to the product of the other and its external segment.*

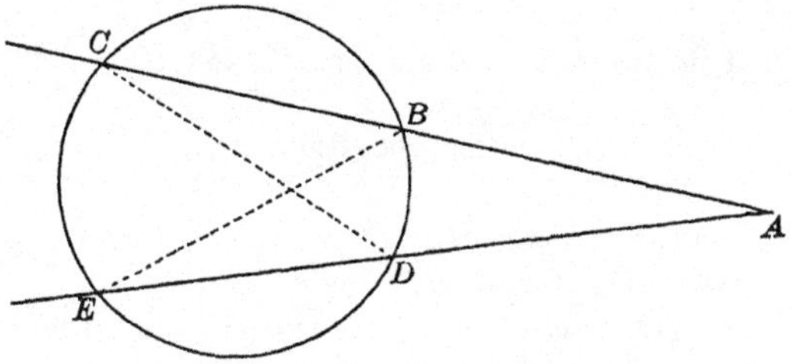

Hyp. Two secants AC and AE cut the circle in B, C, and D, E, respectively.

To prove $AC \times AB = AE \times AD$.

[The proof is left to the student.]

Ex. 637. If, in the diagram for Prop. XXXIV, $CB = 16$, $BA = 2$, $DA = 4$ find DE.

Ex. 638. If, in the same diagram, AE should revolve about A, until AE becomes a tangent, what would $AB \times AC$ be equal to.

Proposition XXXV. Theorem

314. *If, from a point without a circle, a tangent and a secant be drawn, the tangent is the mean proportional between the secant and its external segment.*

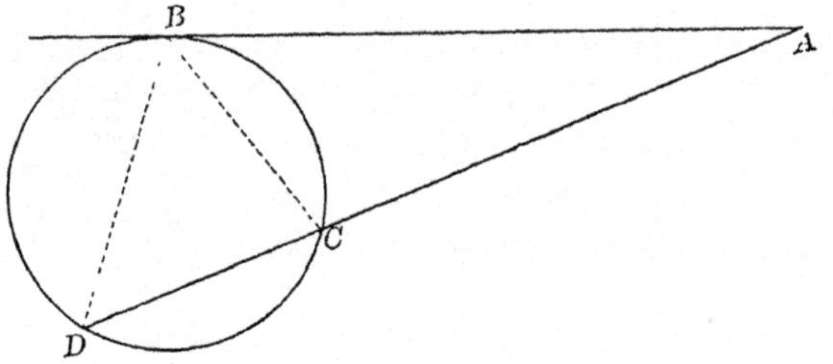

Hyp. The tangent AB touches the $\odot BDC$ in B, and the secant AD cuts the circle in C and D.

To prove $AD : AB = AB : AC.$

[The proof is left to the student.]

Ex. 639. If, in the diagram for Prop. XXXV, $DC = 5$, $AC = 4$, find AB.

Ex. 640. Construct the mean proportional between two given lines by means of Prop. XXXV.

Ex. 641. Tangents to two intersecting circles, drawn from any point in the common chord produced, are equal.

Ex. 642. If the diameter of the earth is equal to 8000 mi., how far can you see from a lighthouse 100 ft. high ?

Ex. 643. If two circles are tangent externally and from any point in the common internal tangent secants are drawn to the two circles, the products of the secants and their external segments will be equal.

* **Ex. 644.** To construct a circle passing through two given points and touching a given line.

315. Def. A straight line is said to be divided in **extreme and mean ratio,** when it is divided into two segments, such that *the greater is the mean proportional between the smaller and the whole line.*

Thus, AB is divided by C in extreme and mean ratio, if

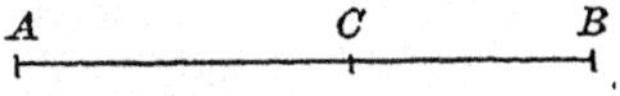

$$AB : AC = AC : CB.$$

Proposition XXXVI. Problem

316. *To divide a line in extreme and mean ratio.*

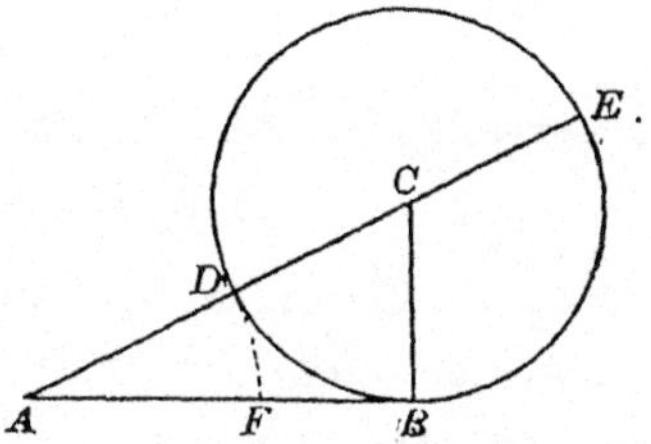

Given. Line AB.

Required. To divide AB in extreme and mean ratio.

Construction. At B, draw $BC = \frac{1}{2} AB$ and $\perp$ to AB.

From C as a center, with a radius CB, describe a circumference.

Draw AC meeting the circumference in D and E.

On AB, lay off $AF = AD$.

Then AB is divided as required.

Proof. $$AE : AB = AB : AD.$$

$$AE - AB : AB = AB - AD : AD. \qquad (271)$$

But $$AB = DE \text{ and } AD = AF.$$

$$\therefore AF : AB = FB : AF,$$

or $$AB : AF = AF : FB.$$

Q.E.D.

317. Def. **The projection of a point upon a line is the foot of the perpendicular from the point to the line.**

318. Def. The **projection of one line upon another** is the length between the projections of the extremities of the first line upon the second.

Ex. 645. If the side of an equilateral triangle equals 10 in., what is the length of the projection of one side upon another?

Ex. 646. Find the projection of AB upon a line XY, if AB and XY include an angle of 45°, and $AB = 2$.

Ex. 647. Find the projection of AB upon XY, if $AB = m$, and the two lines include an angle (a) of 60°, (b) of 30°. (Ex. 645.)

319. Note. — $\triangle abc$ denotes a triangle whose sides are a, b, and c. The other notations used in the following Propositions are in accordance with (230).

Proposition XXXVII. Theorem

320. *In any triangle, the square of a side opposite an acute angle is equal to the sum of the squares of the other two sides diminished by twice the product of one of those sides and the projection of the other side upon it.*

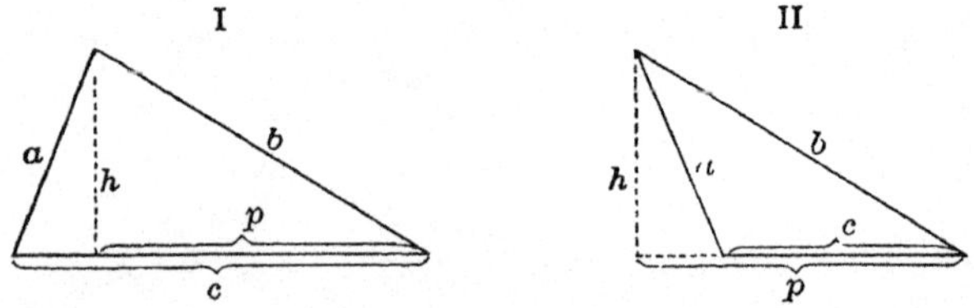

Hyp. In $\triangle abc$, p is the projection of b upon c, and the angle opposite a is an acute angle.

To prove $a^2 = b^2 + c^2 - 2\,cp$.

Proof. Denote the perpendicular upon c by h.

In the figure on the left

$$a^2 = h^2 + (c - p)^2,$$

but $$h^2 = b^2 - p^2,$$

$$\therefore a^2 = \quad ?$$

[To be completed by the student.]
Similarly for Fig. II.

321. Remark. The equation, $a^2 = b^2 + c^2 - 2\,cp$, contains four quantities. Therefore **any one of them may be found by algebraical methods if the other three are given.** (Similarly in the following propositions.)

Ex. 648. The sides of a triangle are 13, 14, and 15. Find the projection of 13 upon 14.

Proposition XXXVIII. Theorem

322. *In any obtuse triangle, the square of the side opposite the obtuse angle is equal to the sum of the squares of the other two sides, increased by twice the product of one of these sides and the projection of the other side upon it.*

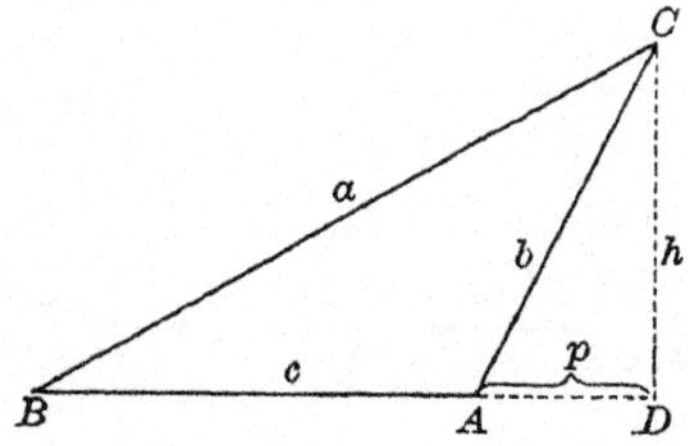

Hyp. In $\triangle abc$, p is the projection of b upon c, and the angle opposite a is obtuse.

To prove $a^2 = b^2 + c^2 + 2cp$.

Proof. $a^2 = h^2 + (c + p)^2$.

but $h^2 = b^2 - p^2$.

$\therefore\ a^2 = ?$

[To be completed by the student.]

Ex. 649. If two sides of a triangle equal 15 and 25, respectively, and the projection of 15 upon 25 equals 9, what is the value of the third side.

Ex. 650. Two sides of a triangle are 16 and 12 inches, respectively, and include an angle of 60°. Find the third side.

Ex. 651. Two sides of a triangle are 20 and 30, respectively, and include an angle of 45°. Find the third side.

323. SCHOLIUM. If we consider a projection of one side of a triangle upon another as positive when the projection lies on that line, but as negative when it lies on the prolongation, Props. XXXVIII and XXXII become special cases of Prop. XXXVII, and we have always:

$$a^2 = b^2 + c^2 - 2\,cp.$$

To compute the projections of sides of a triangle whose angles are not known, always apply this equation. If the result is negative, the triangle is obtuse.

324. COR. 1. In $\triangle abc$, if p_c denotes the projection of b upon c,

$$p_c = \frac{b^2 + c^2 - a^2}{2\,c}.$$

325. COR. 2. If h_c denotes the altitude upon c,

$$h_c = \sqrt{b^2 - p_c^{\,2}} = \sqrt{b^2 - \left(\frac{b^2 + c^2 - a^2}{2\,c}\right)^2}.$$

[This expression can be simplified by algebraical operations:

$$h_c^{\,2} = b^2 - \left(\frac{b^2 + c^2 - a^2}{2\,c}\right)^2 = \left(b + \frac{b^2 + c^2 - a^2}{2\,c}\right)\left(b - \frac{b^2 + c^2 - a^2}{2\,c}\right)$$

$$= \frac{(2\,bc + b^2 + c^2 - a^2)(2\,bc - b^2 - c^2 + a^2)}{4\,c^2}$$

$$= \frac{[(b + c)^2 - a^2][a^2 - (b - c)^2]}{4\,c^2}$$

$$= \frac{(a + b + c)(b + c - a)(a - b + c)(a + b - c)}{4\,c^2}.$$

Let $a + b + c = 2s$, *i.e.* let s denote half the perimeter.

$$b + c - a = 2(s - a).$$

$$a - b + c = 2(s - b).$$

$$a + b - c = 2(s - c).$$

$$h_c^2 = \frac{4s(s-a)(s-b)(s-c)}{c^2}$$

or

$$h_c = \frac{2}{c}\sqrt{s(s-a)(s-b)(s-c)}.$$

Ex. 652. In $\triangle abc$, $a = 20$, $b = 15$, and $c = 7$. Find the projection of b upon c. Is the triangle obtuse or acute?

Ex. 653. The sides of a triangle are 4, 13, and 15, respectively. Find the three altitudes.

Ex. 654. The sides of a triangle are 13, 14, and 15, respectively. Find the three altitudes.

Proposition XXXIX. Theorem

326. *In any triangle, the square of one side plus four times the square of the corresponding median is equal to twice the sum of the squares of the other sides.*

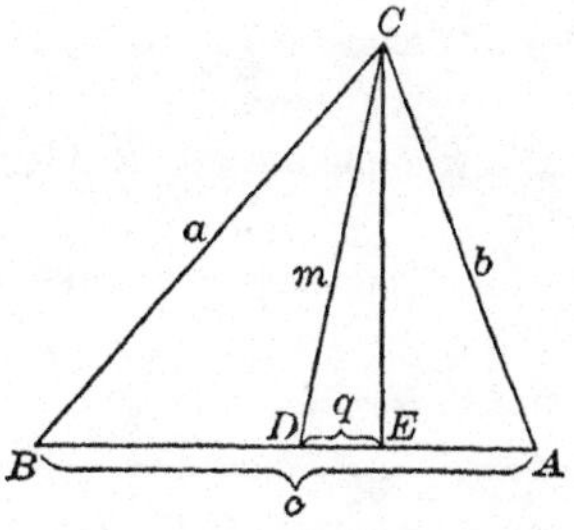

Hyp. In $\triangle ABC$, m_c is the median to c.

To prove $c^2 + 4m_c^2 = 2a^2 + 2b^2$.

Proof. Draw $CE \perp AB$, and suppose E to fall between A and D. Let $DE = q$.

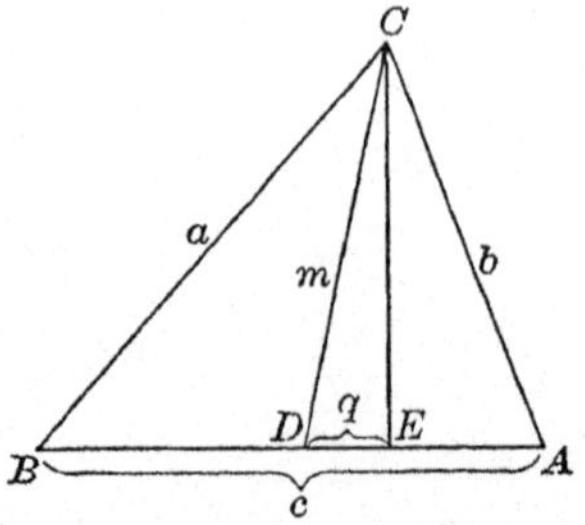

$$a^2 = \left(\frac{c}{2}\right)^2 + m^2 + 2\left(\frac{c}{2}\right)q.$$

$$b^2 = \left(\frac{c}{2}\right)^2 + m^2 - 2\left(\frac{c}{2}\right)q.$$

$$a^2 + b^2 = 2\left(\frac{c}{2}\right)^2 + 2m^2. \qquad \text{(Ax. 2.)}$$

or $$2a^2 + 2b^2 = c^2 + 4m^2. \qquad \text{(Ax. 6.)}$$ Q.E.D.

327. COR. In $\triangle abc$, if m_c denotes the median drawn to c, $m_c = \frac{1}{2}\sqrt{2a^2 + 2b^2 - c^2}$.

Ex. 655. The sides of a triangle are 9, 10, and 17, respectively. Find the three altitudes.

Ex. 656. The sides of a triangle are 11, 25, and 30, respectively. Find the three altitudes.

Ex. 657. The sides of a triangle are 7, 8, and 9, respectively. Find the length of the median to 8.

Ex. 658. The sides of a triangle are 7, 4, and 9, respectively. Find the length of the median to 9.

Ex. 659. In $\triangle abc$, $a = 8$, $b = 11$, and $m_c = 8\frac{1}{2}$. Find c.

Ex. 660. In $\triangle abc$, $a = 28$, $c = 32$, and $m_c = 38$. Find b.

Ex. 661. The sides of a triangle are 10, 5, and 9, respectively. Find the length of the median to 9.

Ex. 662. The sides of a triangle are 22, 20, and 18, respectively. Find the length of the median to 18.

Ex. 663. The sum of the squares of the four sides of a parallelogram is equal to the sum of the squares of the diagonals.

PROPOSITION XL. THEOREM

328. *In any triangle, the product of two sides is equal to the square of the bisector of the included angle plus the product of the segments of the third side.*

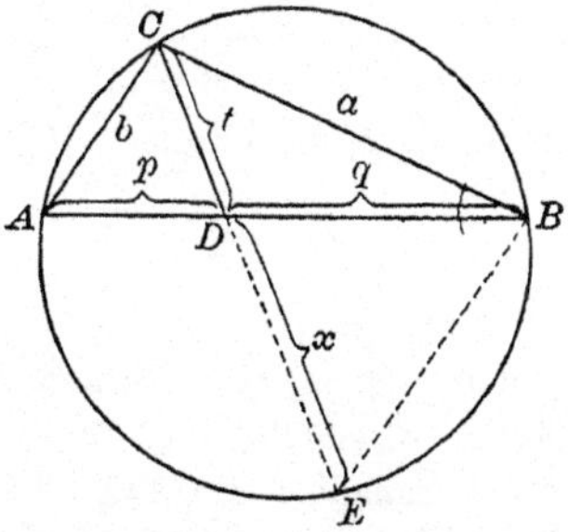

Hyp. In $\triangle abc$, the bisector t divides c into the segments, p and q.

To prove $ab = t^2 + pq$.

Proof. Circumscribe a circle about $\triangle abc$.

Produce the bisector CD to meet the circumference in E. Draw EB, and let $DE = x$.

$$\angle ACD = \angle ECB. \qquad \text{(Hyp.)}$$

$$\angle A = \angle E. \qquad (220)$$

$$\triangle ACD \sim \triangle EBC. \qquad (291)$$

$$\therefore b : t + x = t : a,$$

or
$$ab = t(t + x).$$

$$ab = t^2 + tx.$$

But
$$tx = pq. \qquad (312)$$

$$\therefore ab = t^2 + pq.$$

Q.E.D.

329. COR. $t^2 = ab - pq.$

By applying (287), it may be proven that

$$p = \frac{bc}{a+b},$$

$$q = \frac{ac}{a+b},$$

or

$$t^2 = ab - \frac{abc^2}{(a+b)^2}.$$

By using the method and notations of (325), we may obtain

$$t = \frac{2}{a+b}\sqrt{abs(s-c)}.$$

Ex. 664. The sides of a triangle are 18, 9, and 21, respectively. Find the length of the bisector corresponding with 21.

Ex. 665. The sides of a triangle are 21, 14, and 25, respectively. Find the length of the bisector corresponding with 25.

Ex. 666. The sides of a triangle are 22, 11, and 21, respectively. Find the length of the bisector corresponding with 21.

Ex. 667. The sides of a triangle are 6, 3, and 7, respectively. Find the length of the bisector corresponding with 7.

PROPOSITION XLI. THEOREM

330. *In any triangle the product of two sides is equal to the altitude upon the third side, multiplied by the diameter of the circumscribed circle.*

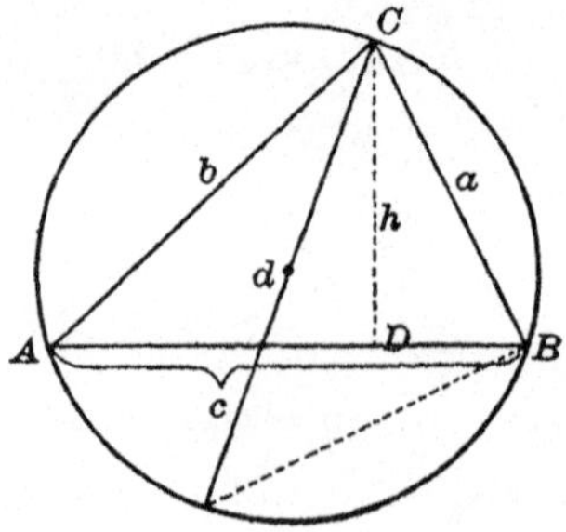

Hyp. d is a diameter of the circle circumscribed about $\triangle abc$, and h is the altitude upon c.

To prove $ab = hd$.

[The proof is left to the student.]

331. Cor. The diameter of the circumscribed circle of any triangle is equal to the product of two sides divided by the altitude upon the third side.

$$\left(d = \frac{ab}{h} \text{ or } d = \frac{abc}{2\sqrt{s(s-a)(s-b)(s-c)}}\right).$$

Ex. 668. $\triangle ABC$ is inscribed in a circle of radius $= 5$ inches. Find the altitude to BC if $AB = 4$, and $AC = 5$.

Ex. 669. The three sides of a triangle are 4, 13, and 15, respectively. Find the radius of the circumscribed circle.

Ex. 670. In $\triangle abc$, $a = 20$, $b = 15$, and the projection of b upon c equals 9. Find the radius of the circumscribed circle.

Ex. 671. In $\triangle abc$, $a = 9$ and $b = 12$. Find c if the diameter of the circumscribed circle equals 15.

PROBLEMS OF COMPUTATION

Ex. 672. A straight line $AB = 4$, is divided externally in the ratio $5:4$. Find the segments.

Ex. 673. The shadow of a church steeple upon level ground is 60 ft., while a pole 10 ft. high casts a shadow 3 ft. long. How high is the steeple?

Ex. 674. The arms of a right triangle are 8 and 15 respectively. Compute the hypotenuse and the altitude upon the hypotenuse.

Ex. 675. In $\triangle abc$, $a = 9$, $b = 15$, and $c = 17$. Is the triangle obtuse, right, or acute?

Ex. 676. The arms of a right triangle are $m^2 - n^2$ and $2mn$ respectively Find the hypotenuse.

Ex. 677. The distance of a chord 14 inches long from the center is 12 inches. Find the radius of the circle.

Ex. 678. The distance of a chord 24 inches long from the center is 5 inches. Find the distance of a chord 10 inches long from the center.

Ex. 679. In $\triangle abc$, $a = 5$, $b = 8$, and the angle opposite to c equals $60°$. Find c.

Ex. 680. In $\triangle abc$, $a = 8$, $b = 15$, and the angle opposite to c equals $60°$. Find c.

Ex. 681. In $\triangle abc$, $a = 3$, $b = 5$, and the angle opposite to c equals $120°$. Find c.

Ex. 682. In $\triangle abc$, $a = 7$, $b = 8$, and the angle opposite to c equals $120°$. Find c.

Ex. 683. In $\triangle abc$, $a = 10$, $b = 17$, $c = 21$. Find the altitude upon 10.

Ex. 684. The line of centers of two circles is equal to 30. Find the length of the common chord if the radii are 8 and 26 respectively.

Ex. 685. The base of an isosceles triangle is 48 inches. Find the altitude if each arm equals 50 inches.

Ex. 686. The diagonal of a square is 20 inches. Find the side.

Ex. 687. The sides of a rectangle are 16 and 30 respectively. Find the diagonal.

Ex. 688. The diameter AB of a circle is produced to C, and from C a tangent is drawn to the circle. Find the length of the tangent if $AB = 30$ and $BC = 2$.

Ex. 689. In $\triangle abc$, $a = 11$, $b = 9$, and $c = 16$. Find the median to 16.

Ex. 690. In $\triangle abc$, $a = 16$, $b = 18$, and $c = 22$. Find the median to b.

Ex. 691. The base of an isosceles triangle is 4, and the arm 7. Find the median to one of the arms.

Ex. 692. A ladder 17 ft. long reaches a window 15 ft. high. How far is the lower end of the ladder from the house?

Ex. 693. In $\triangle abc$, $a = 18$, $b = 23$, and $c = 9$. Find the bisector of the angle opposite b.

Ex. 694. In $\triangle abc$, $a = 50$, $b = 32$, and $c = 78$. Find the radius of the circumscribed circle.

Ex. 695. The base of an isosceles triangle is b, and each arm a. Find the altitude.

Ex. 696. The non-parallel sides AB and CD of a trapezoid are produced till they meet in E. Find AE and BE if $AB = 7$ and the bases are 5 and 3 respectively.

Ex. 697. The altitude of a trapezoid is h, the bases a and b respectively. Find the altitudes of the two triangles formed by producing the non-parallel sides until they meet.

Ex. 698. From a point 24 ft. above sea-level the visible horizon has a radius of 6 miles. Find the diameter of the earth.

***Ex. 699.** In a quadrilateral $ABCD$ $AB = 10$, $BC = 17$, $CD = 13$, $DA = 20$, and $AC = 21$. Find the diagonal BD.

***Ex. 700.** Find the greatest segment of a line 10 inches long, when it is divided in extreme and mean ratio.

PROBLEMS OF CONSTRUCTION

Ex. 701. In a given line AB, to find a point C such that $AC : BC = 1 : \sqrt{2}$.

Ex. 702. To divide a given line into segments proportional to any number of given lines by means of Prop. XXIV.

Ex. 703. Divide any side of a triangle into two parts proportional to the other two sides.

To construct a triangle, having given: (230)

Ex. 704. a, b, $(b : c)$.

Ex. 705. a, $(a : b)$, and $(a : c)$.

Ex. 706. a, $b + c$, and $(b : c)$.

Ex. 707. A, $(b : c)$, t_c.

Ex. 708. To construct the fourth proportional to three given lines by means of Prop. XXXIV.

Ex. 709. From a given rectangle to cut off a similar rectangle by a line parallel to one of its sides.

Ex. 710. In a given circle, to inscribe a triangle similar to a given triangle.

Ex. 711. About a given circle, to circumscribe a triangle, similar to a given triangle.

Ex. 712. Construct a triangle similar to a given triangle and having a given altitude.

Ex. 713. To inscribe a square in a given triangle. (Ex. 610.)

Ex. 714. In a given triangle, ABC, to inscribe a parallelogram similar to a given parallelogram, so that one side lies in AB, and the other two vertices lie in BC and AC respectively. (Ex. 610.)

Ex. 715. To draw a parallel to one side of a triangle, cutting off another triangle of given perimeter.

Ex. 716. From a point without a circle, to draw a secant whose external segment is equal to one-half the secant.

Ex. 717. To construct a circle, touching a given circle in a given point, and touching a given line.

Ex. 718. To construct a circle, touching two parallel lines and passing through a given point.

Ex. 719. About a given circle, to circumscribe a rhombus, having given an angle.

Ex. 720. In a given circle, to inscribe a rectangle, having given the ratio of two sides.

***Ex. 721.** To divide a trapezoid into two similar trapezoids by a line parallel to the base.

Ex. 722. In the prolongation of the side AB of the triangle ABC to find a point X such that $AX \times BX = CX^2$.

Ex. 723. Through a given point P, to draw a line such that its distances from two other given points, R and S, shall have a given ratio.

***Ex. 724.** Through a given point within a circle, to draw a chord so that its segments have a given ratio.

***Ex. 725.** To construct a circle passing through a given point and touching two given lines.

***Ex. 726.** Through a point of intersection of two circles, to draw a line forming chords which have a given ratio.

Ex. 727. To construct two lines, having given their mean proportional and their difference.

THEOREMS

Ex. 728. If a chord is bisected by another, either segment of the first is a mean proportional between the segments of the other.

Ex. 729. If in the triangle ABC the altitudes BD and AE meet in F, and $AB = BC$, then

$$BC : AF = BD : CD.$$

Ex. 730. Two triangles are similar if an angle of the one is equal to an angle of the other, and the altitudes corresponding with the other angles are proportional.

Ex. 731. If, between two parallel tangents, a third tangent is drawn, the radius is the mean proportional between the segments of the third tangent.

Ex. 732. If two circles are tangent externally, and through the point of contact a secant is drawn, the chords formed are proportional to the radii.

Ex. 733. If C is the midpoint of the arc AB, and a chord CD meets the chord AB in E, then

$$CE : CA = CA : CD.$$

Ex. 734. If two circles intersect, their common chord produced bisects the common tangents.

Ex. 735. If an isosceles triangle is inscribed in a circle, the tangents drawn at the vertices form another isosceles triangle.

Ex. 736. The tangents drawn at the vertices of an inscribed rectangle enclose a rhombus.

Ex. 737. Two parallelograms are similar when they have an angle of the one equal to an angle of the other, and the including sides proportional.

Ex. 738. Two rectangles are similar if two adjacent sides are proportional.

Ex. 739. A circumference described upon the arm of an isosceles triangle as a diameter bisects the base.

Ex. 740. Three times the sum of the squares of the sides of a triangle is equal to four times the sum of the squares of the medians.

Ex. 741. If in the parallelogram

$$ABCD\ \angle A = 60^\circ,\ \overline{AC}^2 = \overline{AB}^2 + \overline{BC}^2 + AB \times BC.$$

Ex. 742. If the altitude upon the hypotenuse of a right triangle divides the hypotenuse in extreme and mean ratio, the smaller arm is equal to the non-adjacent segment of the hypotenuse.

Ex. 743. If in a triangle the squares of two unequal sides have the same ratio as their projections, upon the third side, the triangle is a right triangle.

Ex. 744. If from a point O, OA, OB, OC, and OD are drawn so that the angle AOB is equal to the angle BOC, and the angle BOD equal to a right angle, any line intersecting OA, OB, OC, and OD is divided harmonically. (287, 288.)

Ex. 745. The sum of the squares of the four sides of any quadrilateral is equal to the sum of the square diagonals plus four times the square of the line joining the midpoints of the diagonals. (326.)

* **Ex. 746.** If, from a point within the triangle ABC the perpendiculars OX, OY, and OZ be drawn upon AB, BC, and CA, respectively,

$$\overline{AX}^2 + \overline{BY}^2 + \overline{CZ}^2 = \overline{BX}^2 + \overline{YC}^2 + \overline{ZA}^2.$$

* **Ex. 747.** State and prove the converse of the preceding exercise.

* **Ex. 748.** If two circles are tangent externally, the common tangent is a mean proportional between the diameters.

Ex. 749. If through a vertex C of triangle ABC a line CE is drawn parallel to AB, and any point H in CE is joined to F, the midpoint of AB, then FH is divided harmonically by BC and AC produced. (300.)

BOOK IV

AREAS OF POLYGONS

332. Def. The **unit of surface** is a square whose side is the unit of length, *e.g.* a square inch, a square meter, etc.

333. The **area of a surface** is the number of units of surface it contains.

334. Two figures are **equivalent** if their areas are equal.

Proposition I. Theorem

335. *Rectangles having equal altitudes are to each other as their bases.*

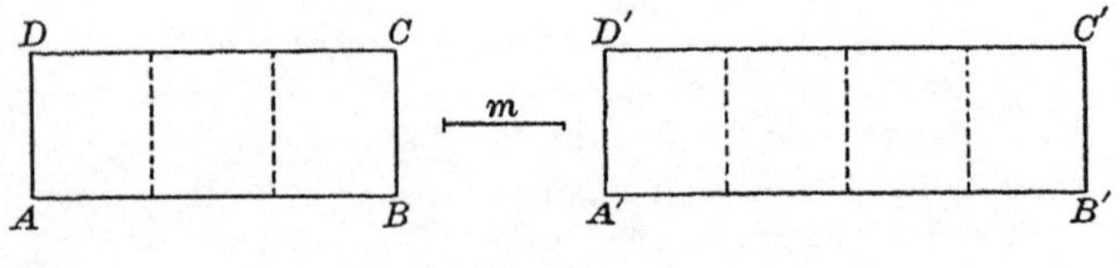

Hyp. In rectangles $ABCD$ and $A'B'C'D'$, altitude $AD =$ altitude $A'D'$.

To prove $$\frac{ABCD}{A'B'C'D'} = \frac{AB}{A'B'}.$$

Proof. Case I. AB and $A'B'$ are commensurable.

Let m be a common measure contained in AB 3 times, and in $A'B'$ 4 times.

[To be completed by the student. Compare (280).]

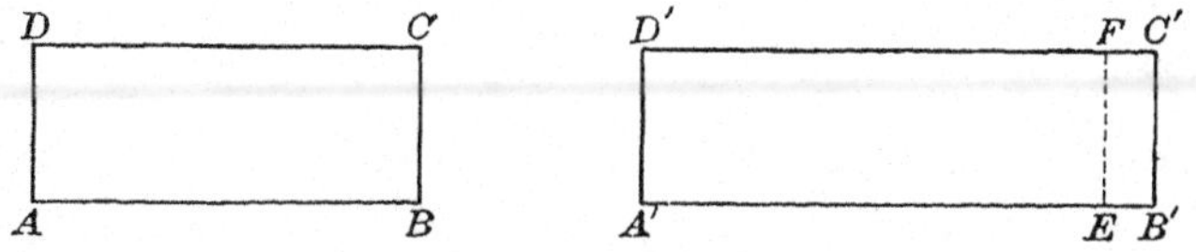

CASE II. AB and $A'B'$ are incommensurable.

Divide AB into any number of equal parts, and lay off one of these parts on $A'B'$ as often as possible.

As AB and $A'B'$ are incommensurable, there must be a remainder, EB', less than one of the equal parts.

Draw $EF \perp A'B'$.

As AB and $A'E$ are commensurable,

$$\frac{ABCD}{A'EFD'} = \frac{AB}{A'E}.$$

By increasing the number of parts into which AB is divided, we can diminish the length of these parts, and, therefore, the length of EB' indefinitely.

Hence, $A'E$ approaches $A'B'$ as a limit, and $A'EFD'$ approaches $A'B'C'D'$ as a limit.

The variables $\frac{ABCD}{A'EFD'}$ and $\frac{AB}{A'E}$ being always equal, must have equal limits.

Whence $$\frac{ABCD}{A'B'C'D'} = \frac{AB}{A'B'}.$$ Q.E.D.

336. COR. Rectangles having equal bases are to each other as their altitudes.

337. NOTE. — As a polygon can be measured by the unit of surface only, the words "triangle," "quadrilateral," etc., are frequently used for the area of a triangle, area of a quadrilateral, etc.

Ex. 750. From a given rectangle, to cut off another rectangle whose area is $\frac{4}{5}$ of the given one.

Ex. 751. To construct a rectangle, which is to a given one as $m : n$, when m and n are two given lines.

PROPOSITION II. THEOREM

338. *The areas of two rectangles are to each other as the products of their bases and altitudes.*

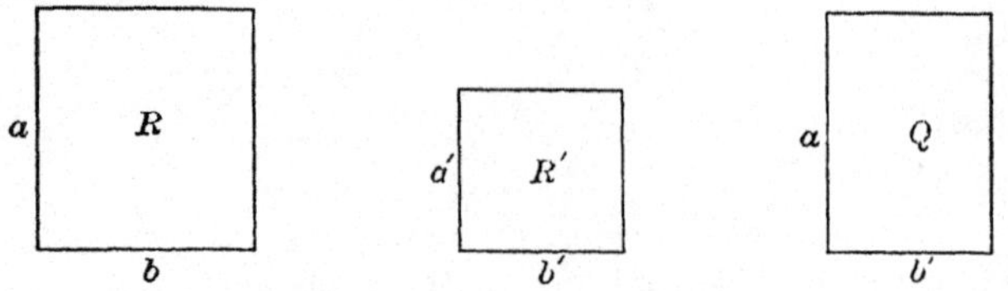

Hyp. Rectangles R and R' have the bases b and b' and the altitudes a and a' respectively.

To prove $$\frac{R}{R'} = \frac{a \times b}{a' \times b'}.$$

Proof. Construct the rectangle Q, having the same base as R', and the same altitude as R.

$$\frac{R}{Q} = \frac{b}{b'}. \qquad (335)$$

$$\frac{Q}{R'} = \frac{a}{a'}.$$

Multiplying member by member,

$$\frac{R}{R'} = \frac{a \times b}{a' \times b'}.$$

Q.E.D

Ex. 752. Find the ratio of a rectangle 4 by 5 feet, and a square having a side of 10 feet.

Ex. 753. The diagonal of a rectangle is 26 inches long, and one of its sides 24 inches. The diagonal of another rectangle is 25 inches long, and one of its sides 20 inches. Find the ratio of the areas of the two rectangles.

PROPOSITION III. THEOREM

339. *The area of a rectangle is equal to the product of its base and altitude.*

Hyp. R is a rectangle with base b and altitude a.

To prove area of $R = a \times b$.

Proof. Let U be the unit of surface.

Then $$\frac{R}{U} = \frac{a \times b}{1 \times 1}. \quad (338)$$

But $\frac{R}{U}$ is the area of R. (333)

$\therefore$ area of $R = a \times b$. Q.E.D.

340. SCHOLIUM. If the base and altitude of a rectangle are expressed by integral numbers, Prop. III may be proved by dividing the figure into squares.

Thus, if the base contains 5 and the altitude 3 linear units, the figure can be divided into fifteen squares, each being the unit of surface.

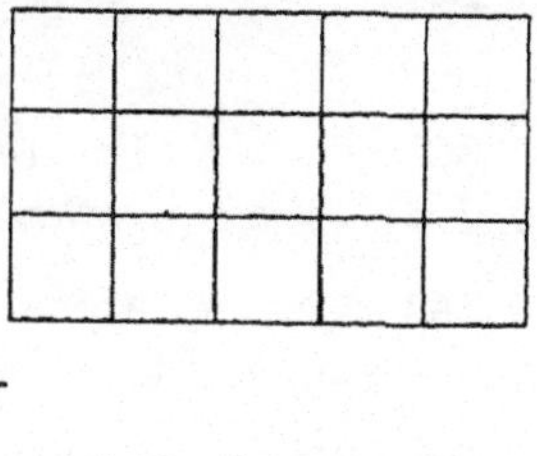

Ex. 754. A rectangular field is 24 yds. long and 15 yds. wide. Find the area.

Ex. 755. The area of a rectangle is 360 square inches, and its base is 4 yds. Find the altitude.

Proposition IV. Theorem

341. *The area of a parallelogram is equal to the product of its base and its altitude.*

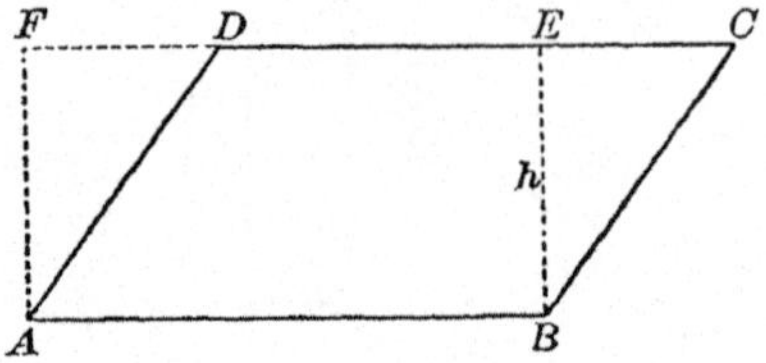

Hyp. $ABCD$ is a parallelogram, having the base $AB = b$, and the altitude $BE = h$.

To prove area of $ABCD = b \times h$.

Proof. Draw $AF \perp AB$, meeting CD produced in F.

Then $ABEF$ is a rectangle, having its base b and its altitude h.

$$ABCF = ABCF,$$
(identical).

But $$\triangle AFD = \triangle BEC,$$
(hy. arm = hy. arm).

Subtracting member from member,

$$\text{area } ABCD = \text{area } ABEF.$$

But $$\text{area } ABEF = b \times h. \qquad (339)$$

$$\therefore \text{area } ABCD = b \times h.$$ Q.E.D.

342. Cor. 1. Parallelograms having equal bases and equal altitudes are equivalent.

343. Cor. 2. Any two parallelograms are to each other as the products of their bases and altitudes.

344. Cor. 3. Parallelograms having equal bases are to each other as their altitudes.

345. Cor. 4. Parallelograms having equal altitudes are to each other as their bases.

Ex. 756. Find the area of a parallelogram whose base is 15 inches and whose altitude is 2 feet.

Ex. 757. Two sides of a parallelogram are 15 and 20 respectively, and include an angle of 30°. Find the area.

Ex. 758. The sides of a parallelogram are 5 and 8 respectively, and the projection of 5 upon 8 is 3. Find the area.

Ex. 759. To divide a parallelogram into three equivalent parts.

PROPOSITION V. THEOREM

346 *The area of a triangle is equal to one-half the product of its base and altitude.*

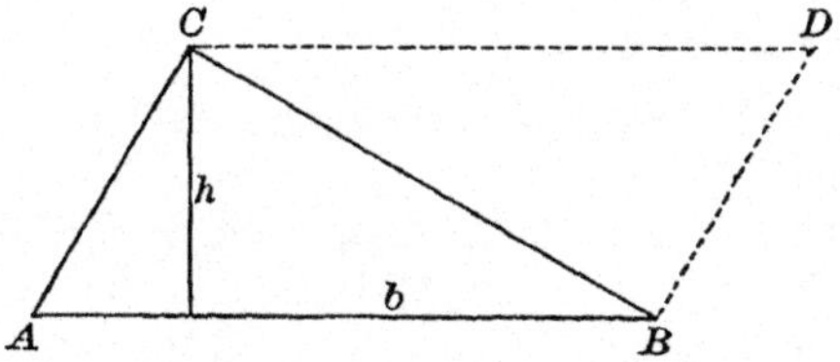

Hyp. In $\triangle ABC$, the base is b, and the altitude h.

To prove area of $ABC = \frac{1}{2} b \times h$.

Proof. Construct the parallelogram $ABDC$.

The diagonal of a parallelogram divides it into two equal triangles.

$$\therefore \triangle ABC \approx \tfrac{1}{2} \square ABDC.$$

But area $ABDC = b \times h$.

$$\therefore \text{area } ABC = \tfrac{1}{2} b \times h.$$

Q.E.D.

347. COR. 1. Triangles having equal bases and equal altitudes are equivalent.

348. COR. 2. Any two triangles are to each other as the products of their bases and altitudes.

349. COR. 3. Triangles having equal bases are to each other as their altitudes.

350. Cor. 4. Triangles having equal altitudes are to each other as their bases.

351. Def. To **transform a figure** means to find another figure equivalent to the given one.

Ex. 760. If A denotes the area and $2s$ the perimeter of $\triangle abc$, then $A = \sqrt{s(s-a)(s-b)(s-c)}$.

Ex. 761. To transform a given triangle into another one, having the same base.

Ex. 762. What is the locus of the vertices of all the equivalent triangles constructed on the same base?

Ex. 763. To transform a given triangle into a right triangle.

Ex. 764. To transform $\triangle ABC$ into an isosceles triangle, having its base equal to AB.

Ex. 765. To transform $\triangle ABC$ into an isosceles triangle, having an arm equal to AB.

Ex. 766. In Exs. 761–765, how many parts of the given triangle are not altered by the transformation?

Ex. 767. To transform a given triangle into another one, having given one side.

Ex. 768. To transform a given triangle into a right triangle, having the hypotenuse equal to one side of the given triangle.

Ex. 769. To transform a given triangle into another one, having its vertical angle equal to a given angle.

352. Remark. — *Cor. 1 of Prop. V is the principal means for* **transformation of figures.** *Frequently two transformations are necessary.*

Ex. 770. To transform $\triangle ABC$ into another triangle, having two of its sides respectively equal to two given lines m and n.

Ex. 771. To transform $\triangle ABC$ into another triangle, having a side equal to a given line m, and one adjacent angle equal to a given angle A.

Ex. 772. To transform a given triangle into a right triangle, having a given arm.

Ex. 773. To transform $\triangle ABC$ into a right triangle, having the hypotenuse equal to a given line m.

Ex. 774. To transform $\triangle ABC$ into an isosceles triangle, having the base equal to a given line m.

Ex. 775. To transform $\triangle ABC$ into an isosceles triangle, having an arm equal to a given line m.

Ex. 776. To transform $\triangle ABC$ into a triangle having one side equal to a given line m, and the opposite angle equal to a given angle A.

Ex. 777. To transform a parallelogram into a rectangle.

Ex. 778. To transform a parallelogram into a rhombus.

Ex. 779. To transform a parallelogram into another one having a given side.

Ex. 780. To transform a parallelogram into another one containing a given angle.

Ex. 781. To transform a parallelogram into a rectangle, having one side equal to a given line.

Ex. 782. To divide a triangle ABC into three equivalent parts by two lines passing through A.

Ex. 783. To divide a given parallelogram into two equivalent parts by a line parallel to the bases.

Ex. 784. To divide a given parallelogram into two equivalent parts by a line perpendicular to the bases.

Ex. 785. To divide a given parallelogram into two equivalent parts by a line parallel to a given line.

Ex. 786. To transform a given quadrilateral $ABCD$ into another one, having three of its vertices in A, B, and C, respectively.

Ex. 787. To transform a given quadrilateral into a triangle.

Ex. 788. To transform $\triangle ABC$ into a triangle containing the angle A, and having one vertex in D, if D is a given point in AC.

HINT. — Consider $ABCD$ a quadrilateral, and apply Ex. 787.

Ex. 789. The diagonals divide a parallelogram into four equivalent triangles.

Ex. 790. Two triangles are equivalent if two sides of the one are respectively equal to two sides of the other, and the included angles are supplementary.

Ex. 791. The lines joining the midpoint of a diagonal of a quadrilateral with the opposite vertices divide the figure into two equivalent parts.

Ex. 792. To divide a quadrilateral into three equivalent parts.

Ex. 793. The area of a triangle is 600 square inches, and the altitude is 20 inches. Find the base.

Ex. 794. Two sides of a triangle are 5 and 8, respectively, and include an angle of 30°. Find the area.

Ex. 795. Find the area of a triangle whose sides are respectively (*a*) 13, 14, 15, (*b*) 9, 10, 17, (*c*) 11, 25, 30.

Ex. 796. To construct a triangle three times as large as a given triangle.

PROPOSITION VI. PROBLEM

353. *To transform a polygon into a triangle.*

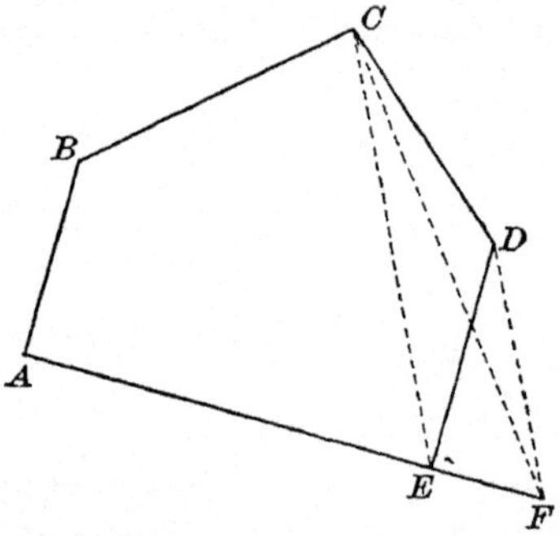

Given. Polygon $ABCDE$.

Required. To construct a triangle equivalent to $ABCDE$.

Construction. Draw CE.
Through D, draw $DF \parallel CE$, meeting AE produced in F.
Then polygon $ABCF \approx ABCDE$, and has one less side.
Continue the process until a triangle is obtained.

[The proof is left to the student.]

Ex. 797. To transform a parallelogram into a right triangle.

Ex. 798. To transform a trapezoid into an isosceles triangle.

PROPOSITION VII. THEOREM

354. *The area of a trapezoid is equal to one-half the product of its altitude and the sum of its bases.*

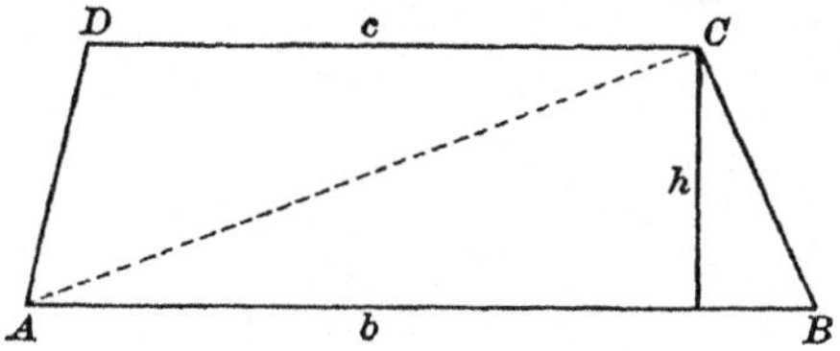

Hyp. Trapezoid $ABCD$ has the bases b and c respectively, and the altitude h.

To prove area $ABCD = \frac{1}{2}h(b+c)$.

Proof. Draw AC.

Area $ABC = \frac{1}{2}h \times b$. (346)

Area $DCA = \frac{1}{2}h \times c$. (346)

$\therefore$ area $ABCD = \frac{1}{2}h(b+c)$. Q.E.D.

Ex. 799. A line joining the midpoints of the bases of a trapezoid divides the trapezoid into two equivalent parts.

Ex. 800. The bases of a trapezoid are 12 and 8 respectively, and the altitude is 5. Find the area.

Ex. 801. In the diagram for Prop. VII, if $b = 6$, $c = 4$, $BC = 10$, and $\angle B = 30°$. Find the area.

Ex. 802. The area of a trapezoid is 200, the bases are 15 and 25 respectively. Find the altitude.

Ex. 803. The area of a trapezoid is 30, the altitude is 5, and one base is 8. Find the other base.

Ex. 804. In the diagram for Prop. VII if E, the midpoint of DA, be joined to C and B, $\triangle CBE$ is one-half of $ABCD$.

Ex. 805. If the three altitudes of a triangle are equal, the triangle is equilateral.

PROPOSITION VIII. THEOREM

355. *The square constructed on the sum of two lines is equivalent to the sum of the two squares constructed on these lines, increased by twice the rectangle of these lines.*

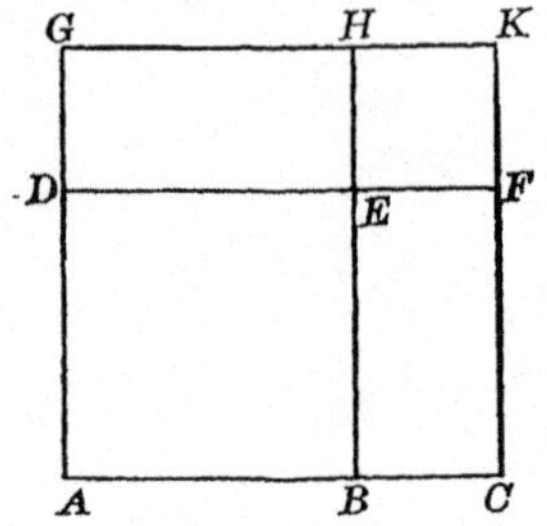

Hyp. Line $AC = AB + BC$.

To prove the square on $AC \Leftrightarrow$ to the sum of the squares on AB and BC increased by twice the rectangle contained by AB and BC.

Proof. On AC, construct the square $ACKG$.

On AG, lay off $AD = AB$.

Through B, draw $BH \parallel AG$, through D, draw $DF \parallel AC$.

The square AK is thus divided into four parts, AE, EK, GE, and BF, all of which are rectangles. (Why ?)

But AE is the square on AB; (Why ?)

EK is the square $\Leftrightarrow$ to the square on BC; (Why ?)

GE is $\Leftrightarrow$ to the rectangle contained by AB and BC; (Why ?)

and

BF is $\Leftrightarrow$ to the rectangle contained by AB and BC. (Why ?)

$\therefore$ the square on AC is $\Leftrightarrow$ to the sum of the squares on AB and BC, increased by twice the rectangle contained by AB and BC. Q.E.D.

356. SCHOLIUM. This theorem illustrates geometrically the algebraic formula,

$$(a + b)^2 = a^2 + 2\,ab + b^2.$$

Proposition IX. Theorem

357. *The square constructed on the difference of two lines is equivalent to the sum of the two squares constructed on these lines, diminished by twice the rectangle of these lines.*

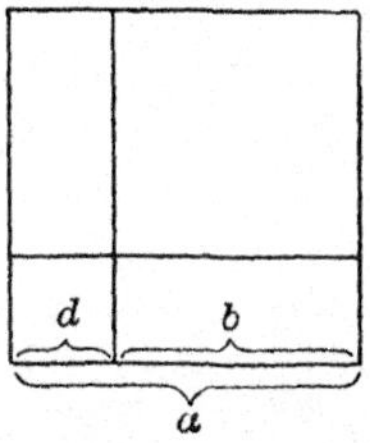

Hint. — Demonstrate that d^2 is obtained when we construct geometrically

$$a^2 - ab + b^2 - ab.$$

358. Scholium 1. This theorem illustrates geometrically the algebraic formula,

$$(a - b)^2 = a^2 - 2\,ab + b^2.$$

359. Scholium 2. Any homogeneous equation of the second degree can be illustrated geometrically.

Ex. 806. Prove, geometrically, the algebraic formula,

$$a(b + c) = ab + ac.$$

Ex. 807. Prove, geometrically, the algebraic formula,

$$a(b - c) = ab - ac.$$

Ex. 808. Prove, geometrically, the algebraic formula,

$$a^2 - b^2 = (a + b)(a - b).$$

Ex. 809. Prove, geometrically, the algebraic formula,

$$(a + b)^2 - (a - b)^2 = 4\,ab.$$

Ex. 810. Prove, geometrically, the algebraic formula,

$$(a + b)(c + d) = ac + bc + ad + bd.$$

PROPOSITION X. PROBLEM

360. *To transform a rectangle into a square.*

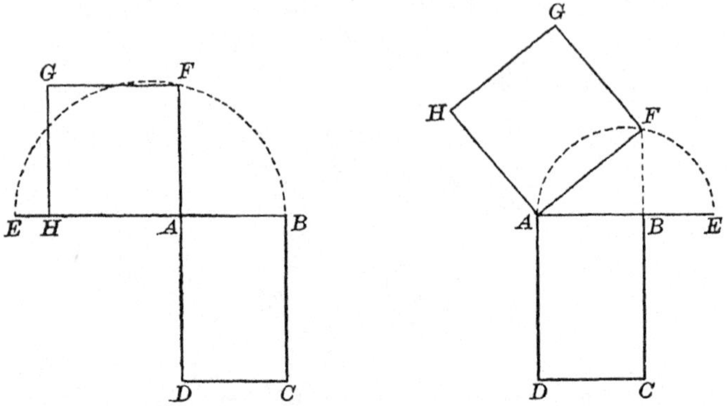

Given. Rectangle $ABCD$.

Required. A square ⇔ to $ABCD$.

Construction. Construct a mean proportional AF between AB and AD. The square on AF is the required square.

Proof. $AB : AF = AF : AD.$ (Con.)

$\therefore AB \times AD = \overline{AF}^2,$ (262)

or area $ABCD =$ area AG. Q.E.F.

361. SCHOLIUM. The mean proportional may be constructed by any of the methods of Book III. The two indicated in the diagram, however, are the most useful ones.

Ex. 811. To transform a parallelogram into a square.

Ex. 812. To transform a triangle into a parallelogram.

Ex. 813. To transform a triangle into a square.

Ex. 814. To transform a pentagon into a square.

Ex. 815. To construct a square equivalent to one-third of a given triangle.

PROPOSITION XI. PROBLEM

362. *To construct a square that shall be any given part of a given square.*

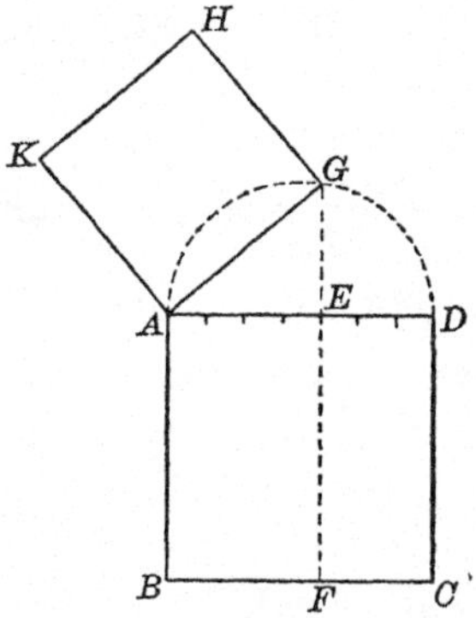

Given. Square AC.

Required. A square equivalent to $\frac{4}{7}$ of AC.

Construction. On AD, lay off $AE = \frac{4}{7} AD$.

Draw $EF \perp AD$, and transform rectangle $AEFB$ into a square, AH. (360)

AH is the required square.

Proof. $AH \approx AF$. (Con.)

$AF \approx \frac{4}{7} AC$. (335)

$\therefore AH \approx \frac{4}{7} AC$. Q.E.F.

363. SCHOLIUM. If, instead of $\frac{4}{7}$, the ratio $\frac{m}{n}$ is given, where m and n are two given lines, make AE the fourth proportional to n, m, and AD.

Ex. 816. To construct a square three times as large as a given square.

Ex. 817. In a given square $ABCD$, to construct four squares, all containing the angle A, and having areas respectively equal to $\frac{1}{5}$, $\frac{2}{5}$, $\frac{3}{5}$, and $\frac{4}{5}$ of $ABCD$.

Ex. 818. In the diagram for Prop. XI, find two lines whose squares have the ratio $AE : ED$.

Ex. 819. In the same diagram, find a rectangle equivalent to the square on GD.

Ex. 820. In the same diagram, find a square equivalent to the sum of the squares on AG and GD.

Proposition XII. Theorem

364. *In a right triangle the sum of the squares of the arms is equivalent to the square of the hypotenuse.*

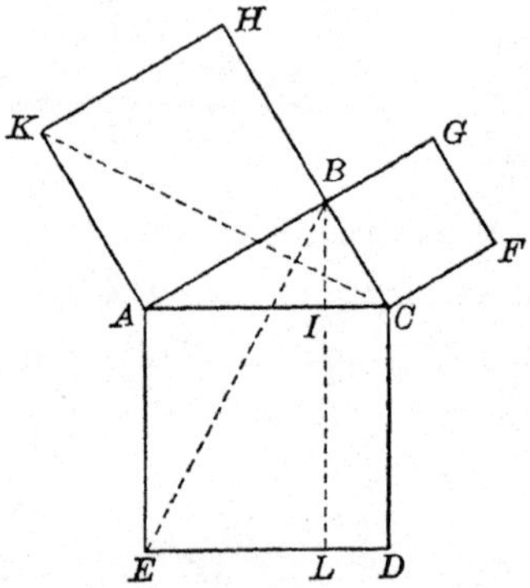

Hyp. AH and BF are squares on the arms, AD the square on the hypotenuse of the right triangle ABC.

To prove $AD \Leftrightarrow AH + BF$.

Proof. From B draw $BL \parallel AE$, intersecting AC and ED in I and L, respectively.

Draw BE and KC.

$$\angle HBC = \text{st.} \angle,$$

or HBC is a straight line.

In $\triangle$ *AKC* and *BAE*,

$$KA = AB \text{ and } AC = AE.$$

$$\angle KAC = \angle BAE. \quad \text{(Why ?)}$$

$$\therefore \triangle KAC = \triangle BAE. \quad \text{(Why ?)}$$

But $\quad \text{area } AH = 2 \text{ area } KAC, \quad$ (Why ?)

and $\quad \text{area } EI = 2 \text{ area } BAE.$

$$\therefore \text{area } AH = \text{area } EI.$$

In like manner,

$$\text{area } BF = \text{area } CL.$$

$$\therefore \text{area } AH + \text{area } BF = \text{area } AD. \quad \text{Q.E.D.}$$

365. Cor. The square of an arm of a right triangle is equivalent to the difference between the squares of the hypotenuse and the other arm.

366. Note. — The above theorem was first demonstrated by Pythagoras about 550 B.C., and is called after him the **Pythagorean Theorem.**

Ex. 821. The square on the difference of the arms of a right triangle increased by twice the rectangle contained by them, is equivalent to the square on the hypotenuse.

Ex. 822. The square on the sum of the arms of a right triangle, increased by the square on their difference, is equivalent to twice the square on the hypotenuse.

Proposition XIII. Problem

367. *To construct a square equivalent to the sum of two given squares.*

[The solution is left to the student.]

Proposition XIV. Problem

368. *To construct a square equivalent to the difference of two given squares.*

[The solution is left to the student.]

Ex. 823. In a right triangle, the rectangle contained by the arms is equivalent to the rectangle contained by the hypotenuse and the altitude upon the hypotenuse.

Ex. 824. To construct a square equivalent to the sum of three given squares.

Proposition XV. Theorem

369. *The areas of two triangles which have an angle of the one equal to an angle of the other are to each other as the products of the sides including the equal angles.*

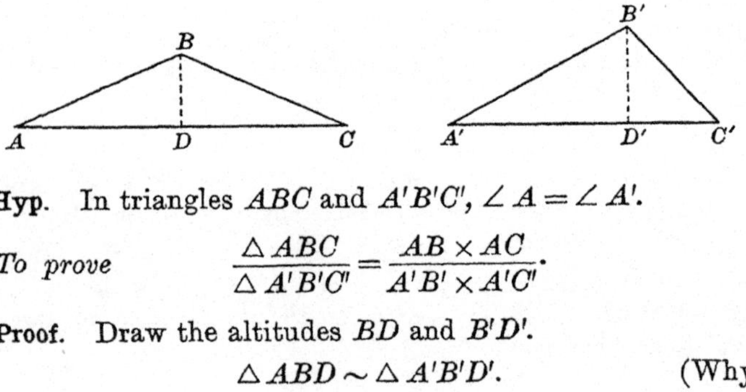

Hyp. In triangles ABC and $A'B'C'$, $\angle A = \angle A'$.

To prove $$\frac{\triangle ABC}{\triangle A'B'C'} = \frac{AB \times AC}{A'B' \times A'C'}.$$

Proof. Draw the altitudes BD and $B'D'$.

$$\triangle ABD \sim \triangle A'B'D'. \qquad \text{(Why ?)}$$

$$\therefore \frac{BD}{B'D'} = \frac{AB}{A'B'}.$$

But $$\frac{\triangle ABC}{\triangle A'B'C'} = \frac{AC \times BD}{A'C' \times B'D'} = \frac{AC}{A'C'} \times \frac{BD}{B'D'} = \frac{AC}{A'C'} \times \frac{AB}{A'B'}$$

$$= \frac{AC \times AB}{A'C' \times A'B'}.$$

Q.E.D.

Ex. 825. In the diagram for Prop. XV, if $\angle A = \angle A'$, $AB = 6$ in., $AC = 9$ in., $A'B' = 1$ ft., $A'C' = 2$ ft. Find the ratio of $\triangle ABC$ to $\triangle A'B'C'$.

Ex. 826. If $\triangle ABC \backsim \triangle A'B'C'$, and $\angle A = \angle A'$, then

$$AB : A'B' = A'C' : AC.$$

Proposition XVI. Theorem

370. *Similar triangles are to each other as the squares of their homologous sides.*

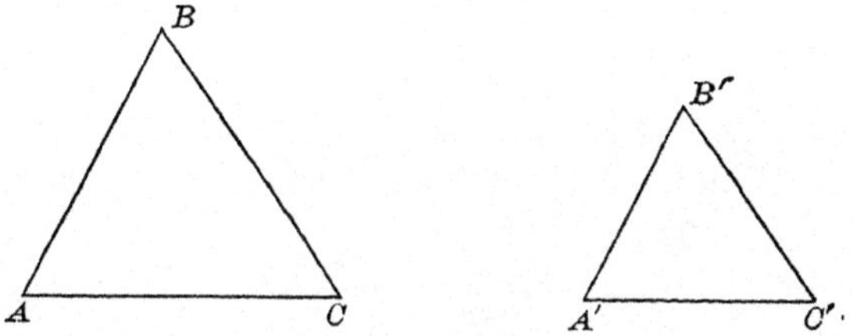

Hyp. $\triangle ABC \sim \triangle A'B'C'$.

To prove $\dfrac{\triangle ABC}{\triangle A'B'C'} = \dfrac{\overline{AB}^2}{\overline{A'B'}^2}$.

Proof.

$$\frac{\triangle ABC}{\triangle A'B'C'} = \frac{AB \times AC}{A'B' \times A'C'} = \frac{AB}{A'B'} \times \frac{AC}{A'C'} = \frac{AB}{A'B'} \times \frac{AB}{A'B'} = \frac{\overline{AB}^2}{\overline{A'B'}^2}.$$

Q.E.D.

Ex. 827. The sides of a triangle are 4, 7, and 8. What are the sides of a similar triangle, whose area is four times as large?

Ex. 828. Similar triangles are to each other as the squares of homologous altitudes.

Ex. 829. Find the ratio of two similar triangles, two of whose homologous sides are 2 and 3 in. respectively.

Ex. 830. Construct a triangle nine times as large as a given triangle and similar to it.

Ex. 831. The areas of two similar triangles are to each other as 4 : 5. Find the ratio of their homologous sides.

PROPOSITION XVII. THEOREM

371. *Similar polygons are to each other as the squares of their homologous sides.*

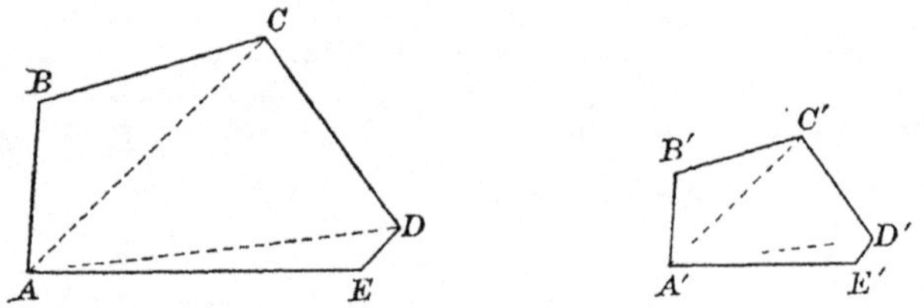

Hyp. AB and $A'B'$ are homologous sides of two similar polygons BE and $B'E'$, whose areas are S and S' respectively.

To prove $$\frac{S}{S'}=\frac{\overline{AB}^2}{\overline{A'B'}^2}.$$

Proof. Draw AC, AD, $A'C'$, and $A'D'$.

Then $\triangle ABC \sim \triangle A'B'C'$. (Why ?)

Whence $$\frac{\triangle ABC}{\triangle A'B'C'}=\frac{\overline{AB}^2}{\overline{A'B'}^2}.$$

In like manner $$\frac{\triangle ACD}{\triangle A'C'D'}=\frac{\overline{DC}^2}{\overline{D'C'}^2}=\frac{\overline{AB}^2}{\overline{A'B'}^2}.$$

Similarly for $\triangle$s ADE and $A'D'E'$.

$$\therefore \frac{\triangle ABC}{\triangle A'B'C'}=\frac{\triangle ACD}{\triangle A'C'D'}=\frac{\triangle ADE}{\triangle A'D'E'}=\frac{\overline{AB}^2}{\overline{A'B'}^2}.$$

$$\therefore \frac{\triangle ABC+\triangle ACD+\triangle ADE}{\triangle A'B'C'+\triangle A'C'D'+\triangle A'D'E'}=\frac{\overline{AB}^2}{\overline{A'B'}^2}. \quad (273)$$

$$\frac{S}{S'}=\frac{\overline{AB}^2}{\overline{A'B'}^2}.$$ Q.E.D.

372. Scholium. If a and a' are homologous sides of two similar polygons whose areas are S and S' respectively, and K represents a constant, such that $S = Ka^2$, then $S' = Ka'^2$. (268)

Ex. 832. Two homologous sides of two similar polygons are 4 in. and 3 in. respectively, the area of the greater polygon is 1 sq. ft. What is the area of the smaller polygon?

Ex. 833. On a map whose scale is 1 : 1,000,000, how many square kilometers are represented by a polygon equivalent to 2 sq cm.?

Ex. 834. Two similar parallelograms are to each other as the products of their diagonals.

373. Remark. *As similar polygons (and triangles) are to each other as the* **squares of their homologous sides,** *Prop. XI may be used to draw a polygon that shall be* **any given part of a given polygon, and similar to it.**

Ex. 835. Construct a triangle equivalent to two-sevenths of a given triangle, and similar to it.

Ex. 836. Construct a triangle similar to a given triangle and three times as large.

Ex. 837. To divide a triangle into two equivalent parts by a line parallel to one of its sides.

Ex. 838. To divide a triangle into five equivalent parts by lines parallel to one of its sides.

Ex. 839. Construct a polygon equivalent to three-fifths of a given polygon and similar to it.

Ex. 840. Construct a pentagon equivalent to three times a given pentagon and similar to it.

Ex. 841. To divide a regular (*i.e.* equiangular and equilateral) pentagon into two equivalent parts so that one part is again a regular pentagon.

* **Ex. 842.** To construct two lines having the same ratio as the areas of two similar given triangles.

PROPOSITION XVIII. THEOREM

374. *If similar polygons are constructed on the three sides of a right triangle, the sum of the polygons on the arms is equivalent to the polygon on the hypotenuse.*

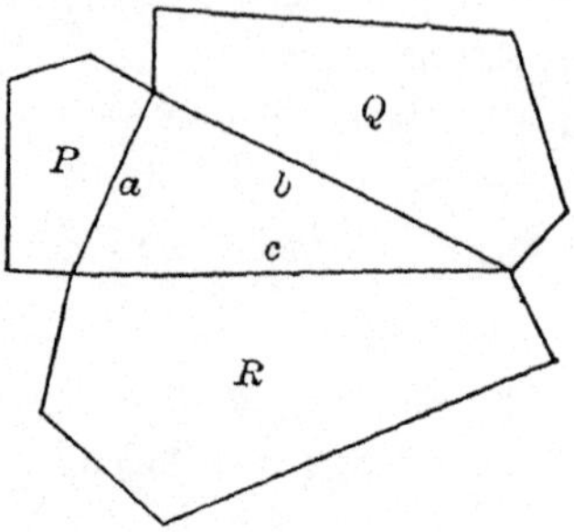

Hyp. P, Q, and R are the areas of similar polygons constructed, respectively, on the arms a and b and the hypotenuse c of the rt. $\triangle\, abc$.

To prove $P + Q = R$.

Proof.

Suppose that $R = Kc^2$, where K denotes a number,

then $P = Ka^2$, (372)

$Q = Kb^2$.

$\therefore\ P + Q = K(a^2 + b^2) = Kc^2 = R$. Q.E.D.

375. COR. If similar polygons are constructed on the three sides of a right triangle, the polygon on one arm is equivalent to the difference of the polygons on the hypotenuse and the other arm.

Ex. 843. Construct an equilateral triangle equivalent (a) to the sum of two given equilateral triangles; (b) to the difference of two given equilateral triangles.

PROPOSITION XIX. PROBLEM

376. *To construct a polygon similar to two given similar polygons and equivalent to their sum.*

[The solution is left to the student.]

PROPOSITION XX. PROBLEM

377. *To construct a polygon similar to two given similar polygons and equivalent to their difference.*

[The solution is left to the student.]

Ex. 844. To construct an isosceles right triangle equivalent to the difference of two isosceles right triangles.

Ex. 845. Construct a triangle similar to two given similar triangles and equivalent to their sum.

PROPOSITION XXI. PROBLEM

378. *To construct a rectangle equivalent to a given square, having the sum of its base and altitude equal to a given line.*

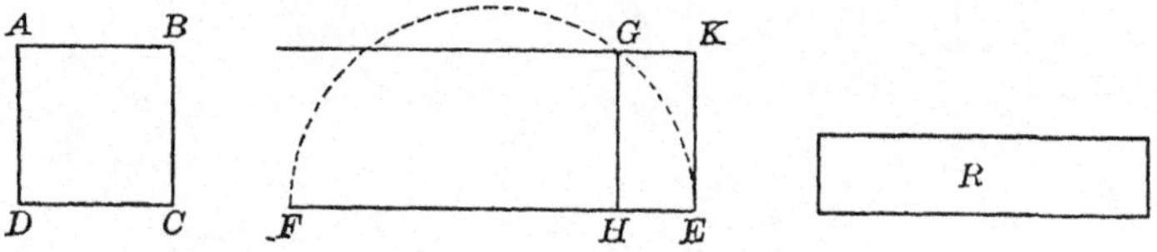

Given. A square $ABCD$ and a line EF.

Required. To construct a rectangle equivalent to $ABCD$, having the sum of its base and altitude equal to EF.

Construction. On EF as a diameter construct a semicircle. Draw EK perpendicular to EF and equal to AB.

Draw $KG \parallel EF$, intersecting the semicircle in G. Draw $GH \perp EF$.

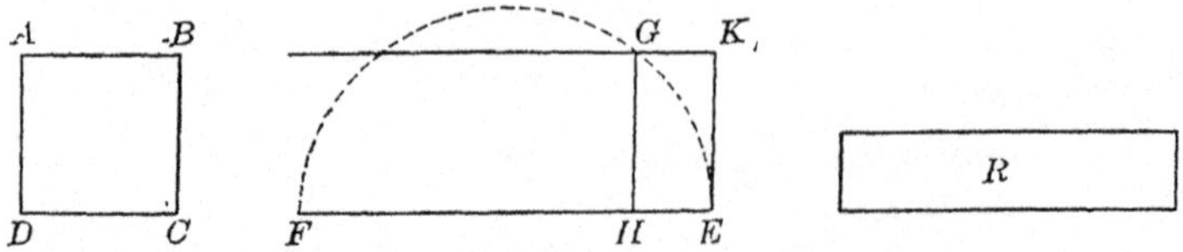

The rectangle R, constructed with FH as the base and HE as the altitude, is the required one.

Proof. $$FH : GH = GH : HE. \qquad \text{(Why?)}$$

$$\therefore \overline{GH}^2 = FH \times HE.$$

But $$\overline{GH}^2 = \overline{KE}^2 = \text{area } AC,$$

and $$FH \times HE = \text{area } R$$

$$\therefore \text{area } AC = \text{area } R,$$

and $$FH + HE = FE.$$ Q.E.F.

Proposition XXII. Problem

379. *To construct a rectangle equivalent to a given square, having the difference of its base and altitude equal to a given line.*

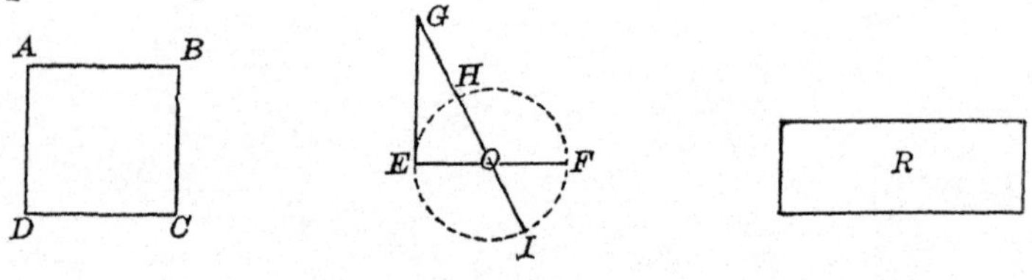

Given. A square $ABCD$ and a line EF.

Required. To construct a rectangle equivalent to $ABCD$, and having the difference of its base and altitude equal to EF.

Construction. On EF as a diameter draw a circle. Draw EG perpendicular to EF and equal to AB. Through the center O draw GO, intersecting the circumference in H and I.

Then the rectangle R, constructed with the base and altitude equal to GI and GH respectively, is the required one.

Proof. $GI - GH = IH = EF.$ (Why?)

$\overline{GE}^2 = GI \times GH,$ (314)

or area $ABCD$ = area R. Q.E.F.

PROPOSITION XXIII. PROBLEM

380. *To construct a polygon similar to a given polygon and equivalent to another polygon.*

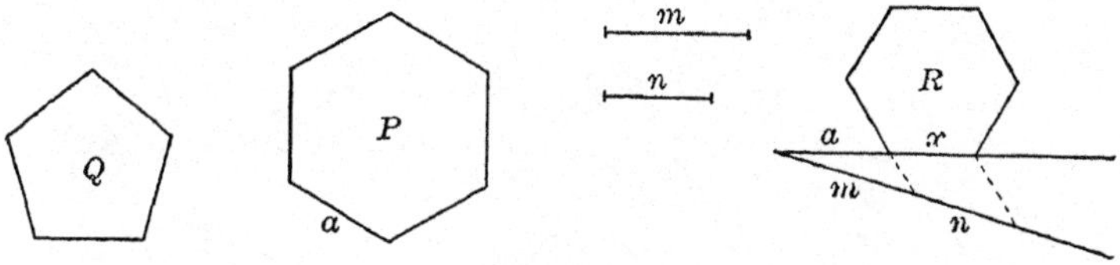

Given. Two polygons, P and Q, a being a side of P.

Required. To construct a polygon equivalent to Q and similar to P.

Construction. Construct squares equivalent to P and Q, and denote their sides by m and n respectively. Construct x, the fourth proportional, to m, n, and a.

Upon x, homologous to a, construct the polygon R similar to P.

R is the required polygon.

Proof. $m : n = a : x,$ (Con.)

$m^2 : n^2 = a^2 : x^2,$ (276)

or $P : Q = a^2 : x^2.$

But $P : R = a^2 : x^2.$ (371)

$\therefore$ area Q = area R. (263) Q.E.F.

PROPOSITION XXIV. PROBLEM

381. *To transform the triangle ABC so that the angle A is not altered, and the side opposite the angle A becomes parallel to a given line MN.*

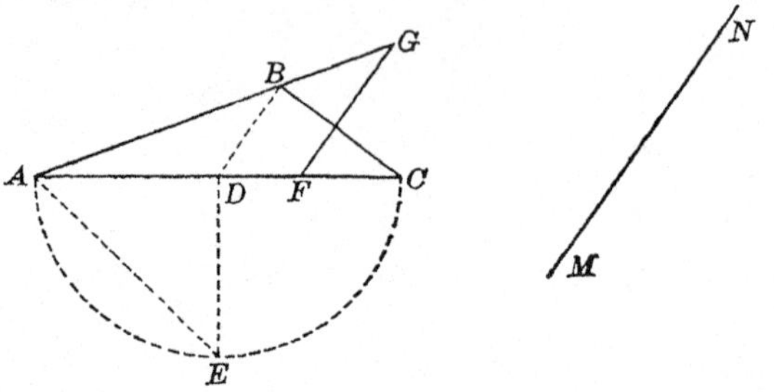

Construction. Draw $BD \parallel MN$, intersecting AC in D.

Construct AE, a mean proportional between AD and AC.

On AC lay off $AF = AE$.

Draw $FG \parallel MN$, meeting AB produced in G.

AFG is the required triangle.

Proof.

$$\frac{\triangle ABC}{\triangle AGF} = \frac{AC \times AB}{AF \times AG} = \frac{AC}{AF} \times \frac{AB}{AG} = \frac{AC}{AF} \times \frac{AD}{AF} = \frac{AC \times AD}{\overline{AF}^2} = 1.$$

$\therefore$ area ABC = area AGF. Q.E.F.

Ex. 846. To transform a given triangle into another one containing two given angles.

Ex. 847. To transform a given triangle into an isosceles triangle, having a given vertical angle.

Ex. 848. To transform a given triangle into an equilateral triangle.

Ex. 849. To transform a given parallelogram into an equilateral triangle.

Ex. 850. Construct an equilateral triangle equivalent to one-half of a given square.

Ex. 851. Transform a given triangle into a right triangle containing a given acute angle.

382. REMARK. — **To divide a polygon into n equivalent parts by lines passing through a point P,** *one of the following two methods is usually used:*

(*a*) Transform the figure into a triangle having one vertex at P. Divide the triangle into n equal parts, and transform the parts thus obtained so as to form parts of the original figure.

or, (*b*) Divide the figure into n parts by any method, and transform the parts thus obtained so as to fulfil the given conditions.

The two methods are illustrated by the following exercises:

Ex. 852. To divide quadrilateral $ABCD$ into three equal parts by straight lines passing through A.

(*a*) Transform $ABCD$ into $\triangle ADE$.

Divide $\triangle ADE$ into the three equivalent parts ADF, AFG, and AGE.

As the last two parts do not lie entirely in the given quadrilateral, draw $GH \parallel CA$.

Then $AFG \approx AFCH$, and AF and AH are the required lines,

or, (*b*) Trisect DB. Draw AF, AE, CF, and CE.

Then the broken lines AFC and AEC divide the figure into three equivalent parts. To transform these parts so as to fulfil the conditions, draw FH and EK parallel to AC.

AH and AK are the required lines.

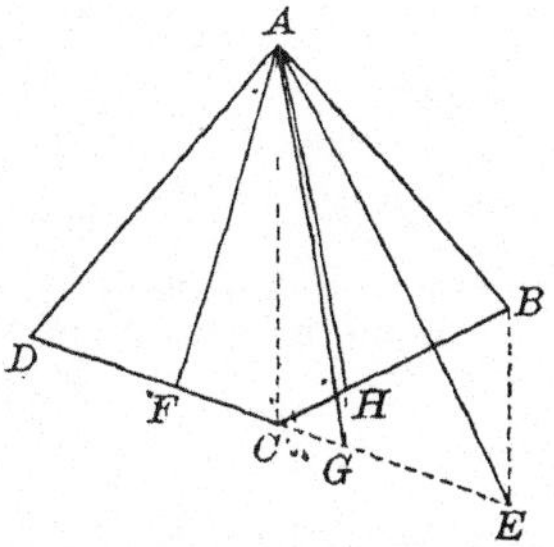

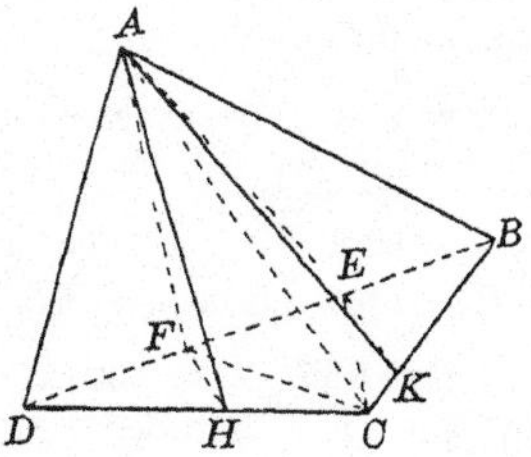

Ex. 853. To bisect a trapezoid by a line drawn from a point P in the smaller base.

Ex. 854. To bisect a triangle by a line drawn from a point P in the base.

Ex. 855. To bisect a triangle by a line drawn perpendicular to the base.

Ex. 856. To bisect a triangle by a line parallel to a given line.

MISCELLANEOUS EXERCISES

THEOREMS

Ex. 857. A straight line passing through the midpoint of a diagonal of a parallelogram divides the figure into two equivalent parts.

Ex. 858. The area of a circumscribed polygon is equivalent to one-half the product of perimeter and radius.

Ex. 859. If through any point in a diagonal of a parallelogram, parallels are drawn to the sides, four parallelograms are formed, of which the two which do not contain the diagonal are equivalent.

Ex. 860. If any point within a parallelogram be joined to the four vertices, the sum of either pair of opposite triangles is equivalent to one-half the parallelogram.

Ex. 861. The lines joining the midpoints of the sides of a quadrilateral in succession form a parallelogram equivalent to one-half the quadrilateral.

Ex. 862. The areas of two similar triangles are to each other as the squares of any two homologous bisectors.

Ex. 863. The areas of two similar triangles are to each other as the squares of any two homologous medians.

Ex. 864. If E is any point in the diagonal AC of the parallelogram $ABCD$, prove that $\triangle AEB \approx \triangle ADE$.

* **Ex. 865.** A quadrilateral is equivalent to a triangle if its diagonals and the angle included between them are respectively equal to two sides and the included angle of the triangle.

* **Ex. 866.** Two quadrilaterals are equivalent if the diagonals and the included angle of one are equal, respectively, to the diagonals and the included angle of the other. (Ex. 865.)

* **Ex. 867.** If two chords intersect within a circle at right angles, the sum of the squares upon their segments is constant.

Ex. 868. In any quadrilateral the sum of the squares of the four sides is equal to the sum of the squares of the diagonals, increased by four times the square of the line joining the midpoints of the diagonals.

Ex. 869. If, on two sides of triangle ABC, parallelograms DB and BG are constructed, their sides DE and GH be produced to meet in F, and on AC a parallelogram be constructed, having AK equal and parallel to FB, then the parallelogram AL is equivalent to parallelogram AE plus parallelogram BG. (Pappus's Theorem.)

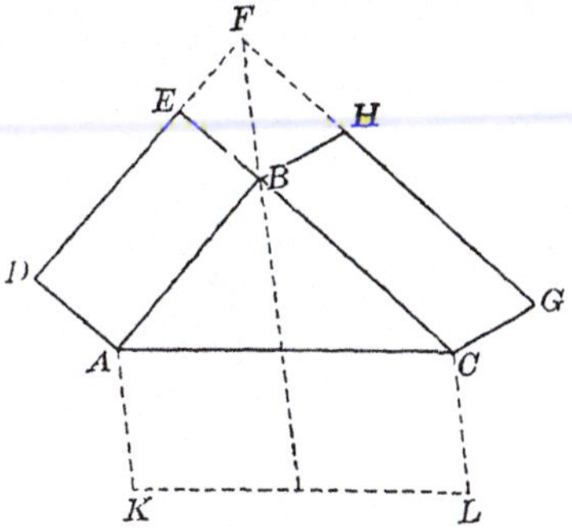

Ex. 870. Find a similar proposition for triangles constructed on the three sides of a given triangle.

Ex. 871. Prove the Pythagorean Theorem by means of Ex. 869.

PROBLEMS OF COMPUTATION

Ex. 872. The side of an equilateral triangle is 10 in. Find the area.

Ex. 873. Find the area of an isosceles triangle if the base is 6 and the arm 5.

Ex. 874. Find the area of a trapezoid whose bases are 9 and 11 respectively, and whose altitude is 12 ft.

Ex. 875. Find the area of a rhombus whose diagonals are 9 and 10 ft. respectively.

Ex. 876. Find the area of quadrilateral $ABCD$ if $AB = 39$, $BC = 52$, $CD = 25$, $AD = 60$, diagonal $AC = 65$.

Ex. 877. A side of equilateral triangle ABC is 8. Find the side of an equilateral triangle equivalent to three times triangle ABC.

Ex. 878. The perimeter of a rectangle is 20 m., one side is 6 m. Find the area.

Ex. 879. What is the side of a square whose area is 900 sq. m.?

Ex. 880. The area of a rhombus is equal to m, and one diagonal is equal to d. Find the other diagonal.

Ex. 881. The area of a trapezoid is 400 sq. m., its altitude is 8 m. Find the length of the line joining the midpoints of the non-parallel sides.

Ex. 882. The hypotenuse of a right triangle is 20, and the projection of one arm upon the hypotenuse is 4. What is its area?

Ex. 883 The base and altitude of a triangle are 12 and 20 respectively. At a distance of 6 from the base, a parallel is drawn to the base. Find the areas of the two parts of the triangle.

Ex. 884. Find the area of a rectangle having one side equal to 6 and a diagonal equal to 10.

Ex. 885. Find the area of a polygon whose perimeter equals 20 ft., circumscribed about a circle whose radius is 3 ft.

Ex. 886. Find the side of an equilateral triangle equivalent to a parallelogram, whose base and altitude are 10 and 15 respectively.

Ex. 887. The sides of two equilateral triangles are 13 and 12 respectively. Find the side of an equilateral triangle equivalent to their difference.

Ex. 888. Two similar polygons have two homologous sides equal to 7 and 24 respectively. Find the homologous side of a third polygon, similar to the given polygons and equivalent to their sum.

Ex. 889. The sides of a triangle are as $8:15:17$. Find the sides if the area is 480 sq. ft.

Ex. 890. The sides of a triangle are 8, 15, and 17. Find the radius of the inscribed circle.

Ex. 891. The sides of a triangle are 6, 7, and 8 ft. Find the areas of the two parts into which the triangle is divided by the bisector of the angle included by 6 and 7.

Ex. 892. Find the area of an equilateral triangle whose altitude is equal to h.

PROBLEMS OF CONSTRUCTION

Ex. 893. To construct a triangle equivalent to the sum of two given triangles.

Ex. 894. To transform a rectangle into another one, having given one side.

Ex. 895. To construct a triangle equivalent to the difference of two given parallelograms.

Ex. 896. To transform a square into an isosceles triangle, having a given base.

Ex. 897. To transform a rectangle into a parallelogram, having a given diagonal.

Ex. 898. To divide a triangle into three equivalent parts by lines drawn through a point in one of the sides.

Ex. 899. To bisect a parallelogram by a line perpendicular to a side.

Ex. 900. To bisect a parallelogram by a line perpendicular to a given line.

Ex. 901. To divide a parallelogram into three equivalent parts by lines drawn through a vertex.

Ex. 902. To bisect a trapezoid by a line drawn through a vertex.

* **Ex. 903.** Divide a triangle into three equivalent parts by lines drawn from a point P within the triangle.

Ex. 904. Divide a pentagon into four equal parts by lines drawn through one of its vertices.

Ex 905. Divide a quadrilateral into four equal parts by lines drawn from a point in one of its sides.

Ex. 906. Find a point within a triangle such that the lines joining the point to the vertices shall divide the triangle into three equivalent parts.

Ex. 907. Construct a square that shall be to a given triangle as 5 is to 4.

Ex. 908. Construct a square that shall be to a given triangle as m is to n, when m and n are two given lines.

Ex. 909. Construct an equilateral triangle, that shall be to a given rectangle as 4 is to 5.

Ex. 910. Find a point within a triangle such that the lines joining the point with the vertices shall form three triangles, having the ratio 3 : 4 : 5.

Ex. 911. Divide a given line into two segments such that one segment is to the line as $\sqrt{2}$ is to $\sqrt{5}$.

Ex. 912. To transform a triangle into a right isosceles triangle.

Ex. 913. Construct a triangle similar to a given triangle and equivalent to another given triangle.

* **Ex. 914.** Bisect a trapezoid by a line parallel to the bases.

BOOK V

REGULAR POLYGONS. MEASUREMENT OF THE CIRCLE

REGULAR POLYGONS

PROPOSITION I. THEOREM

382′. DEF. A **regular polygon** is one which is equiangular and equilateral.

383. *A circle can be circumscribed about any regular polygon.*

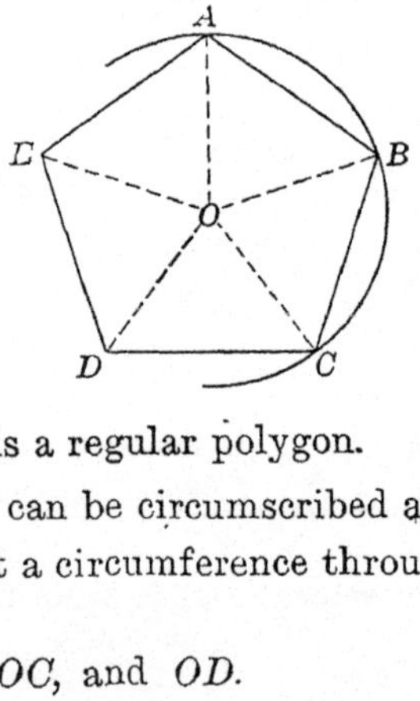

Hyp. $ABCDE$ is a regular polygon.

To prove a circle can be circumscribed about $ABCDE$.

Proof. Construct a circumference through A, B, and C, and let O be its center.

Draw OA, OB, OC, and OD.

$\angle ABC = \angle BCD.$ (Why?)

$\angle OBC = \angle OCB.$ (Why?)

$\therefore \angle OBA = \angle OCD.$

But $OC = OB$ and $AB = CD.$ (Why?)

$\triangle OBA = \triangle OCD.$ (Why?)

$\therefore OD = OA.$

∴ the circumference passes through D.

In like manner, it may be proven that the circumference passes through the remaining vertices of the polygon.

∴ a circle can be circumscribed about the given polygon. Q.E.D.

PROPOSITION II. THEOREM

384. *A circle can be inscribed in any regular polygon.*

HINT. — Circumscribe a circle about the given polygon and prove that the center is equidistant from the sides.

385. DEF. The **center of a regular polygon** is the common center of the circumscribed and inscribed circles of the polygon.

386. DEF. The **radius of a regular polygon** is the radius of the circumscribed circle.

387. DEF. The **central angle** is the angle between two radii drawn to the ends of a side.

388. DEF. The **apothem of the polygon** is the radius of the inscribed circle.

389. COR. 1. The angle at the center of a regular polygon of n sides is equal to $\frac{4}{n}$ right angles.

390. COR. 2. The angle formed by an apothem to a side of a regular polygon, and a radius to an extremity of that side, is equal to $\frac{2}{n}$ right angles.

Ex. 915. The lines joining opposite vertices of a regular hexagon pass through the center.

Ex. 916. A triangle is regular if the centers of the circumscribed and inscribed circles coincide.

Ex. 917. A polygon is regular if the centers of the circumscribed and inscribed circles coincide.

Ex. 918. The angle at the center of a regular polygon is the supplement of an angle of the polygon.

Proposition III. Theorem

391. *If the circumference of a circle is divided into any number of equal parts:*

(1) *The chords joining the points of division successively form a regular inscribed polygon.*

(2) *Tangents drawn at the points of division form a regular circumscribed polygon.*

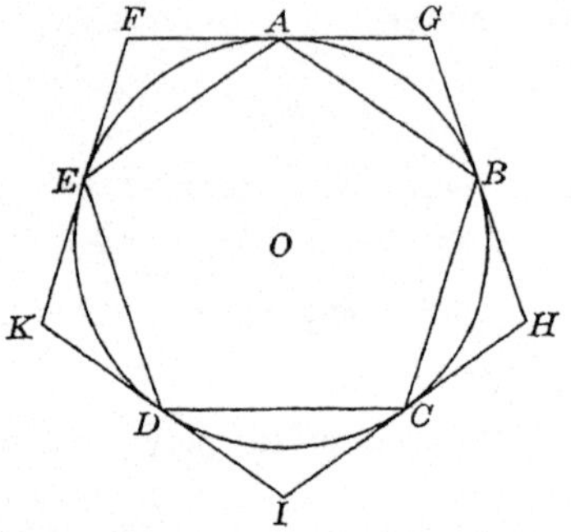

Hyp. The circumference ACE is divided into the equal arcs AB, BC, CD, etc.

(1) *To prove* $ABCDE$ is a regular polygon.

Proof. arc $AB =$ arc $BC =$ arc CD, etc. (Why ?)

$\therefore$ arc $AC =$ arc $BD =$ arc CE, etc. (Ax. 2.)

$\therefore \angle EAB = \angle ABC = \angle BCD$, etc. (Why ?)

But $AB = BC = CD$. (Why ?)

$\therefore$ polygon $ABCDE$ is regular. (Why ?)

(2) *To prove* tangents drawn at A, B, C, etc., form the regular circumscribed polygon $FGHIK$.

Proof. $\angle GAB = \angle GBA = \angle CBH = \angle HCB$, etc., (Why ?)

and $AB = BC = CD$, etc. (Why ?)

∴ △ ABG, CBH, CDI, etc., are equal and isosceles. (Why?)

$$\therefore \angle G = \angle H = \angle I, \text{ etc.},$$

and $AG = GB = BH = HC$, etc.

Whence $GH = HI = IK$, etc. (Ax. 2.)

∴ circumscribed polygon $FGHIK$ is regular. Q E D

392. Cor. 1. The perimeter of a regular inscribed polygon is less than the perimeter of a regular inscribed polygon of double the number of sides.

393. Cor. 2. The perimeter of a regular circumscribed polygon is greater than the perimeter of a regular circumscribed polygon of double the number of sides.

Ex. 919. An equilateral polygon inscribed in a circle is regular.

Ex. 920. An equilateral polygon circumscribed about a circle is regular if the number of its sides is odd.

Ex. 921. An equiangular polygon circumscribed about a circle is regular.

Proposition IV. Problem

394. *To inscribe a square in a given circle.*

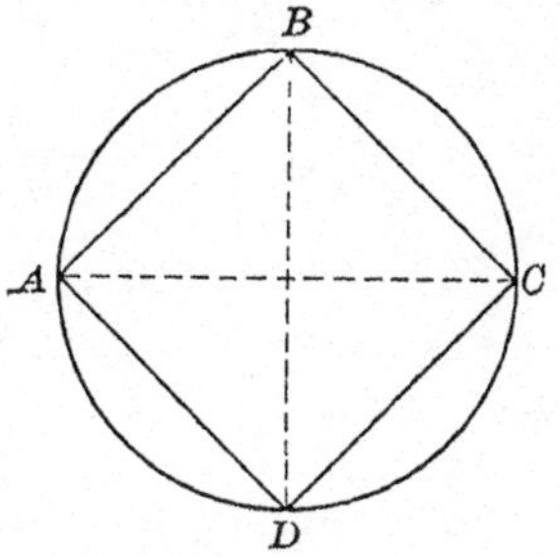

Construction. In the given circle ABC, draw diameters AC and BD perpendicular to each other, and join AB, BC, CD, and DA.

Then $ABCD$ is the required square.

[The proof is left to the student.]

395. Cor. 1. By bisecting the central angles, the arcs AB, BC, etc., will be bisected, and a polygon of eight sides may be inscribed in the circle. By repeating the process, polygons of 16, 32, $\cdots$, 2^n sides may be constructed.

396. Cor. 2. By drawing tangents at A, B, C, and D, a square may be circumscribed about the circle.

Ex. 922. To circumscribe an octagon about a given circle.

Ex. 923. To construct a regular octagon, having a given side.

Ex. 924. The side of an inscribed square is equal to the radius multiplied by $\sqrt{2}$.

Ex. 925. Find the area of a square, if its radius is equal to r.

Proposition V. Problem

397. *To inscribe a regular hexagon in a given circle.*

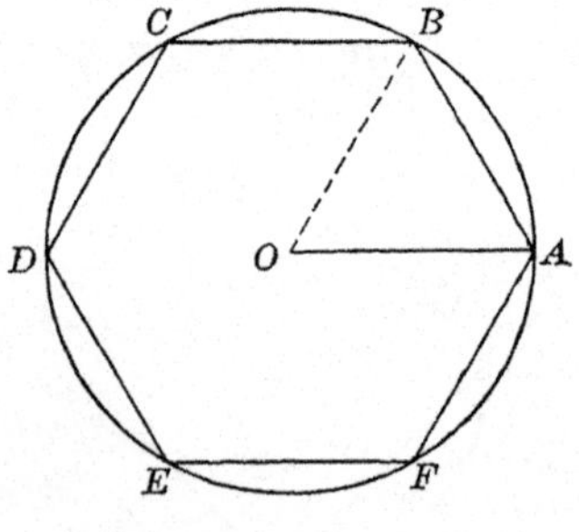

Construction. In the given circle ACD, draw the radius AO.

From A as a center, with a radius equal to OA, draw an arc meeting the circumference in B. Draw AB.

AB is the side of a regular hexagon.

By applying the radius six times as a chord, the regular hexagon $ABCDEF$ is formed.

Proof. $\triangle ABO$ is equilateral. (Why?)

$\therefore \angle O = \frac{2}{3}$ rt. or $\frac{1}{6}$ of 2 st. $\angle$s. (Why?)

$\therefore$ arc $BA = \frac{1}{6}$ of the circumference,

and $ABCDEF$ is a regular hexagon. Q.E.F.

398. Cor. 1. By joining the alternate vertices of an inscribed regular hexagon, an inscribed equilateral triangle is formed.

399. Cor. 2. Polygons of 3, 6, 12, 24, etc., sides may be inscribed in and circumscribed about a given circle.

Ex. 926. The side of an equilateral triangle is equal to its radius multiplied by $\sqrt{3}$.

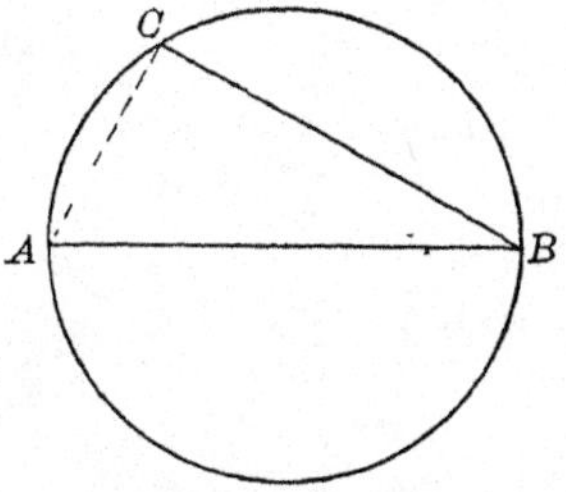

Ex. 927. If the radius of a circle is r, the apothem of the inscribed hexagon is equal to $\frac{r}{2}\sqrt{3}$.

Ex. 928. To circumscribe a regular hexagon about a given circle.

Ex. 929. To construct a regular polygon of twelve sides, having given a side.

Ex. 930. In Ex. 929 how many degrees are in the angle at the center?

Ex. 931. Find the area of a regular hexagon if its radius is equal to r.

Ex. 932. The apothem of an equilateral triangle is equal to one-half its radius.

Ex. 933. The area of an inscribed equilateral triangle is equal to one-half the area of the inscribed regular hexagon.

Ex. 934. The areas of triangles inscribed in equal circles are to each other as the products of their three sides.

Ex. 935. A square constructed on a diameter of a circle is equivalent to twice the area of the inscribed square.

Proposition VI. Problem

400. *To inscribe a regular decagon in a given circle.*

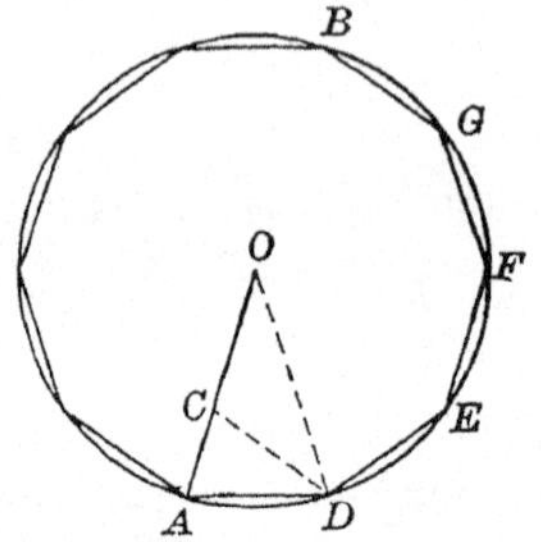

Construction. In the given circle AB, draw any radius OA, and divide it in extreme and mean ratio, so that

$$OA : OC = OC : CA.$$

Then OC is the side of the required decagon, and by applying it ten times as a chord, a regular decagon $ADEF$, etc., is formed.

Proof. Draw DO and DC.

$$AO : OC = OC : CA,$$

or since $$OC = AD,$$

$$AO : AD = AD : AC.$$

$$\therefore \triangle OAD \sim \triangle CAD. \quad \text{(Why ?)}$$

$$\therefore \angle O = \angle CDA. \quad \text{(Why ?)}$$

But $\triangle OAD$ being isosceles, the similar triangle DAC must be isosceles,

or, $$CD = AD = CO,$$

whence $$\angle O = \angle CDO. \quad \text{(Why ?)}$$

But $$\angle O = \angle CDA.$$

$$\therefore 2\angle O = \angle ODA, \quad \text{(Ax. 2.)}$$

and $$2\angle O = \angle DAO.$$

But $\angle O + \angle ODA + \angle DAO = 2$ rt. $\angle$s.

$\therefore 5\angle O = 2$ rt. $\angle$s,

or $\angle O = \frac{2}{5}$ rt. $\angle$, or $\frac{1}{10}$ of 4 rt. $\angle$s.

$\therefore$ arc AD is $\frac{1}{10}$ of the circumference, and AD is the side of an inscribed regular decagon. Q.E.F.

401. Cor. 1. By joining the alternate vertices A, E, G, etc., an inscribed regular pentagon is formed.

402. Cor. 2. Polygons of 5, 10, 20, etc., sides may be inscribed in and circumscribed about a given circle.

Proposition VII. Problem

403. *To construct the side of a regular polygon of fifteen sides inscribed in a given circle.*

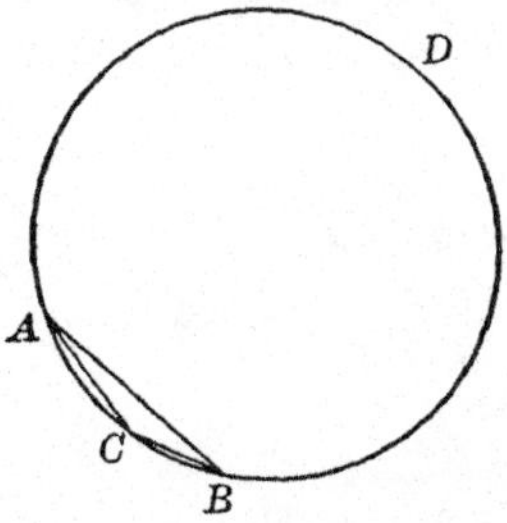

Construction. In the given circle ABD, draw the chord AB equal to the radius, and chord AC equal to the side of the inscribed regular decagon.

Then CB is the side of the inscribed regular polygon of fifteen sides.

Proof. Arc $AB = \frac{1}{6}$ of the circumference.

Arc $AC = \frac{1}{10}$ of the circumference.

Arc $CB = \frac{1}{6} - \frac{1}{10}$, or $\frac{1}{15}$ of the circumference,

and CB is the side of the required polygon. Q.E.F.

404. Cor. Polygons of 15, 30, 60, etc., sides may be inscribed in, and circumscribed about a given circle.

Ex. 936. To construct an angle of 36°.

Ex. 937. To construct a regular decagon, having given a side.

Ex. 938. To divide a right angle into five equal parts.

Ex. 939. The diagonals of a regular pentagon are equal.

Ex. 940. The diagonals of a regular pentagon divide each other into extreme and mean ratio.

Ex. 941. The area of a regular inscribed hexagon is a mean proportional between the areas of the inscribed and circumscribed equilateral triangles.

Ex. 942. Any radius of a regular polygon bisects an angle of the polygon.

Ex. 943. The diagonals drawn from a vertex of a regular decagon divide that angle into eight equal parts.

Ex. 944. To construct a regular pentagon, having given a side.

Ex. 945. An equiangular polygon inscribed in a circle is regular if the number of its sides is odd.

Proposition VIII. Theorem

405. *Regular polygons of the same number of sides are similar.*

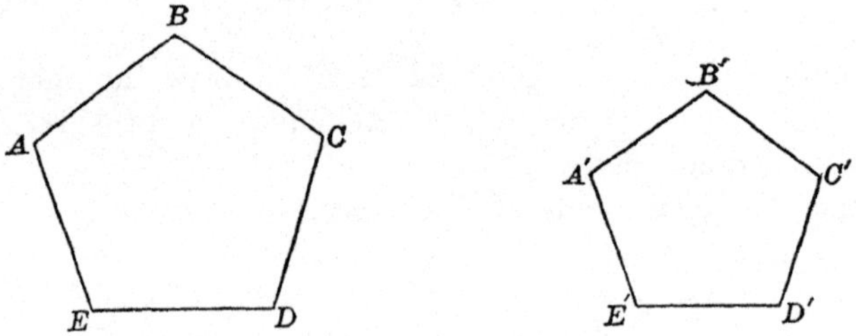

Hyp. $ABCDE$ and $A'B'C'D'E'$ are regular polygons of n sides.

To prove $ABCDE \sim A'B'C'D'E'$.

Proof. Every angle of both polygons is equal to $\frac{n-2}{n}$ st. $\angle$s. (Why ?)

$\therefore$ polygons $ABCDE$ and $A'B'C'D'E'$ are mutually equiangular.

Since $AB = BC$, etc., and $A'B' = B'C'$, etc., (Why ?)

$$AB : A'B' = BC : B'C' = \text{etc.}$$

$\therefore$ $ABCDE$ and $A'B'C'D'E'$ are similar. Q E D

Ex. 946. To construct a regular hexagon equivalent to one-half of a given regular hexagon.

Ex. 947. Construct a regular pentagon equivalent to the sum of two given regular pentagons.

PROPOSITION IX. THEOREM

406. *The perimeters of regular polygons of the same number of sides have the same ratio as their radii or as their apothems.*

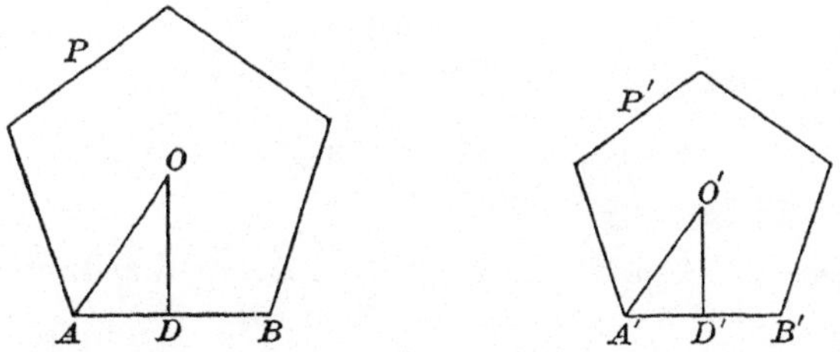

Hyp. P and P' are the perimeters of two regular polygons having the radii OA and $O'A'$, and the apothems OD and $O'D'$, respectively.

To prove $P : P' = OA : O'A' = OD : O'D'$.

Proof. $\angle AOD = \angle A'O'D'$. (390)

$\therefore \triangle OAD \sim \triangle O'A'D'$. (Why ?)

Hence $AD : A'D' = OD : O'D' = AO : A'O'$. (Why ?)

But $P : P' = AB : A'B' = AD : A'D'$. (Why ?)

$\therefore P : P' = OD : O'D' = AO : A'O'$. Q.E.D.

407. Cor. The areas of regular polygons of the same number of sides are to each other as the squares of their radii or apothems.

Ex. 948. The lines joining the midpoints of the radii of a regular pentagon enclose a regular pentagon equivalent to one-fourth the given pentagon.

Proposition X. Theorem

408. *If the number of sides of a regular polygon is increased indefinitely, the apothem will increase toward the radius as a limit.*

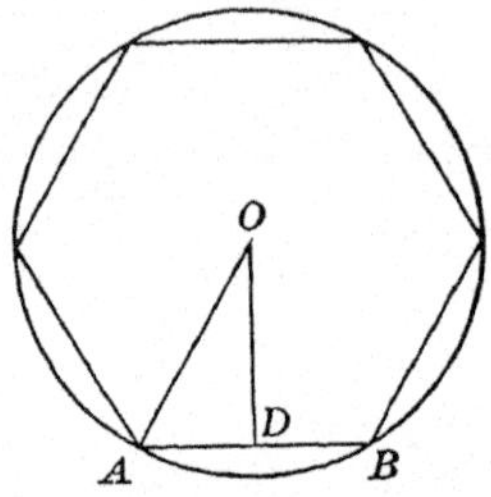

Hyp. AB is a side, OA the radius, and OD the apothem of a regular polygon of n sides.

To prove. If n increases indefinitely, the limit of OD equals OA.

Proof. In $\triangle AOD$, the difference of two sides is less than the third.

$$\therefore AO - OD < AD.$$

By increasing the number of sides indefinitely, AB, and therefore AD, can be made smaller than any imaginable quantity.

$\therefore$ the difference between AO and OD can be made less than any imaginable quantity.

$\therefore$ OA is the limit of OD. Q.E.D.

PROPOSITION XI. THEOREM

409. *The circumference of a circle is greater than the perimeter of any inscribed polygon.*

[The proof is left to the student.]

PROPOSITION XII. THEOREM

410. *The circumference of a circle is less than the perimeter of any enveloping line.*

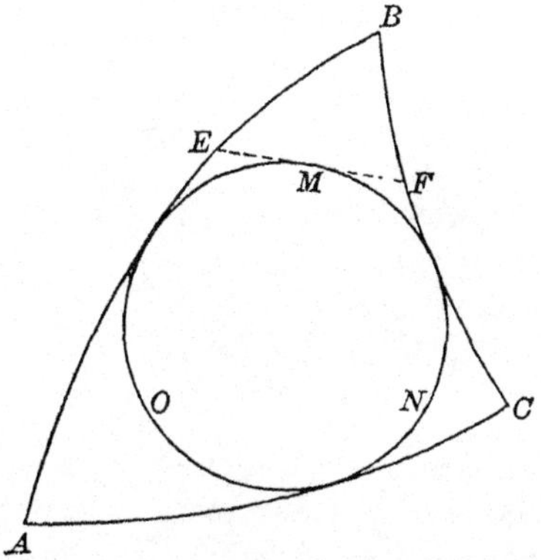

Hyp. *MNO* is a circumference, and any line *ABC* envelopes it.

To prove $MNO < ABC.$

Proof. Among all the lines enclosing *MNO*, there must be one which is the shortest.

Draw tangent *EF*, intersecting *ABC* in *E* and *F*.

$EF < EB + BF.$ (Ax. 10.)

$\therefore AEFC < ABC.$ (Why?)

$\therefore ABC$ is not the shortest line.

Similarly no other enveloping line can be the least.

$\therefore$ *OMN* is the least line. Q.E.D.

411. COR. The circumference of a circle is less than the perimeter of any circumscribed polygon.

PROPOSITION XIII. THEOREM

412. *If the number of sides of an inscribed regular or circumscribed regular polygon is increased indefinitely:*

(1) *Their perimeters approach the circumference as a common limit.*

(2) *Their areas approach the circle as a common limit.*

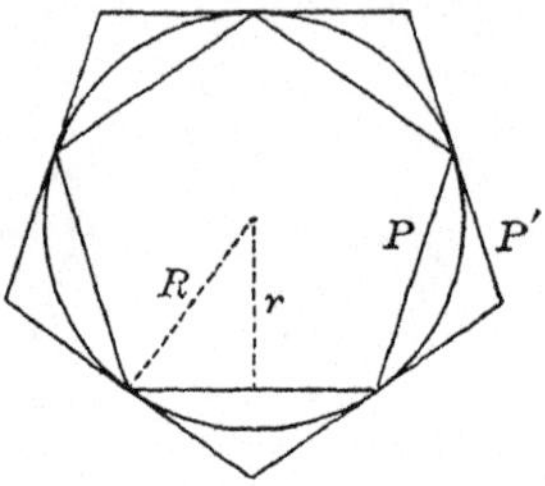

Hyp. P and P' are the perimeters, A and A' the areas of two regular polygons of n sides respectively inscribed in and circumscribed about a circle S, whose circumference is C.

To prove. If n increase indefinitely,

(1) P' and P approach C as a limit.

(2) A' and A approach S as a limit.

Proof. Let R be the apothem of P' and radius of P, and let r be the apothem of P.

$$(1)\quad \frac{P'}{P}=\frac{R}{r}. \qquad \text{(Why ?)}$$

$$\frac{P'-P}{P}=\frac{R-r}{r}. \qquad \text{(Why ?)}$$

By increasing n indefinitely, $R-r$ can be made smaller than any imaginable quantity. Hence $P'-P$ will become indefinitely small.

Since $P' > C > P$. (Why?)

The difference between C and either P' or P is less than $P' - P$, and can be made indefinitely small.

$\therefore$ P' and P approach C as a limit.

$$(2) \quad \frac{A'}{A} = \frac{R^2}{r^2}. \qquad \text{(Why?)}$$

$$\frac{A' - A}{A} = \frac{R^2 - r^2}{r^2}. \qquad \text{(Why?)}$$

By increasing n indefinitely,

$R^2 - r^2$ becomes indefinitely small.

$\therefore$ $A' - A$ becomes indefinitely small.

And since $A' > S > A$,

the difference between S and either A' or A is smaller than $A' - A$, and can be made indefinitely small.

$\therefore$ A and A' approach S as a limit. Q.E.D.

PROPOSITION XIV. THEOREM

413. *Circumferences are to each other as their radii.*

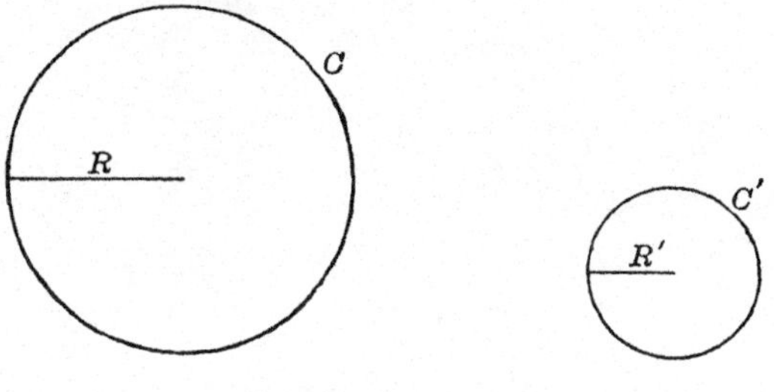

Hyp. C and C' are two circumferences with radii R and R' respectively.

To prove $C : C' = R : R'$.

Proof. Inscribe regular polygons of n sides in each circumference having the perimeters P and P' respectively.

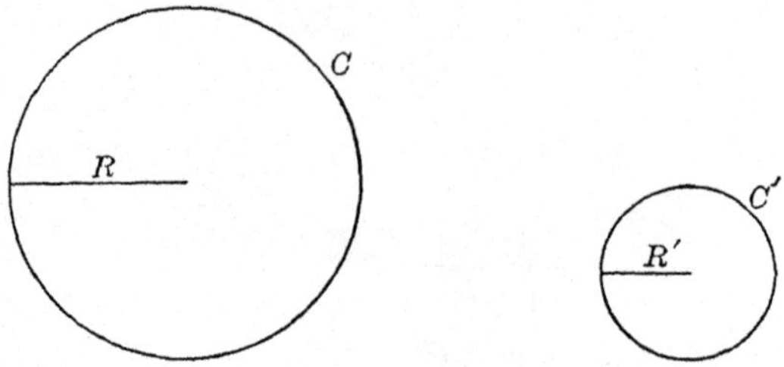

$$\frac{P}{P'} = \frac{R}{R'}, \qquad \text{(Why ?)}$$

or
$$\frac{P}{R} = \frac{P'}{R'}. \qquad \text{(Why ?)}$$

Increase n indefinitely, then the equal variables $\frac{P}{R}$ and $\frac{P'}{R'}$ must have equal limits.

But C and C' are the limits of P and P' respectively.

$$\therefore \frac{C}{R} = \frac{C'}{R'},$$

or
$$C : C' = R : R'. \qquad \text{Q.E.D.}$$

414. Cor. 1. Circumferences are to each other as their diameters.

415. Cor. 2. The ratio of the circumference to its diameter is constant.

Since
$$C : C' = 2\,R : 2\,R',$$

$$\frac{C}{2\,R} = \frac{C'}{2\,R'}.$$

416. SCHOLIUM. The constant value of $\frac{C}{2R}$ is denoted by the Greek letter π. It is an incommensurable number, which can only be found approximately, but with any degree of precision.

Since $$\frac{C}{2R} = \pi \quad C = 2\pi R.$$

Ex. 949. Find a circumference equal to the sum of two given circumferences.

Ex. 950. Draw three circumferences which are to each other as 4 : 5 : 6.

PROPOSITION XV. THEOREM

417. *The area of a regular polygon is equal to one-half the product of its apothem and perimeter.*

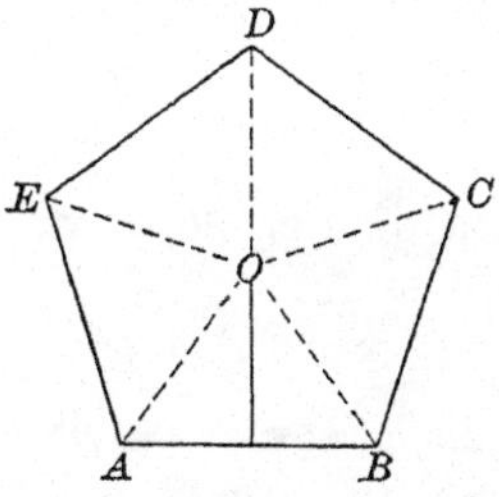

Hyp. S is the area, P the perimeter, and R the apothem of the polygon $ABCDE$, having n sides.

To prove $$S = \tfrac{1}{2} P \times R.$$

Proof. Draw radii OA, OB, etc., dividing the polygon into n equal triangles. (Why?)

$$\text{area } ABO = \tfrac{1}{2} AB \times R. \qquad \text{(Why?)}$$

$$S = n \times \text{area } ABO = \tfrac{1}{2} n \times AB \times R.$$

But $$n \times AB = P,$$

or $$S = \tfrac{1}{2} P \times R. \qquad \text{Q.E.D.}$$

P

418. Def. A **sector** of a circle is a figure bounded by two radii and their intercepted arc.

419. Def. **Similar arcs, segments,** and **sectors** are those which correspond to equal central angles.

Proposition XVI. Theorem

420. *The area of a circle is equal to one-half the product of its circumference and radius.*

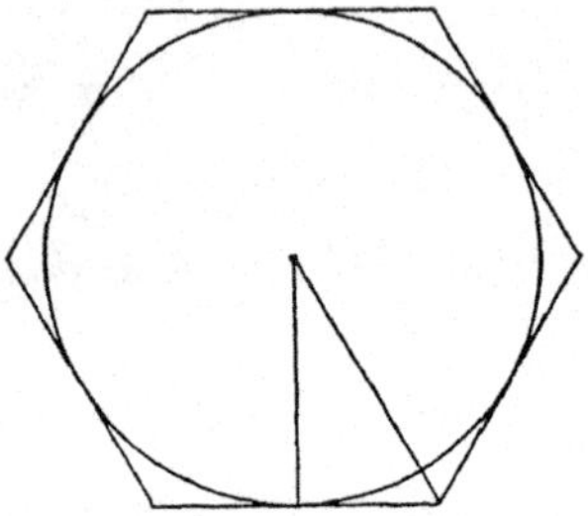

Hyp. S is the area of a circle of radius R and circumference C.

To prove $S = \frac{1}{2} C \times R$.

Proof. Circumscribe a regular polygon about the circle, and let P be its perimeter, A its area.

$$A = \tfrac{1}{2} P \times R.$$

Let the number of sides be increased indefinitely, then

A approaches S as a limit. (Why ?)

P approaches C as a limit. (Why ?)

$\therefore S = \frac{1}{2} C \times R$. (213) Q.E.D.

421. Cor. 1. The area of a circle is equal to π times the square of its radius.

For $S = \frac{1}{2} C \times R$ and $C = 2\pi R$.

$$\therefore S = \pi R^2.$$

422. Cor. 2. The areas of circles are to each other as the squares of their radii.

423. Cor. 3. The area of a sector is equal to one-half the product of its radius and arc.

424. Cor. 4. Similar sectors are to each other as the squares of their radii.

425. Remark. — **Since the areas of circles have the same ratio as the squares of their radii,** *a great many constructions referring to squares may be applied to circles.*

Ex. 951. Construct a circle equivalent to three-fifths of a given circle.

Ex. 952. Construct a circle equivalent to three times a given circle.

Ex. 953. Construct a semicircle equivalent to a given circle.

Ex. 954. Construct a circle equivalent to the sum of two given circles.

Ex. 955. Construct a circle equivalent to the difference of two given circles.

Ex. 956. Construct a circle equivalent to the area bounded by two concentric circumferences.

* **Ex. 957.** If upon the three sides of a right triangle semicircles be drawn as indicated in the diagram, the area of the right triangle is equivalent to the sum of the two crescent-shaped areas, bounded by the semicircles. (Hippocrates' Theorem.)

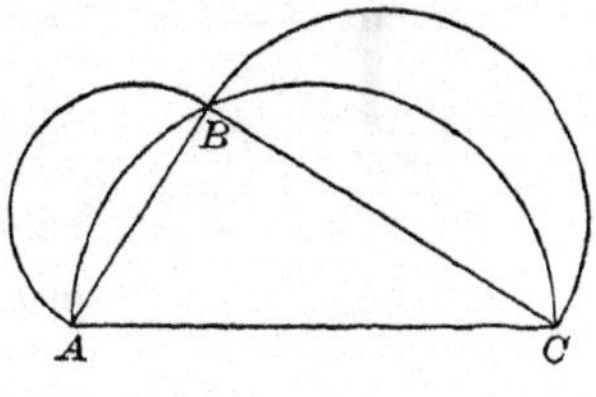

* **Ex. 958.** If C is a point in the straight line AB, the three semicircles, drawn respectively upon AB, AC, and CB as diameters, bound an area equivalent to a circle whose diameter is the perpendicular CD, D being in the largest semicircle.

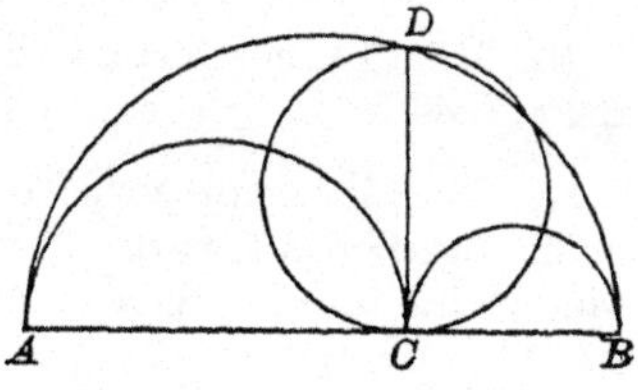

PROPOSITION XVII. THEOREM

426. *Similar segments are to each other as the squares of their radii.*

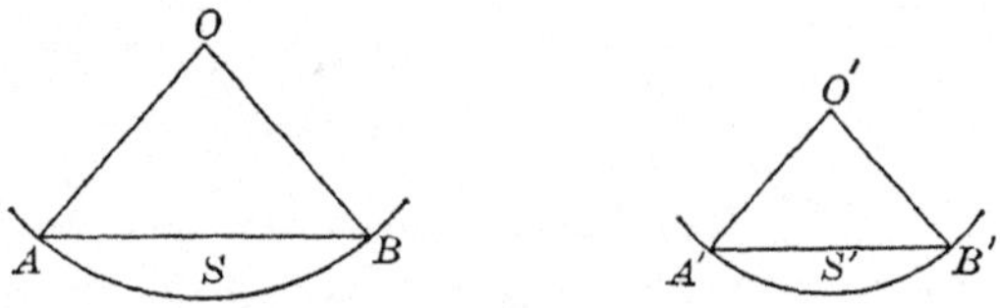

Hyp. S and S' are two similar segments, having the radii R and R' respectively.

To prove $S : S' = R^2 : R'^2$.

Proof. Sector AOB : sector $A'O'B' = R^2 : R'^2$. (Why?)

Or, if K and m denote two constant numbers, and sector $AOB = K \cdot R^2$, then sector $A'O'B' = K \cdot R'^2$. (268)

Similarly, if $\triangle AOB = m \cdot R^2$, then $\triangle A'O'B' = m \cdot R'^2$. (372)

Subtracting member from member,

$$\text{segment } S = (K - m)R^2 \text{ and segment } S' = (K - m)R'^2.$$

By dividing the last two equations,

$$\frac{\text{segment } S}{\text{segment } S'} = \frac{R^2}{R'^2}.$$

Q.E.D.

Ex. 959. To construct a segment similar to two given similar segments and equivalent to their sum.

Ex. 960. If upon three sides of a right triangle arcs of 90° be drawn, so that those upon the arms lie within, and that upon the hypotenuse lies without the triangle, then the area of the right triangle is equal to the area bounded by the three arcs.

PROPOSITION XVIII. PROBLEM

427. *The side and the radius of a regular polygon being given, to compute the side of a regular polygon of the same radius and double the number of sides.*

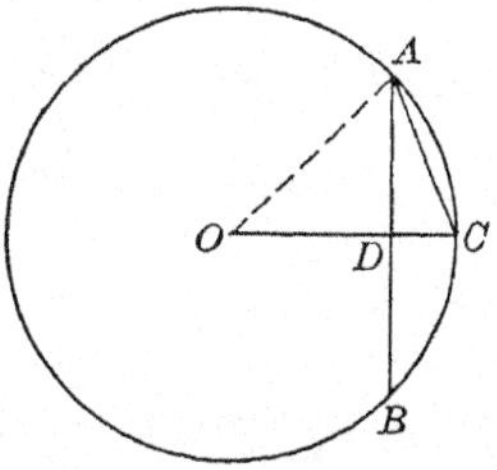

Given. AB equal to a side s of a regular polygon with radius $OA = R$.

Required. To compute AC, the side of a regular polygon of double the number of sides, inscribed in the same circle.

Construction. Draw AO and CO meeting AB in D.

OC bisects AB at right angles. (Why ?)

$$\overline{AC}^2 = \overline{OA}^2 + \overline{OC}^2 - 2\,OC \times OD,$$

(*square of a side opposite an acute* ∠, etc.),

or $$\overline{AC}^2 = 2\,R^2 - 2\,R \times OD.$$

But $$\overline{OD}^2 = R^2 - \left(\frac{s}{2}\right)^2. \qquad \text{(Why ?)}$$

$$\therefore\ OD = \sqrt{R^2 - \frac{s^2}{4}},$$

or $$OD = \tfrac{1}{2}\sqrt{4\,R^2 - s^2}.$$

Hence $$\overline{AC}^2 = 2\,R^2 - R\sqrt{4\,R^2 - s^2}.$$

$$\therefore\ AC = \sqrt{2\,R^2 - R\sqrt{4\,R^2 - s^2}}.$$

428. Cor. If $R = 1$, and the side of a polygon of n sides equals s_n, then

$$s_{2n} = \sqrt{2 - \sqrt{4 - s_n^2}}.$$

Ex. 961. The side of an inscribed square is equal to $R\sqrt{2}$. Compute the side of the regular octagon.

Proposition XIX. Problem

429. *To compute the ratio of the circumference of a circle to the diameter.*

Let $R = 1$, then $s_6 = 1$,

and according to (428), we have

	Length of Side	Length of Perimeter
$s_{12} = \sqrt{2 - \sqrt{4 - 1^2}}$	$= 0.51764$	6.21166
$s_{24} = \sqrt{2 - \sqrt{4 - (0.51764)^2}}$	$= 0.26105$	6.26526
$s_{48} = \sqrt{2 - \sqrt{4 - (0.26105)^2}}$	$= 0.13081$	6.27870
$s_{96} = \sqrt{2 - \sqrt{4 - (0.13081)^2}}$	$= 0.06534$	6.28206
$s_{192} = \sqrt{2 - \sqrt{4 - (0.06534)^2}}$	$= 0.03272$	6.28291
$s_{384} = \sqrt{2 - \sqrt{4 - (0.03272)^2}}$	$= 0.01636$	6.28312
$s_{768} = \sqrt{2 - \sqrt{4 - (0.01636)^2}}$	$= 0.00818$	6.28317

Taking the last perimeter as the approximate value of the circumference,

$$\pi = \frac{C}{2R} = \frac{6.28317}{2} = 3.14159.$$

430. Remark. — **Most computations referring to the circle involve the use of the two formulæ $C = 2\pi R$ and $S = \pi R^2$. If R is not given, compute R first.** (421.)

Ex. 962. Find the circumference of a circle whose radius is 5.

Ex. 963. Find the area of a circle whose radius is 10.

Ex. 964. Find the radius of a circle whose area is 100 sq. in.

Ex. 965. Find the radius of a circle whose circumference equals m inches.

Ex. 966. Find the circumference of a circle whose area equals S.

Ex. 967. Find the area of a circle whose circumference equals C.

Ex. 968. Find the radius of a circle equivalent to a square whose side equals 6.

Ex. 969. The circumference of a circle equals 10. Find the circumference of a circle having twice the area of the given circle.

Ex. 970. Two concentric circles have their circumferences equal to 30 and 40 respectively. Find the area bounded by the two circumferences.

Ex. 971. Find a semicircle equivalent to an equilateral triangle whose side equals 5.

Ex. 972. Find the area of a sector whose radius equals 5, and whose central angle equals 40°.

Ex. 973. A square is inscribed in a circle of radius 10. Find the area of a segment bounded by a side of the square.

Ex. 974. From a point without a circle, two tangents are drawn to a circle whose radius equals 10. Find the area bounded by the two tangents and the circumference, if the tangents include an angle of 120°.

Ex. 975. Find the area bounded by three arcs of 60° and radius 5, if the concave sides of the arcs are turned toward the area.

Ex. 976. Find the area bounded by three arcs of 60° and radius 5, if the convex sides of the arcs are turned toward the area.

Ex. 977. A circle has an area of 60 sq. in. Find the length of an arc of 40°.

Ex. 978. Find the central angle of a sector whose perimeter is equal to the circumference.

Ex. 979. Find the central angle of a sector whose area is equal to the square of the radius.

MISCELLANEOUS EXERCISES

Ex. 980. If in a circle two chords intersect at right angles, and circles are constructed on the segments of the chords as diameters, the area of the given circle is equivalent to the sum of the areas of the four circles.

Ex. 981. The side of the regular circumscribed triangle is equal to twice the side of the regular inscribed triangle.

Ex. 982. The square of a side of an inscribed equilateral triangle is equal to three times the square of the inscribed regular hexagon.

Ex. 983. If the radius of a regular octagon be equal to r, prove that its side equals $r\sqrt{2-\sqrt{2}}$.

Ex. 984. If the radius of a regular decagon be equal to r, prove that its side equals $\frac{r}{2}(\sqrt{5}-1)$.

Ex. 985. If the radius of a regular pentagon be equal to r, prove that its side equals $\frac{r}{2}\sqrt{10-2\sqrt{5}}$.

Ex. 986. If the radius of a regular dodecagon be equal to r, prove that its side equals $r\sqrt{2-\sqrt{3}}$.

Ex. 987. If in a circle a regular decagon and a regular pentagon be inscribed, the side of the decagon increased by the radius is equal to twice the apothem of the pentagon.

* **Ex. 988.** The square of the side of a regular decagon increased by the square of the radius is equal to the square of the side of a regular pentagon, having the same radius.

APPENDIX TO PLANE GEOMETRY

ALGEBRAIC SOLUTIONS OF GEOMETRICAL PROPOSITIONS. MAXIMA AND MINIMA

CONSTRUCTION OF ALGEBRAIC EXPRESSIONS

431. NOTE. — In the following propositions, a, b, c, d, etc., denote given lines, while x, y, z, etc., denote required lines.

432. (1) Construct $x = a + b$.

(2) Construct $x = a - b$.

(3) If m denotes a given number.

Construct $x = m \cdot a$.

(4) Construct $x = \frac{a}{m}$.

(5) Construct $x = \frac{ab}{c}$.

HINT. — x is the fourth proportional to c, a, and b.

(6) Construct $x = \frac{a^2}{b}$.

Ex. 989. Construct $x = \frac{ab}{c}$ by means of (312).

Ex. 990. Construct $x = \frac{a^2}{b}$ by means of (314).

(7) Construct $x = \sqrt{ab}$.

(Three methods: 307 and 314.)

(8) Construct $x = \sqrt{a^2 + b^2}$.

(9) Construct $x = \sqrt{a^2 - b^2}$.

(10) Construct $x = a\sqrt{2}$.

(11) If m denotes a known number.

Construct $x = a\sqrt{m}$.

HINT. $x = \sqrt{a(am)}$.

Ex. 991. Construct $x = \frac{a}{2}\sqrt{3}$ by means of an equilateral triangle.

433. Complex algebraical expressions are constructed by means of the eleven constructions of (432). It is quite often necessary to transform the algebraical expressions, in order to make them special cases of (432). Different algebraical transformations lead to different solutions.

Construct:

Ex. 992.

$$x = \sqrt{3\,ab},$$

$$x = \sqrt{(3\,a)b},$$

i.e. x is the mean proportional between $3\,a$ and b,

or

$$x = \sqrt{ab}\,\sqrt{3},$$

i.e. find $\sqrt{ab}$ by means of (432, 7), and $\sqrt{ab}\,\sqrt{3}$ by means of (Ex. 991).

Ex. 993.

$$x = \frac{a^2 - b^2}{4c} = \frac{(a+b)(a-b)}{4\,c},$$

i.e. find the fourth proportional to $4\,c$, $a + b$, and $a - b$.

Ex. 994. $x = \sqrt{a^2 - ab}$.

HINT. $x = \sqrt{a(a - b)}$.

Ex. 995.

$$x = \sqrt{a^2 - bc} = \sqrt{a^2 - (\sqrt{bc})^2},$$

i.e. find the mean proportional between b and c, and construct a right triangle, having one arm equal to the mean proportional, and the hypotenuse equal to a.

434. Scholium. The expressions to be constructed in Geometry are always homogeneous.

435. The **impossibility of the construction** is indicated if:

(1) The expression is imaginary.

(2) The value of x is not within the limits indicated by the problem.

(3) Sometimes, if the result is negative. If the required line has a certain direction, a negative result would indicate a *line* of opposite direction.

SOLUTION OF PROBLEMS BY MEANS OF ALGEBRAIC ANALYSIS *

436. If a problem requires the construction of lines, it is often possible to state the condition in the form of an equation. The solution of the equation gives the unknown line in an algebraic form, which may be constructed according to (433).

Ex. 996. To divide a line externally so that the small external segment is the mean proportional between the other segment and the given line.

Analysis. Let x be one segment, then $a+x$ is the other, and

$$a+x : x = x : a.$$

$$x^2 = a^2 + ax.$$

The solution of the equation gives

$$x = \frac{a}{2} \pm \sqrt{\frac{a^2}{4} + a^2}.$$

As the minus sign would give a negative answer, *i.e.* an internal segment, we have to construct

$$x = \frac{a}{2} + \sqrt{\left(\frac{a}{2}\right)^2 + a^2}.$$

* The present chapter requires the knowledge of quadratic equations.

Construction.

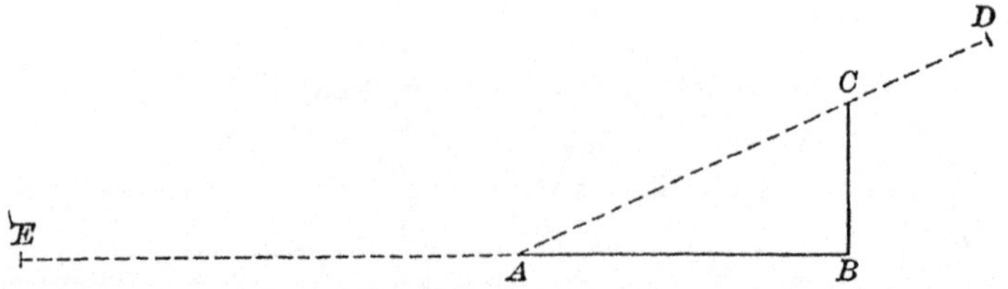

Draw $AB = a$. At B, draw BC perpendicular to AB and equal to $\frac{a}{2}$.

Draw AC, and produce it to D, so that $CD = CB$.

Then $$x = AD.$$

$\therefore$ produce BA to E, so that $AE = AD$.

E is the required point.

437. Scholium 1. The analysis used in this problem is called **algebraic analysis** in distinction from the purely **geometric analysis** given in Book I.

438. Scholium 2. The algebraic analysis contains a proof for the correctness of the construction, although quite often a purely geometric proof may be found.

Ex. 997. From a triangle ABC, to cut off an isosceles triangle AXY, equivalent to one-half of ABC.

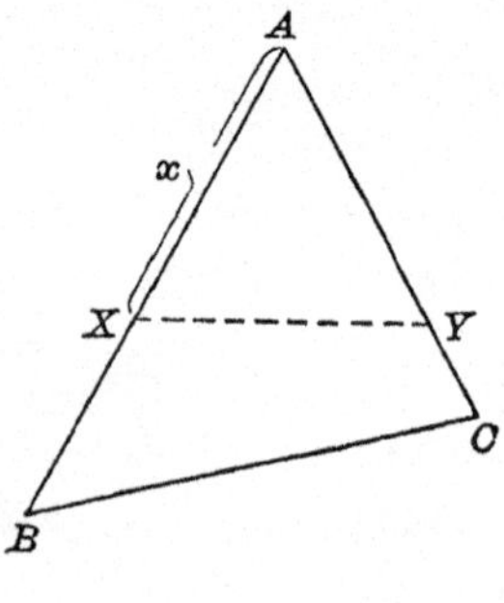

Analysis. Let $AX = AY = x$, $AB = c$, and $AC = b$.

$$\triangle AXY : \triangle ABC = x^2 : bc = 1 : 2. \quad \text{(Why?)}$$

$$\therefore 2x^2 = bc.$$

$$x = \sqrt{\frac{bc}{2}}.$$

Construction. Construct x, a mean proportional between $\frac{c}{2}$ and b, and on AB and AC, respectively, lay off AX and AY equal to x.

Then $\triangle AXY$ is the required one.

Ex. 998. To construct a right triangle, having given an arm a, and the projection p, of the other arm upon the hypotenuse.

Analysis. Let x be the projection of a upon the hypotenuse.

$$x(x+p)=a^2, \qquad \text{(Why?)}$$

or

$$x^2+xp=a^2.$$

$$\therefore x=-\frac{p}{2}\pm\sqrt{\left(\frac{p}{2}\right)^2+a^2}.$$

The minus sign would give a negative answer.

$$\therefore \text{construct } x=-\frac{p}{2}+\sqrt{\left(\frac{p}{2}\right)^2+a^2}.$$

Construction. Draw $AB=a$, and BC perpendicular to AB and equal to $\frac{1}{2}p$. Join

$$AC\left(=\sqrt{\left(\frac{p}{2}\right)^2+a^2}\right).$$

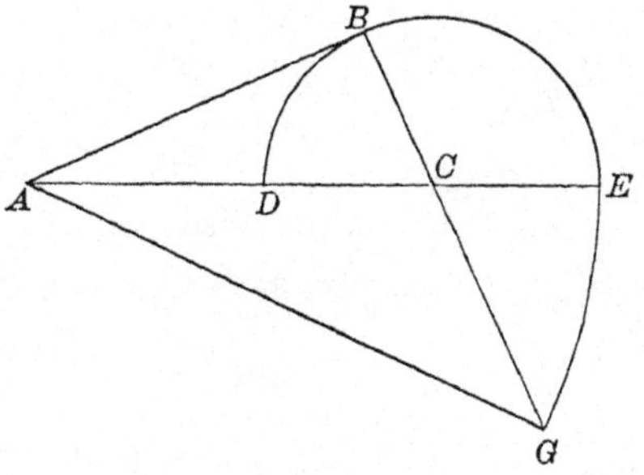

From C as a center, with a radius equal to BC, draw an arc intersecting AC in D and E.

Then $AD=x$,

and

$AE=x+p=$ hypotenuse.

$\therefore$ from A as a center, with a radius equal to AE, draw an arc meeting BC produced in G.

AGB is the required triangle.

Ex. 999. From a given line b, to cut off a part x, that shall be a mean proportional between the remainder $b-x$ and another given line a.

Analysis. $(b-x):x=x:a.$

$$x^2=ab-ax.$$

$$\therefore\ x=-\frac{a}{2}+\sqrt{\frac{a^2}{4}+ab}.$$

Transforming, either

$$x=-\frac{a}{2}+\sqrt{a\left(\frac{a}{4}+b\right)},$$

or

$$x=-\frac{a}{2}+\sqrt{\left(\frac{a}{2}\right)^2+\left(\sqrt{ab}\right)^2}.$$

Using the second method, we have:

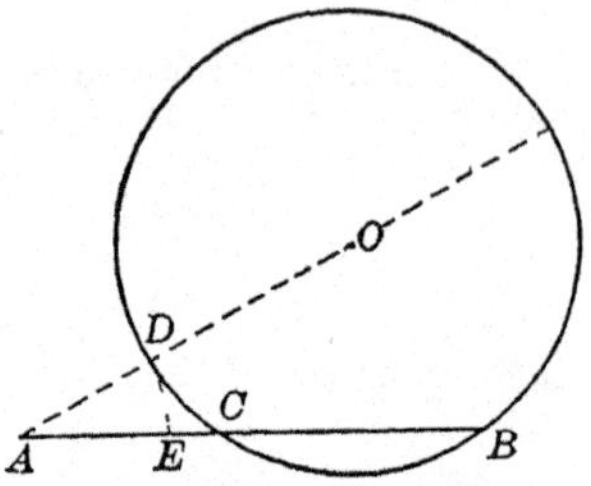

Construction. Draw $AB=a$ and $AC=b$. From C and B as centers, with a radius equal to $\frac{a}{2}$, draw two arcs intersecting in O.

From O as a center and the same radius, describe a circle.

Draw OA, intersecting the circumference in D.

Then $x=AD.$

$\therefore$ on AB, lay off $AE=AD$.

Then E is the required point.

Proof. From A, draw tangent AF.

$$\overline{AF}^2=AC\times AB, \qquad \text{(Why ?)}$$

or

$$AF=\sqrt{ab}.$$

$$\overline{AO}^2=\overline{FO}^2+\overline{AF}^2.$$

$$AO=\sqrt{\left(\frac{a}{2}\right)^2+\left(\sqrt{ab}\right)^2}.$$

$$AD=\sqrt{\left(\frac{a}{2}\right)^2+\left(\sqrt{ab}\right)^2}-\frac{a}{2}.$$

Q.E.D.

In the above construction, the tangent theorem has been used to construct the mean proportional, but any other construction of a mean proportional may be used.

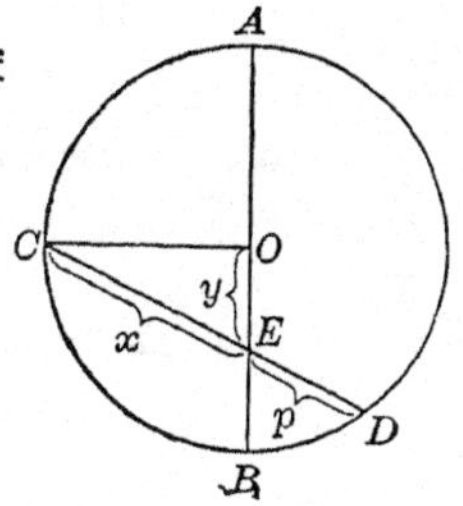

Ex. 1000. Give a purely geometric proof of Ex. 999.

Ex. 1001. In a circle O a diameter $AB = 2r$ is drawn. From the midpoint C of the semicircle AB to draw a chord CD, meeting AB in E, so that DE shall be equal to a given line p.

Analysis. Let $CE = x$, $OE = y$.

$$x^2 = y^2 + r^2, \qquad \text{(Why?)}$$

and
$$x \cdot p = (r + y)(r - y). \qquad \text{(Why?)}$$

Eliminate y and solve. Then

$$x = -\frac{p}{2} \pm \sqrt{\left(\frac{p}{2}\right)^2 + 2r^2}.$$

Only the positive sign gives a line.

Construct
$$x = -\frac{p}{2} + \sqrt{\left(\frac{p}{2}\right)^2 + 2r^2},$$

or transformed,
$$x = \sqrt{\left(\frac{p}{2}\right)^2 + (r\sqrt{2})^2} - \frac{p}{2}.$$

Construction. Draw $CA (= r\sqrt{2})$.

At A, draw AF perpendicular to CA and equal to $\frac{p}{2}$.

Draw CF. On FC, lay off FG equal to FA.

Then $x = CG$.

From C as a center, with a radius equal to CG, draw an arc meeting AB in E. Through E, draw chord CD, which is the required chord.

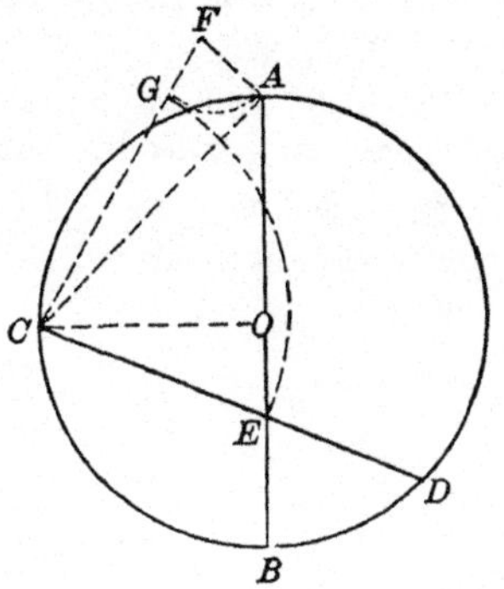

Ex. 1002. Given lines AB and p. In the line AB, to find a point C so that $\overline{AC}^2 - \overline{CB}^2 = p^2$.

Ex. 1003. In the line AB, to find a point C so that $\overline{AC}^2 = 2\,\overline{CB}^2$.

Ex. 1004. In the median AD of $\triangle ABC$, to find a point X such that XD and the perpendiculars dropped from X upon AB and AC divide the figure into three equivalent parts.

Ex. 1005. In a given square $ABCD$, to inscribe an equilateral triangle, having one vertex at A.

Ex. 1006. From a point P without a circumference, to draw a secant which is bisected by the circumference.

Ex. 1007. In a given square, to inscribe another square, having a given side.

Ex. 1008. To inscribe a square in a semicircle.

Ex. 1009. In the triangle ABC, to inscribe a parallelogram having a given area, and having an angle common with $\triangle ABC$.

Ex. 1010. From the midpoint A of arc BC, to draw a chord AD, intersecting chord BC in E, so that DE is equal to a given line p.

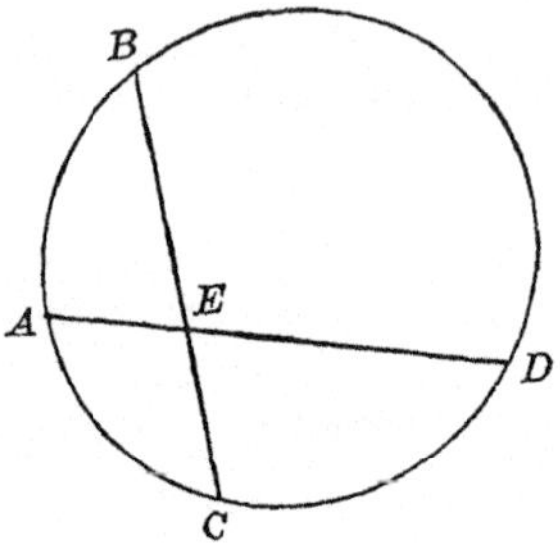

Ex. 1011. To construct a triangle, having given the base, the vertical angle, and the bisector of that angle. (Ex. 1010.)

Ex. 1012. Upon a given line as hypotenuse to construct a right triangle one of whose arms is a mean proportional between the other arm and the hypotenuse.

Ex. 1013. To transform a given square into a rectangle having a perimeter equal to twice the perimeter of the given square.

Ex. 1014. Given two concentric circles. To draw a chord in the larger circle so that it equals twice the chord formed in the smaller circle.

MAXIMA AND MINIMA OF PLANE FIGURES

439. DEF. A **maximum** is the greatest of all magnitudes of the same kind; a **minimum**, the smallest.

440. DEF. **Isoperimetric figures** are those which have **equal perimeters.**

PROPOSITION I. THEOREM

441. *Of all triangles having given two sides, that in which those two sides are perpendicular to each other is a maximum.*

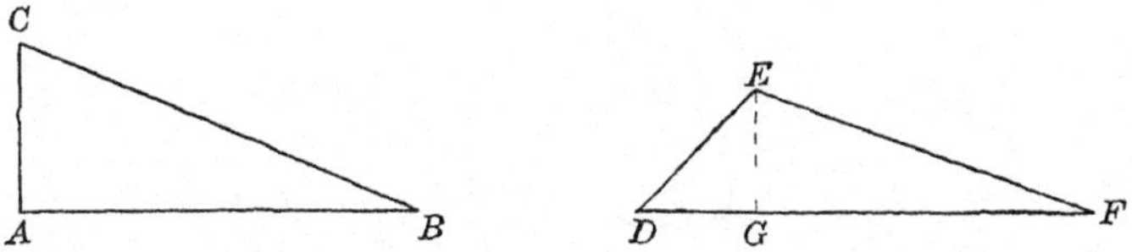

Hyp. In $\triangle$ ABC and DFE, $AB = DF$, $AC = DE$, $\angle A =$ rt. $\angle$, and $\angle D$ is oblique.

To prove area $ABC >$ area DEF.

Proof. From E draw $EG \perp DF$.

Then $EG < DE$, (Why?)

or $EG < AC$.

$\therefore$ $\triangle$ ABC and DEF have equal bases and unequal altitudes.

$\therefore \triangle ABC > \triangle DEF$. (Why?)

$\therefore \triangle ABC$ is a maximum. Q.E.D.

Ex. 1015. To divide a given line into two parts so that the rectangle contained by the segments is a maximum.

Ex. 1016. In the hypotenuse of a right triangle to find a point so that the sum of the squares of perpendiculars drawn from the point upon the arms is a minimum.

PROPOSITION II. THEOREM

442. *Of all triangles having the same base and equal areas, the isosceles triangle has the minimum perimeter.*

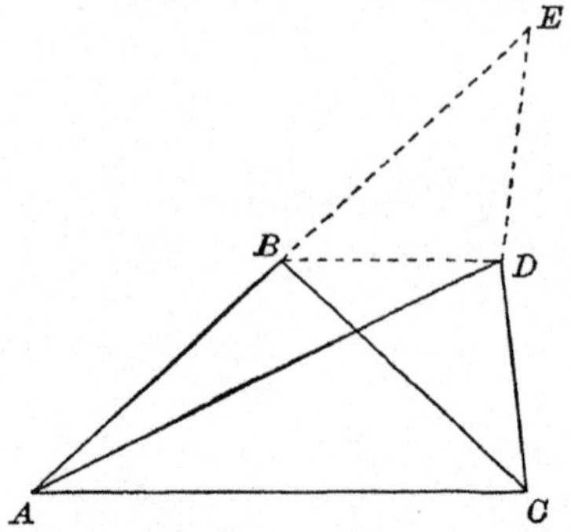

Hyp. $\triangle ADC \approx \triangle ABC$, and $AB = BC$.

To prove. $AB + BC + AC < AD + DC + AC$.

Proof. Produce AB by its own length to E. Join BD and DE.

$$BD \parallel AC,$$

(*for otherwise* $\triangle ABC$ *would not be* $\approx \triangle ADC$).

$\angle EBD = \angle BAC$. (Why?)

$\angle BAC = \angle BCA$. (Why?)

$\angle BCA = \angle CBD$. (Why?)

$\therefore \angle EBD = \angle CBD$. (Ax. 1.)

But $BC = BE$, (Why?)

and BD is common.

$\therefore \triangle BDC = \triangle BDE$.

$\therefore DC = DE$.

But $AD + DE > AB + BE$. (Why?)

$\therefore AD + DC > AB + BC$,

or adding AC,

$AC + AD + DC > AB + BC + AC$. Q.E.D.

Proposition III. Theorem

443. *The maximum of isoperimetric triangles on the same base is the one whose other two sides are equal.*

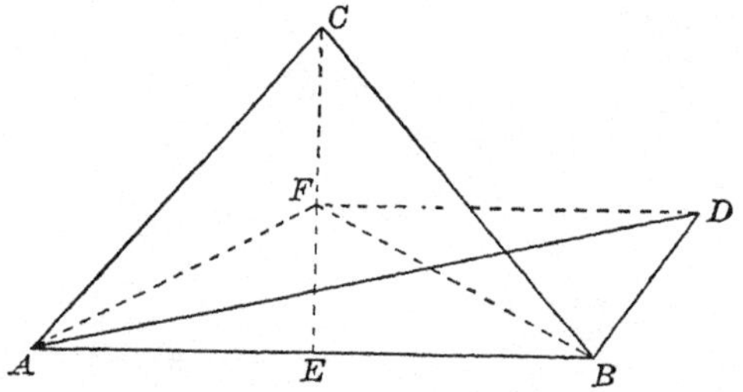

Hyp. $\triangle$s ABC and ABD have equal perimeters, and $AC = CB$.

To prove area $ACB >$ area ADB.

Proof. Draw median CE and $DF \parallel AB$, meeting CE in F. Join FA and FB.

Then CE is the perpendicular bisector of AB. (Why?)

$$\therefore AF = FB.$$

But $\triangle AFB \Bumpeq \triangle ADB$,

(*same base, equal altitudes.*)

$\therefore$ perimeter $AFB <$ perimeter ADB. (442)

$$\therefore AF + FB < AD + DB \text{ or } AC + CB.$$

Hence $AF < AC$.

$\therefore FE < CE$. (311)

$\therefore$ area $AFB <$ area ACB,

(*unequal altitudes*),

or area $ADB <$ area ACB. Q.E.D.

444. Cor. Of all isoperimetric triangles, the equilateral has the maximum area.

PROPOSITION IV. THEOREM

445. *Of all polygons having all sides given but one, the maximum can be inscribed in a semicircle having the undetermined side as diameter.*

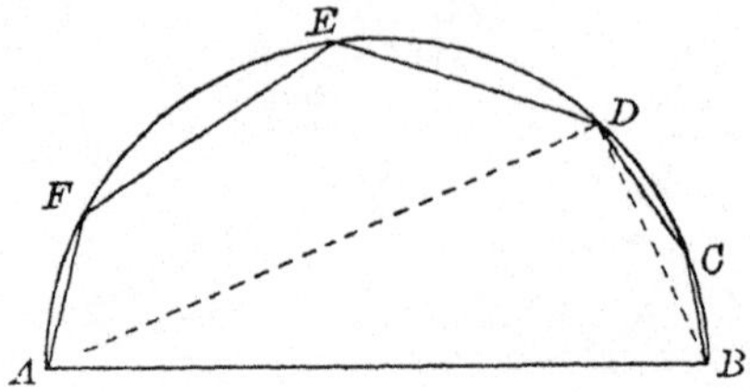

Hyp. Polygon $ABCDEF$ is the maximum of all polygons having given the sides AF, FE, ED, DC, and CB.

To prove $ABCDEF$ can be inscribed in a semicircle whose diameter is AB.

Proof. Join any vertex, as D, with A and B.

Then $\triangle ADB$ must be the maximum of all triangles that can be formed with sides AD and DB. For otherwise by making $\angle ADB$ a right one without changing the sides AD and DB, we could increase $\triangle ADB$ without altering the remaining parts $AFED$ and DCB of the polygon. Or polygon $ABCDEF$ would be increased, which is contrary to the hypothesis, since $ABCDEF$ is a maximum.

$\therefore \triangle ADB$ is the maximum of all triangles having AD and DB given.

Then $\angle ADB$ is a right angle, (Why?)

and D is on a semicircumference that can be constructed on AB.

For the same reason every vertex of the polygon must lie on the semicircumference. Q.E.D.

Ex. 1017. To inscribe an angle in a given semicircle so that the sum of its arms is a maximum.

Proposition V. Theorem

446. *Of all polygons constructed with the same given sides, that which can be inscribed in a circle is the maximum.*

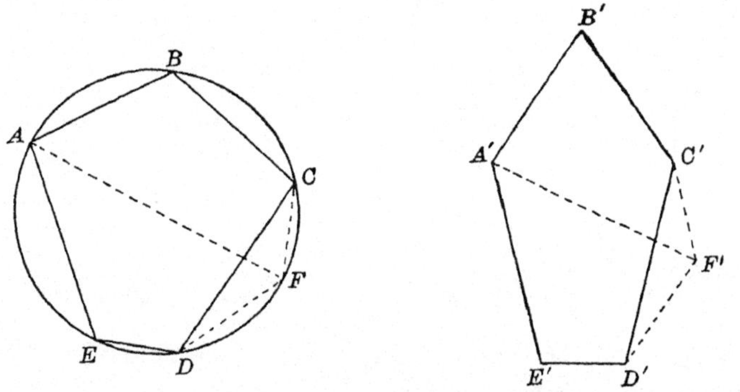

Hyp. Polygon $ABCDE$, which is inscribed in a circle, is mutually equilateral with polygon $A'B'C'D'E'$, which cannot be inscribed in a circle.

To prove area $ABCDE >$ area $A'B'C'D'E'$.

Proof. From A draw diameter AF, and join F to the two nearest vertices C and D.

On $C'D'$, construct $\triangle C'D'F'$ equal to $\triangle CDF$, and join $A'F'$.

Area $ABCF >$ area $A'B'C'F'$. (445)

Area $FDEA >$ area $F'D'E'A'$. (445)

$\therefore$ area $ABCFDE >$ area $A'B'C'F'D'E'$.

But $\triangle CFD = \triangle C'F'D'$.

Then area $ABCDE >$ area $A'B'C'D'E'$. (Ax. 5.) Q.E.D.

PROPOSITION VI. THEOREM

447. *Of all isoperimetric polygons of the same number of sides, the equilateral is the maximum.*

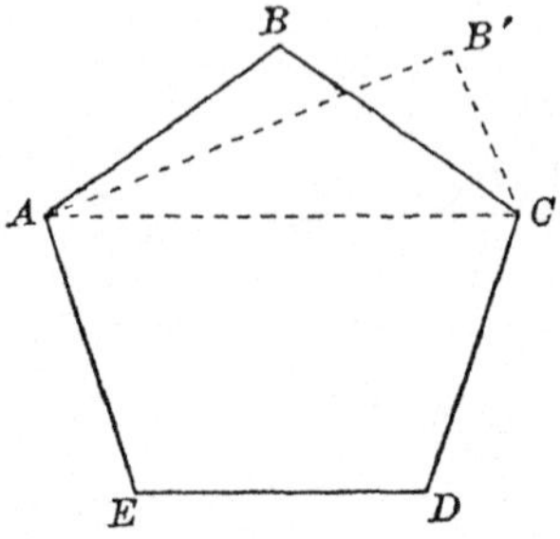

Hyp. $ABCDE$ is the maximum of all isoperimetric polygons of the same number of sides.

To prove $AB = BC = CD = DE = EA$.

Proof. The polygon must be equilateral, for suppose any two sides AB and BC were not equal.

Then construct on the diagonal AC the isosceles triangle $AB'C$, isoperimetric with ABC.

Area $AB'C$ would be greater than area ABC. (443)

$\therefore$ polygon $AB'CDE$ would be greater than the isoperimetric polygon $ABCDE$, which would contradict the hypothesis.

Hence $AB = BC$. Q.E.D.

448. COR. Of all isoperimetric polygons of the same number of sides, **the maximum is regular.**

Ex. 1018. In a given segment to inscribe an angle so that the sum of its arms is a maximum.

Ex. 1019. In a given semicircle to inscribe a trapezoid whose area is a maximum.

PROPOSITION VII. THEOREM

449. *Of two isoperimetric regular polygons, that which has the greater number of sides has the greater area.*

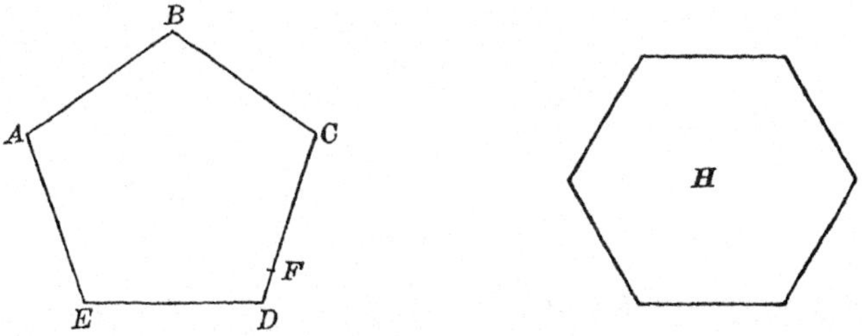

Hyp. Regular pentagon $ABCDE$ is isoperimetric with regular hexagon H.

To prove area $H >$ area $ABCDE$.

Proof. Let F be any point in CD.

$ABCFDE$ may be considered a hexagon, having one of its angles equal to a straight angle.

$\therefore$ area $H >$ area $ABCDE$. (Why?)

Q.E.D.

450. COR. The area of a circle is greater than the area of any polygon whose perimeter equals the circumference of the circle.

Ex. 1020. In a given circle to inscribe a triangle, having a maximum perimeter.

Ex. 1021. The circumscribed regular polygon has a smaller area than any other polygon circumscribed about the same circle.

Ex. 1022. Of all equivalent parallelograms on the same base, which has the minimum perimeter?

PROPOSITION VIII. THEOREM

451. *Of two equivalent regular polygons, that which has the greater number of sides has the smaller perimeter.*

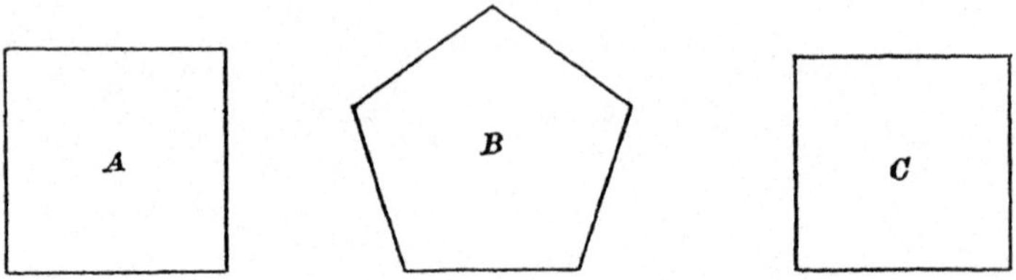

Hyp. Square $A \approx$ regular pentagon B.

To prove perimeter of $A >$ perimeter of B.

Proof. Construct square C isoperimetric with B.

Area $C <$ area B, (449)

or area $C <$ area A.

$\therefore$ perimeter of $C <$ perimeter of A.

$\therefore$ perimeter of $B <$ perimeter of A. Q.E.D.

452. COR. The circumference of a circle is less than the perimeter of any equivalent polygon.

MISCELLANEOUS EXERCISES

Ex. 1023. If two equal lines are divided externally so that the product of the segments of one is equal to the product of the segments of the other, the segments are equal respectively.

* **Ex. 1024.** Two triangles are equal if the base, the opposite angle, and its bisector of one are respectively equal to the base, the opposite angle, and its bisector of the other.

HINT. — Circumscribe circles, produce bisectors until they meet the circumference, and join the point of intersection to one end of the base. (Ex. 733 and Ex. 1023.)

Ex. 1025. If two bisectors of a triangle are equal, the corresponding sides are equal, and the triangle is isosceles. (Ex. 1024.)

* **Ex. 1026.** The square of the side of a regular pentagon increased by the square of one of its diagonals is equal to five times the square of the radius.

* **Ex. 1027.** A regular pentagon is equivalent to a rectangle having one side equal to 5 times the radius, and the other to $\frac{1}{4}$ of a diagonal of the pentagon.

* **Ex. 1028.** The product of the diagonals of an inscribed quadrilateral is equal to the sum of the products of the opposite sides. (Ptolemy's Theorem.)

* **Ex. 1029.** To construct a triangle having given the three feet of the altitudes.

* **Ex. 1030.** If p_n and P_n are respectively the perimeters of an inscribed and circumscribed regular polygon of n sides, and p_{2n} and P_{2n} the perimeters of regular polygons of $2n$ sides, respectively inscribed and circumscribed about the same circle, prove that P_{2n} is the harmonical mean between p_n and P_n; *i.e.*

$$P_{2n} = \frac{2\,p_n\,P_n}{p_n + P_n}.$$

* **Ex. 1031.** Using the notations of Ex. 1030, prove that p_{2n} is the mean proportional between p_n and P_{2n}; *i.e.* $p_{2n} = \sqrt{p_n P_{2n}}$.

Ex. 1032. If circles be circumscribed about the four triangles into which a quadrilateral is divided by its diagonals, their centers form the vertices of a parallelogram.

SOLID GEOMETRY

BOOK VI

LINES AND PLANES IN SPACE—POLYEDRAL ANGLES

Geometry of space or **solid geometry** treats of figures whose elements are not all in the same plane. (15)

453. Def. A **plane** is a surface such that a straight line joining any two points in it lies entirely in the surface. (11)

A plane is determined by given points or lines, if only one plane can be drawn through these points or lines.

Proposition I. Theorem

454. *A plane is determined:*

(1) *By a straight line and a point without the line.*

(2) *By three points not in the same straight line.*

(3) *By two intersecting straight lines.*

(4) *By two parallel straight lines.*

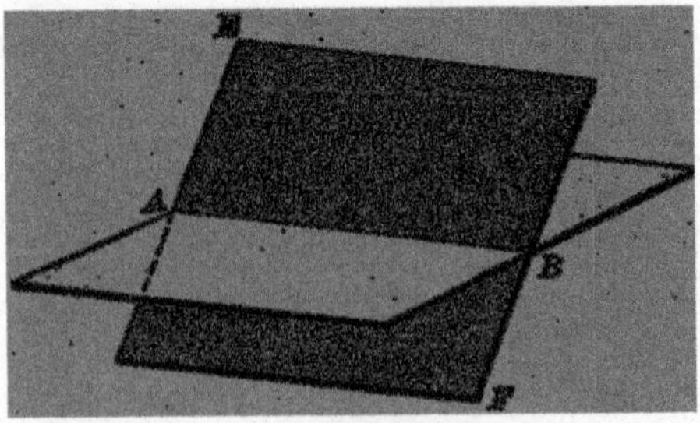

(1) To prove that a plane is determined by a given straight line AB and a given point C.

Turn any plane EF passing through AB about AB as an axis until it contains C.

If the plane, so obtained, be turned in either direction, it would no longer contain C.

Hence, the plane is determined by AB and C.

(2) To prove that three given points A, B, and C determine a plane.

Draw AB. Then AB and C determine the plane. (Case 1.)

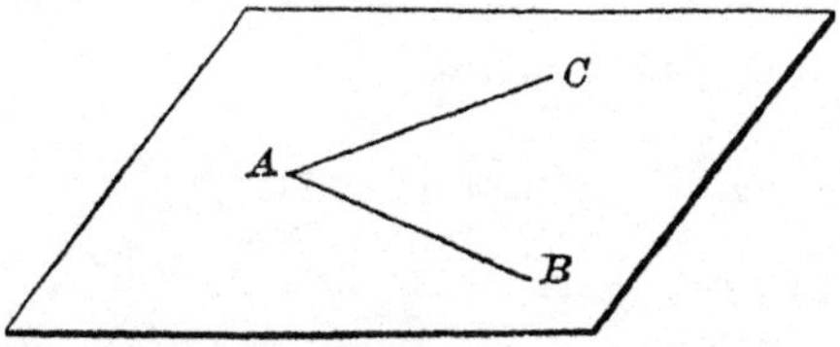

(3) To prove that two intersecting straight lines AB and AC determine a plane.

[Proof by the student.]

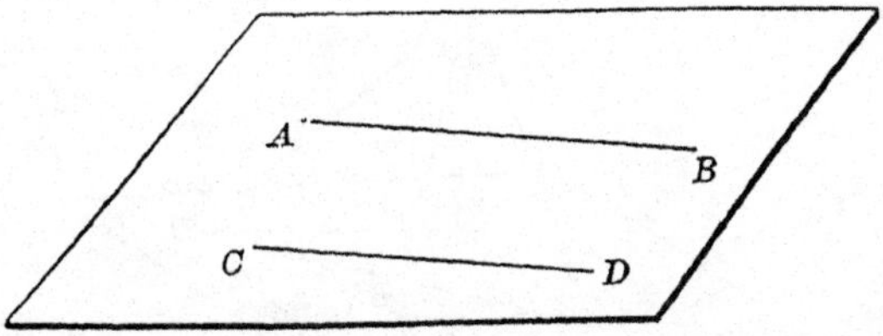

(4) To prove that two parallel lines AB and CD determine a plane.

The parallel lines AB and CD lie in the same plane by definition.

Since AB and the point C determine a plane, the two parallels determine a plane.

455. Cor. The intersection of two planes is a straight line.

For the intersection cannot contain three points not in a straight line, since only one plane can be passed through three such points.

456. Def. The **foot** of the line intersecting a plane is the point of intersection.

457. Def. **A straight line is perpendicular to a plane** if it is perpendicular to every line **drawn through its foot** in the plane.

458. Def. **A plane is perpendicular to a line** if the line is perpendicular to the plane.

459. Def. **A straight line and a plane are parallel** if they do not meet, however far they may be produced.

460. Def. **Two planes are parallel** if they do not meet, however far they may be produced.

Proposition II. Theorem

461. *If a straight line is perpendicular to each of two lines at their point of intersection, it is perpendicular to the plane of those lines.*

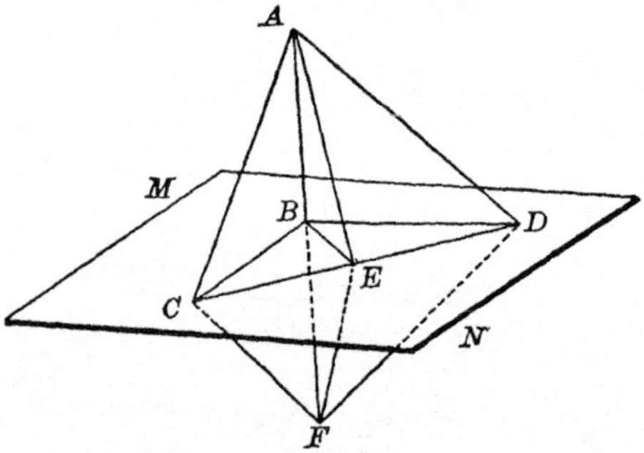

Hyp. AB is $\perp$ to BC and BD at B.

To prove AB is $\perp$ to the plane MN containing BC and BD.

Proof. In plane MN draw any line BE through B.

Draw CD meeting BE in E and produce AB to F, so that $BF = AB$.

Draw AC, AE, AD, CF, EF, and DF.

Since BC is $\perp$ to AF at its midpoint, $AC = FC$.

Similarly $AD = DF$.

CD is common.

Hence $\triangle ACD = \triangle CDF$. (Why ?)

$\therefore \angle ACD = \angle FCD$,

and $\triangle ACE = \triangle FCE$. (Why ?)

$\therefore EA = EF$.

Then, since E and B are respectively equidistant from A and F, BE is $\perp$ to AF. (Why ?)

Hence, AB is $\perp$ to any line passing through its foot in the plane MN, and is therefore $\perp$ to the plane. Q.E.D.

Proposition III. Theorem

462. *All the perpendiculars that can be drawn to a straight line at a given point lie in a plane perpendicular to the line at that point.*

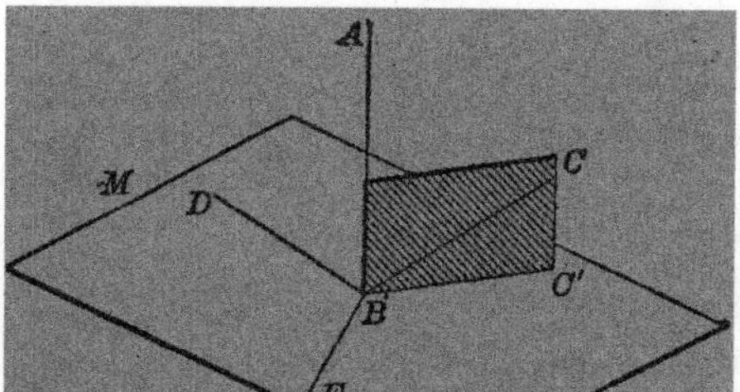

Hyp. AB is $\perp$ to BC, BD, and BE at B.

To prove BC, BD, and BE lie in a plane $\perp$ to AB.

Proof. Pass a plane MN through BD and BE.

Then AB is $\perp$ to MN. (Why ?)

Let the plane containing AB and BC intersect MN in BC'.

Then BC' is $\perp$ to AB. (Why ?)

But BC is $\perp$ to AB.

But in the plane ABC, only one perpendicular can be drawn upon AB at B.

Hence BC and BC' coincide, and BC lies in the plane MN.

Q.E.D

463. Cor. 1. Through a point C, to pass a plane perpendicular to a line AB, draw CD perpendicular to AB in plane ABC, meeting AB in D. At D, in any other plane than ABC, draw DF perpendicular to AB. Then the two perpendiculars DC and DF determine a plane perpendicular to AB.

464. Cor. 2. At a given point in a straight line, one and only one plane can be drawn perpendicular to the straight line.

465. Cor. 3. Through a point without a straight line, one plane and only one can be drawn perpendicular to that line.

Ex. 1033. The locus of a point in space equidistant from the ends of a line is a plane perpendicular to the line at its midpoint.

Proposition IV. Theorem

466. *If three planes intersect each other, the lines of intersection either meet in a point or are parallel.*

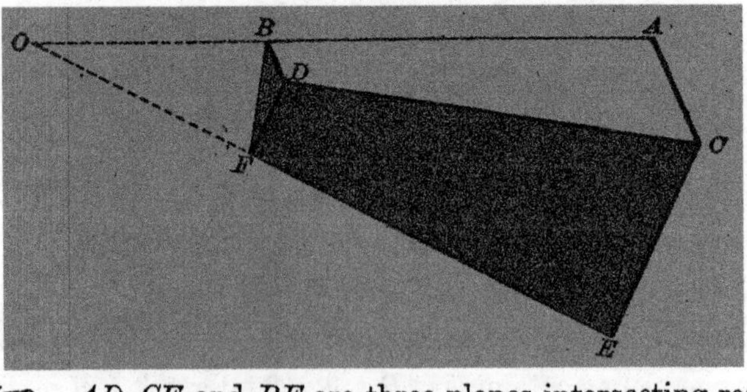

Hyp. AD, CF, and BE are three planes intersecting respectively in CD, AB, and EF.

To prove AB, CD, and EF either meet in a point or are parallel.

Proof. AB and EF lie in a plane.

Hence, they either meet or are parallel.

I. Suppose they meet (produced) in O. O being a point of AB lies in plane AD, and being a point of FE lies in plane CF.

Therefore, O must lie in the intersection of planes AD and CF, or in the line CD; that is, CD (produced) must pass through O.

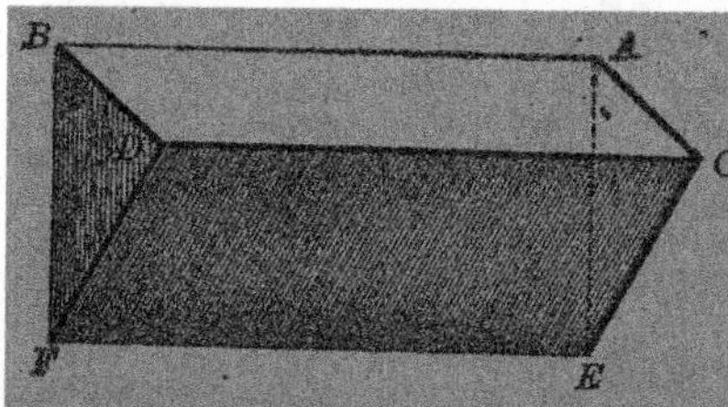

II. Suppose AB and EF are parallel.

Then CD cannot meet AB, for otherwise FE would have to meet AB. (Case I.)

Hence CD and AB, which lie in the same plane, must be parallel.

Similarly CD is parallel to EF. Q.E.D.

467. Cor. If two lines AB and CD are parallel to a third line EF, then AB and CD are parallel.

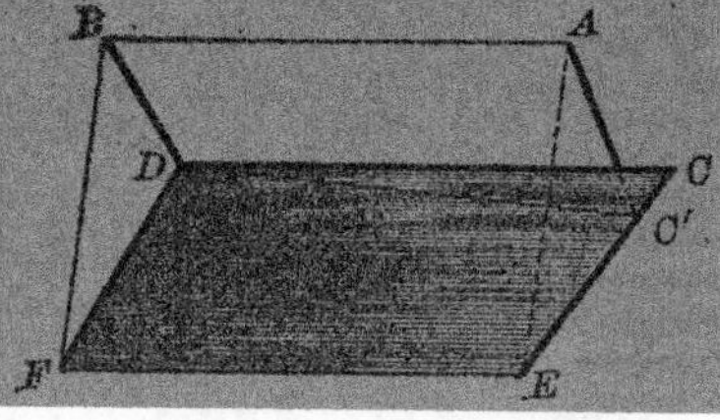

Pass plane AF through AB and EF, plane CF through CD and EF, and plane AD through AB and point D.

Let planes AD and CF intersect in DC'.

Then	DC' is parallel to FE.	(466)
Hence	DC and DC' coincide,	(Ax. 11.)
and	BA and DC are ∥.	Q.E.D.

PROPOSITION V. THEOREM

468. *If two angles, not in the same plane, have their sides parallel and lying in the same direction, they are equal.*

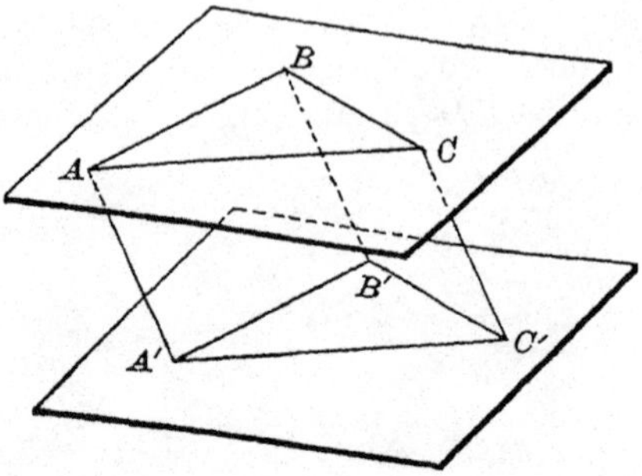

Hyp. The sides of the angles ABC and $A'B'C'$ are respectively parallel and lie in the same direction.

To prove $\angle ABC = \angle A'B'C'$.

Proof. Take $AB = A'B'$ and $BC = B'C'$.

Draw AA', BB', CC', AC, and $A'C'$.

AB' is a parallelogram. (Why?)

BC' is a parallelogram. (Why?)

Hence AA' is equal and parallel to BB',

and BB' is equal and parallel to CC'.

Therefore AA' is equal and parallel to CC'. (Ax. 1.) (467)

Whence $AA'C'C$ is a parallelogram,

and $AC = A'C'$.

$\therefore \triangle ABC = \triangle A'B'C'$. (Why?)

$\therefore \angle B = \angle B'$. Q.E.D.

PROPOSITION VI. THEOREM

469. *If one of two parallel lines is perpendicular to a plane, the other is also perpendicular to the plane.*

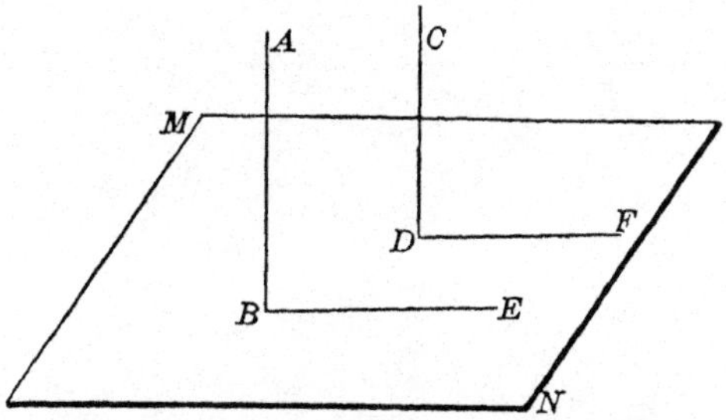

Hyp. AB is $\parallel$ to CD, and AB is $\perp$ to plane MN.

To prove CD is $\perp$ to plane MN.

Proof. In the plane MN through B and D respectively draw two parallel lines BE and DF.

Then $\angle ABE = \angle CDF$. (Why ?)

But $\angle ABE$ is a rt. $\angle$.

Hence $\angle CDF$ is a rt. $\angle$,

and CD is perpendicular to any line drawn through its foot.

Whence CD is $\perp$ to plane MN. Q.E.D.

Ex. 1034. If $ABCD$ is a quadrilateral in space (*i.e.* A, B, C, and D do not lie in the same plane) and $AB = BC$, $CD = DA$, then plane angle A is equal to plane angle C.

Ex. 1035. A line joining the midpoints of two adjacent sides of a quadrilateral in space is equal and parallel to a line joining the midpoints of the other two sides.

PROPOSITION VII. PROBLEM

470. *Through a given point, to draw a line perpendicular to a given plane.*

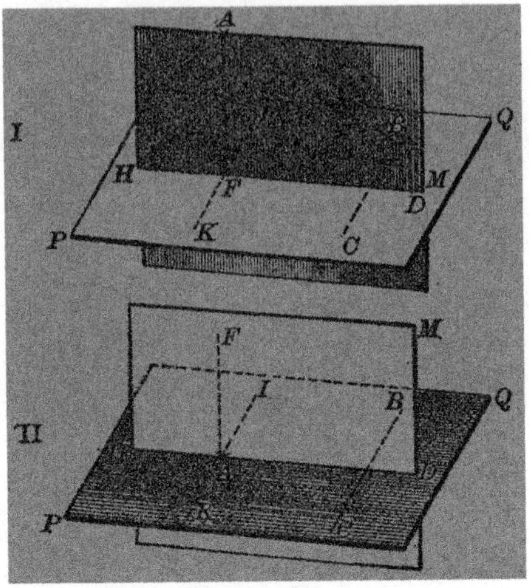

Given. Plane PQ and point A.

Required. Through A, to draw a line $\perp$ to PQ.

Construction. In PQ, draw any line BC.

Through A construct a plane $AM \perp BC$ intersecting PQ in DH.

Through A in the plane AM draw $AF \perp DH$.

AF is the required perpendicular.

Proof. Through the foot of AF (F in Fig. I and A in Fig. II) in plane PQ draw $IK \parallel$ to BC.

Since BC is $\perp$ to plane AM, (Con.)

and BC is $\parallel$ to IK,

$\therefore$ IK is $\perp$ to plane AM. (469)

$\therefore$ FA is $\perp$ to IK. (Why ?)

But FA is $\perp$ to DH. (Con.)

$\therefore$ FA is $\perp$ to plane PQ. (Why ?) Q.E.F.

471. Cor. 1. Through a given point, one and only one perpendicular can be drawn to a given plane.

For if there were two perpendiculars passing through the point, their plane would intersect the given plane in a straight line, perpendicular to both perpendiculars. Or, in a plane, there would be two perpendiculars to a given line, through a given point, which is impossible.

472. Cor. 2. If two lines are perpendicular to the same plane, these lines are parallel.

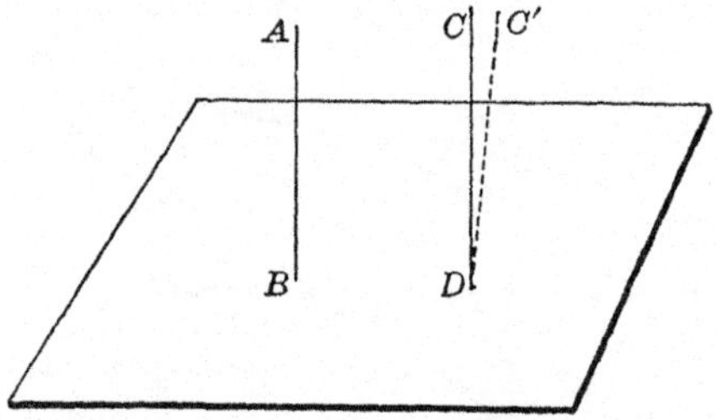

Let AB and CD be $\perp$ to plane MN.

Draw DC' $\parallel$ to AB.

Then DC' is $\perp$ to MN.

But through D only one perpendicular can be drawn upon MN.

Hence DC and DC' coincide, and DC is $\parallel$ to AB. Q.E.D.

Ex. 1036. The lines joining, in succession, the midpoints of the sides of a quadrilateral in space enclose a parallelogram.

Ex. 1037. The lines joining the midpoints of opposite sides of any quadrilateral in space bisect each other.

PROPOSITION VIII. THEOREM

473. *If, from a point without a plane, oblique lines be drawn to the plane:*

(1) *Those meeting the plane at equal distances from the foot of the perpendicular are equal.*

(2) *Of the two lines meeting the plane at unequal distances from the foot of the perpendicular, the more remote is the greater.*

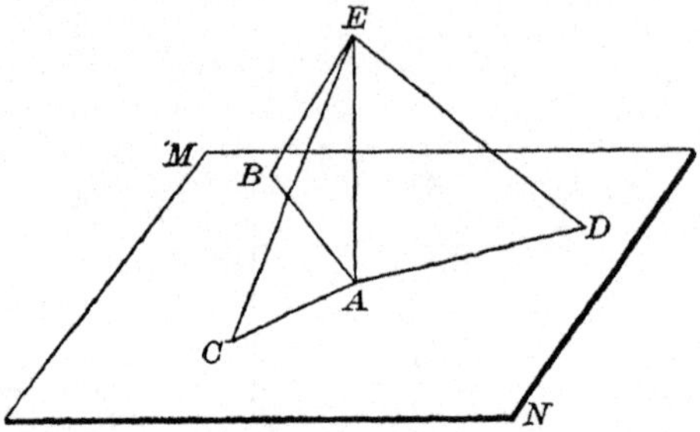

Hyp. EA is $\perp$ to plane MN. Oblique lines EB, EC, and ED are drawn so that $AB = AC$ and $AD > AC$.

To prove (1) $EB = EC$.

(2) $ED > EC$.

HINT.—1. Prove by means of equal triangles

2. $\overline{ED}^2 = \overline{EA}^2 + \overline{AD}^2$, $\overline{EC}^2 = \overline{EA}^2 + \overline{AC}^2$.

474. COR. 1. Conversely, equal oblique lines drawn from a point to a plane meet the plane at equal distances from the foot of the perpendicular from the point to the plane, and of two unequal oblique lines, the greater one meets the plane at a greater distance from the foot.

475. COR. 2. The locus of a point in space that is equidistant from all the points in the circumference of a circle is a straight line perpendicular to the plane of the circle, and passing through the center of the circle.

476. COR. 3. The perpendicular to a plane is the shortest line that can be drawn to the plane from a point without.

477. DEF. The **distance of a point from a plane** is the length of the perpendicular to that plane.

PROPOSITION IX. THEOREM

478. *The intersections of two parallel planes with a third plane are parallel.*

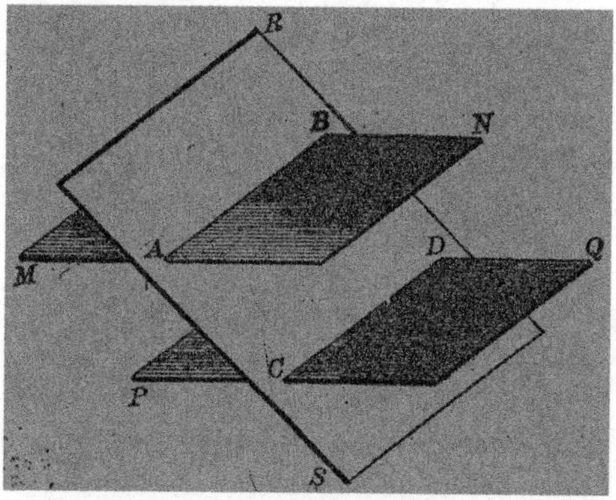

Hyp. The parallel planes MN and PQ are intersected by a third plane RS in AB and CD, respectively.

To prove $AB \parallel CD$.

Proof. AB and CD cannot meet, for otherwise planes MN and PQ would meet.

AB and CD lie in the same plane.

Hence AB is $\parallel$ to CD. Q.E.D.

479. COR. 1. Parallel lines included between parallel planes are equal.

480. COR. 2. Two parallel planes are everywhere equally distant.

Ex. 1038. What is the locus of a point having a given distance from a given plane?

PROPOSITION X. THEOREM

481. *Any plane containing one, and only one, of two parallel lines is parallel to the other line.*

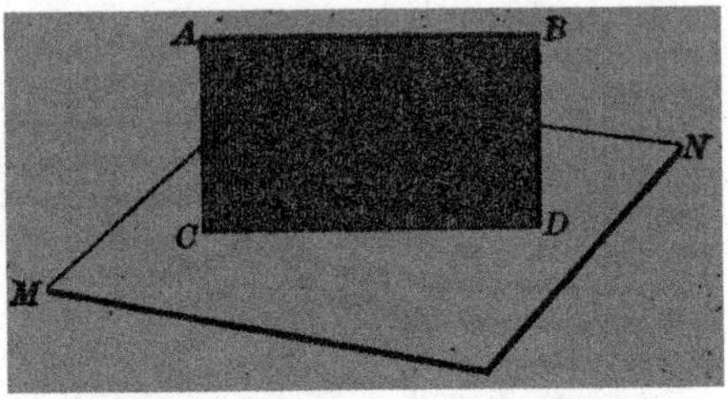

Hyp. AB is ∥ to CD, and plane MN contains CD.

To prove plane MN is ∥ to AB.

Proof. AB and CD determine a plane, intersecting MN in CD.

Hence, if AB meets MN, it must meet MN in CD.

But, since AB ∥ CD, this is impossible.

Hence plane MN is ∥ to AB. Q.E.D.

482. COR. 1. Through a given straight line a plane can be passed parallel to any other given straight line; and if the lines are not parallel, only one such plane can be drawn.

483. COR. 2. Through a given point a plane can be passed parallel to any two given straight lines in space; and if the lines are not parallel, only one such plane can be drawn.

PROPOSITION XI. THEOREM

484. *A line parallel to a plane is parallel to the intersection of this plane with any plane passing through the line.*

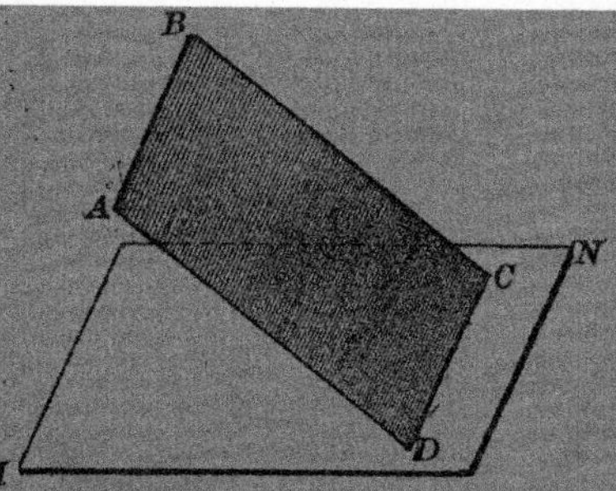

Hyp. Plane MN is ∥ to AB.

To prove any plane AC containing AB intersects MN in CD so that CD is ∥ to AB.

Proof. AB and CD cannot meet, for otherwise AB would meet the plane MN.

AB and CD are in the same plane.

Hence AB is ∥ to CD. Q.E.D.

485. COR. 1. If each of two intersecting lines is parallel to a given plane, their plane is parallel to the given plane.

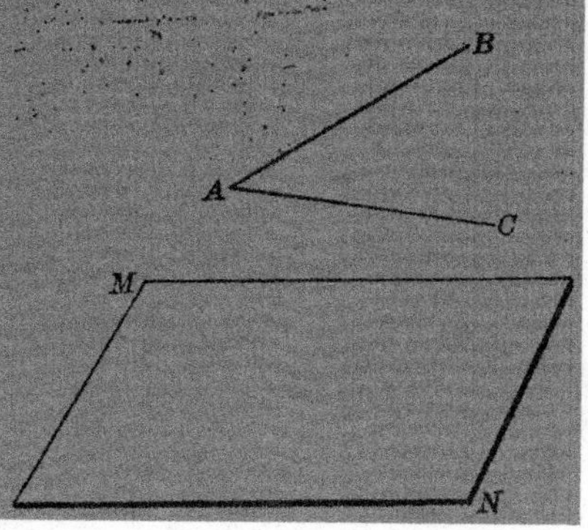

Let AB and AC be parallel to MN. If plane ABC should intersect MN, the intersection would be parallel to both AB and AC, which is impossible.

486. Cor. 2. If two angles ABC and $A'B'C'$ have their sides respectively parallel, their planes are parallel.

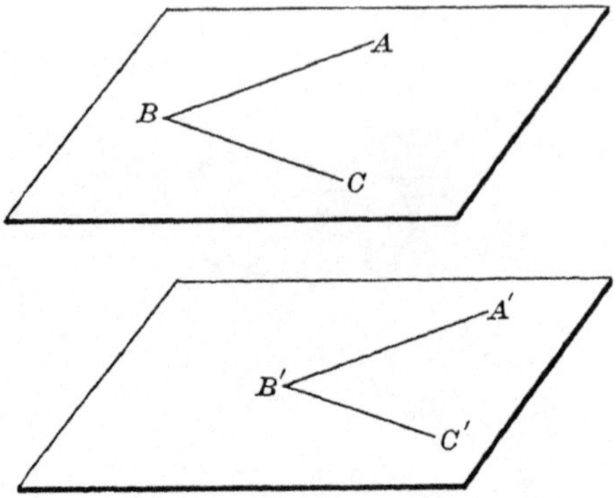

Since AB is ‖ to $A'B'$, plane $A'B'C'$ is ‖ to AB, and, similarly, is ‖ to BC.

Hence planes ABC and $A'B'C'$ are parallel. (Cor. 1.)

Ex. 1039. If a line is parallel to a plane and parallel to another line, the second line is parallel to the plane.

Ex. 1040. Find the locus of a point equidistant from three given points.

Ex. 1041. A line parallel to two intersecting planes is parallel to the intersection of these planes.

Ex. 1042. What is the locus of a point equidistant from two parallel planes?

Ex. 1043. To construct a line parallel to two intersecting planes.

Ex. 1044. The midpoints of two opposite sides of a quadrilateral in space and the midpoints of the two diagonals determine the vertices of a parallelogram.

Ex. 1045. The lines joining the midpoints of the opposite sides of any quadrilateral in space and the line joining the midpoints of the diagonals meet in a point.

Proposition XII. Theorem

487. *If two straight lines are cut by parallel planes, their corresponding segments are proportional.*

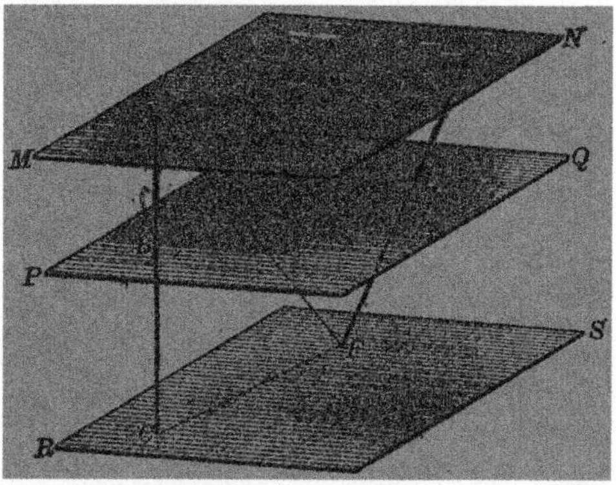

Hyp. Parallel planes MN, PQ, and RS are intersected by two straight lines in A, B, C and D, E, F, respectively.

To prove $AB : BC = DE : EF$.

Proof. Draw AF, and pass a plane through AC and AF, intersecting the planes PQ and RS in BG and CF, respectively.

Then CF is $\parallel$ to BG. (478)

By passing a plane through AF and DF, it follows, similarly,

GE is $\parallel$ to AD.

Hence $$\frac{AB}{BC} = \frac{AG}{GF},$$ (Why?)

and $$\frac{DE}{EF} = \frac{AG}{GF}.$$ (Why?)

$$\therefore \frac{AB}{BC} = \frac{DE}{EF}.$$ (Ax. 1.) Q.E.D.

488. Cor. If through any point straight lines are drawn intersected by two parallel planes, their corresponding segments are proportional.

PROPOSITION XIII. THEOREM

489. *Planes perpendicular to the same straight line are parallel to each other.*

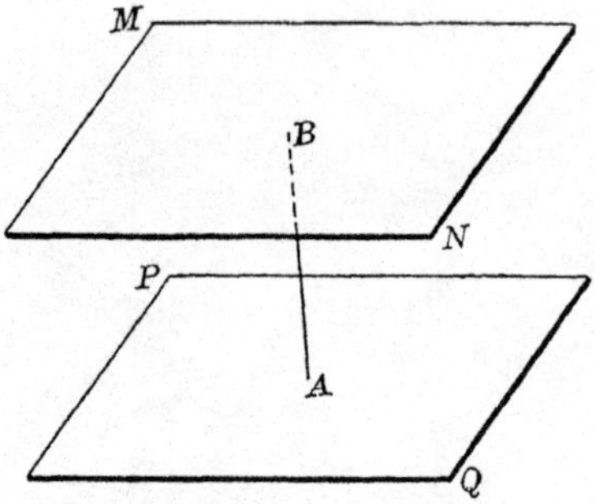

Hyp. Planes MN and PQ are perpendicular to line AB.

To prove MN is $\parallel$ to PQ.

HINT. — Apply the indirect method and (98).

PROPOSITION XIV. THEOREM

490. *A straight line perpendicular to one of two parallel planes is perpendicular to the other also.*

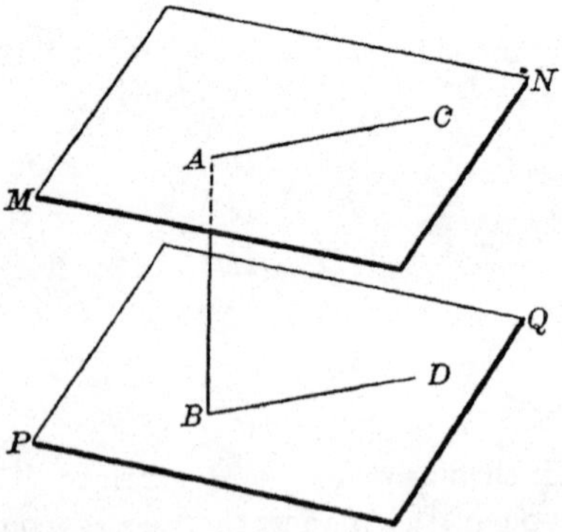

Hyp. Plane MN is $\parallel$ to plane PQ, and AB is $\perp$ to plane MN.

To prove AB is $\perp$ to PQ.

Proof. Through AB pass any plane intersecting MN in AC and PQ in BD, respectively.

AC is $\parallel$ to BD, (478)

and AC is $\perp$ to AB. (461)

Hence BD is $\perp$ to AB. (88)

Therefore AB is perpendicular to any line in PQ passing through B.

Whence AB is $\perp$ to plane PQ. Q E D

491. Cor. Through a given point, one plane and only one may be passed parallel to a given plane.

Ex. 1046. If two planes are parallel to a third plane, they are parallel to each other.

DIEDRAL ANGLES

492. Def. A **diedral angle** is the opening between two planes intersecting in a straight line.

The **edge** is the line of intersection, and the **faces** are the intersecting planes.

Thus, in the diedral angle formed by the planes AC and BE, BC is the edge and AC and BE are the faces.

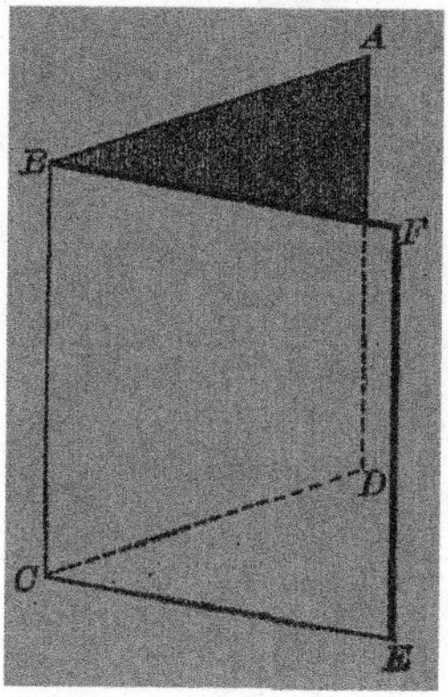

493. A diedral angle may be designated by two letters on its edge; or if several diedral angles have a common edge, by four letters, one on each face, and two on the edge, the letters on the edge being placed between the others.

Thus, the diedral angle in the annexed diagram may be designated by BC or $ABCE$.

494. Def. The **plane angle of a diedral angle** is the angle formed by perpendiculars to the edge at some point, one in each face.

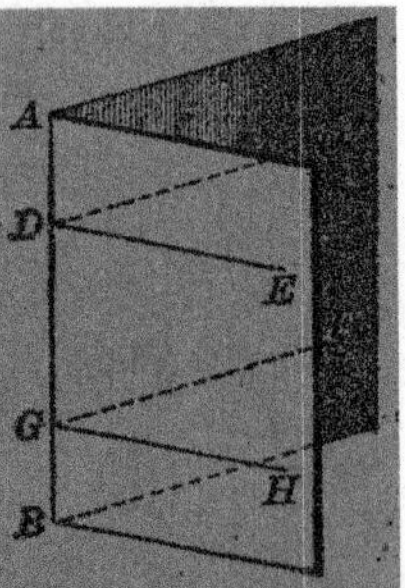

Thus, *CDE* is a plane angle of diedral angle *AB* if *CD* is perpendicular to *AB* and *ED* is perpendicular to *AB*.

Two plane angles of the same diedral angle, as *CDE* and *FGH*, are equal, since their sides are respectively parallel.

Or, the plane angle of a diedral angle is the same, at whatever point of the edge it is drawn.

The size of a diedral angle does not depend upon the extent of its faces, but upon the difference of their positions.

495. Two **diedral angles are equal** when they can be made to coincide.

By superposing two equal diedral angles, their plane angles can be made to coincide.

496. A diedral angle is **acute, right, obtuse,** or **straight,** according as its plane angle is acute, right, obtuse, or straight.

It is evident that the two faces of a straight diedral angle lie in the same plane.

Diedral angles are **complementary, supplementary, adjacent,** etc., according as their plane angles are complementary, supplementary, adjacent, etc.

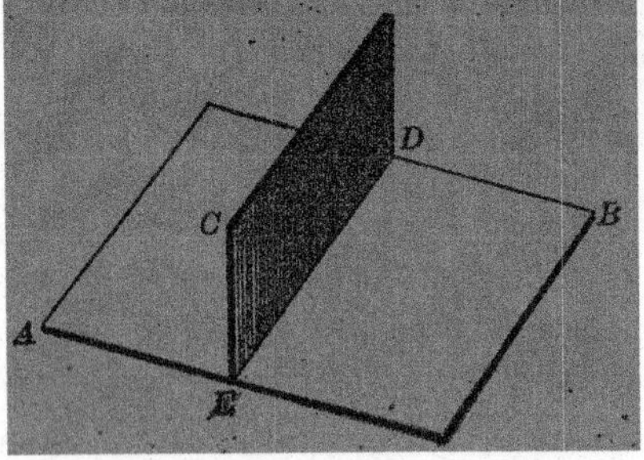

If a plane meets another plane so as to make the two adjacent diedrals equal, each diedral is evidently a right one.

497. Planes forming right diedral angles are said to be **perpendicular to each other.** Thus, plane *CD* is perpendicular to plane *AB* if they form right diedral angles.

498. Def. The **projection** of a point on a plane is the foot of the perpendicular drawn from that point upon the plane.

The **projection of a line** upon a plane is the locus of the projection of all the points of the line.

PROPOSITION XV. THEOREM

499. *Two diedral angles are equal if their plane angles are equal.*

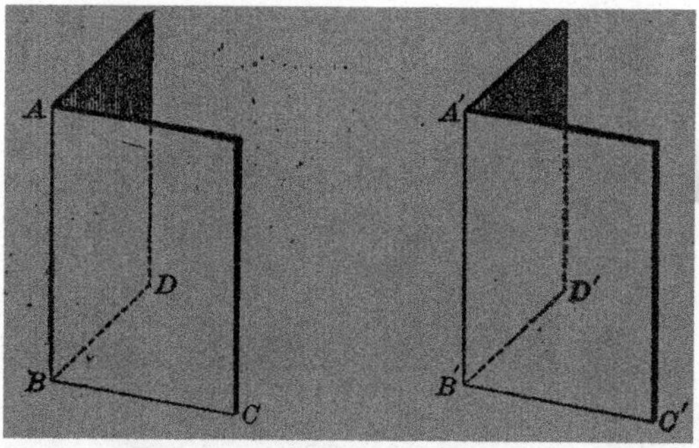

Hyp. The diedral angles AB and $A'B'$ have equal plane angles CBD and $C'B'D'$ respectively.

To prove diedral angle $AB =$ diedral angle $A'B'$.

Proof. Since AB is $\perp$ to BD and $\perp$ to BC,

AB is $\perp$ to plane BDC.

And, similarly, $A'B'$ is $\perp$ to plane $B'D'C'$.

Apply plane $\angle DBC$ to $\angle D'B'C'$.

Then B coincides with B',

and plane BDC coincides with plane $B'D'C'$.

Hence BA coincides with $B'A'$. (471)

Since BA coincides with $B'A'$,

and DB coincides with $D'B'$,

plane AD coincides with plane $A'D'$.

In like manner it may be proved that plane AC coincides with plane $A'C'$.

Hence, diedral angles AB and $A'B'$ coincide. Q.E.D.

Therefore they are equal.

Ex. 1047. Vertical diedral angles are equal.

Ex. 1048. If two parallel planes are intersected by a third one, (*a*) the corresponding diedral angles are equal, (*b*) the alternate diedral angles are equal.

Ex. 1049. State and prove the converse of **Ex.** 1048.

Proposition XVI. Theorem

500. *Two diedral angles have the same ratio as their plane angles.*

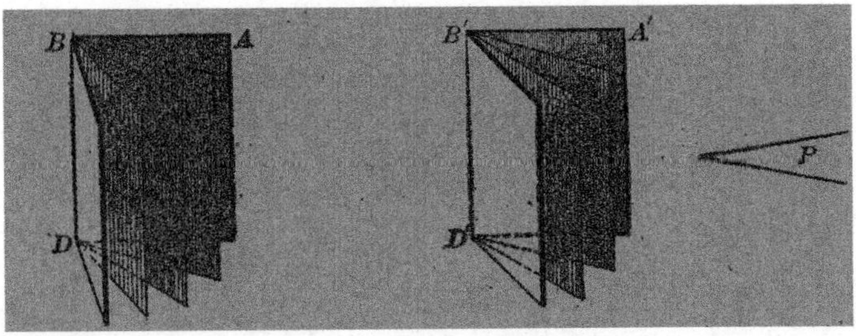

Hyp. $\angle s$ ABC and $A'B'C'$ are respectively the plane angles of diedral angles $ABDC$ and $A'B'D'C'$.

To prove $$\frac{ABDC}{A'B'D'C'}=\frac{\angle ABC}{\angle A'B'C'}.$$

Proof. Case I. Angles ABC and $A'B'C'$ are commensurable.

Apply a common measure P to these angles and let P be contained in ABC m times, and in $A'B'C'$ n times.

Then $$\frac{\angle ABC}{\angle A'B'C'}=\frac{m}{n}.$$

Through the lines of division draw planes which contain BD and $B'D'$, respectively.

These planes divide $ABDC$ into m and $A'B'C'D'$ into n equal parts.

Therefore $$\frac{ABDC}{A'B'D'C'} = \frac{m}{n}.$$

Hence $$\frac{ABDC}{A'B'D'C'} = \frac{\angle ABC}{\angle A'B'C'}.$$ Q.E.D.

Case II. Angles ABC and $A'B'C'$ are incommensurable.

Divide $\angle ABC$ into any number of equal parts and apply one of the parts to $\angle A'B'C'$ as often as possible.

[To be completed by the student, who may compare 280, Bk. III.]

501. Cor. A diedral angle is measured by its plane angle.

Proposition XVII. Theorem

502. *If a straight line is perpendicular to a plane, every plane passed through this line is perpendicular to the plane.*

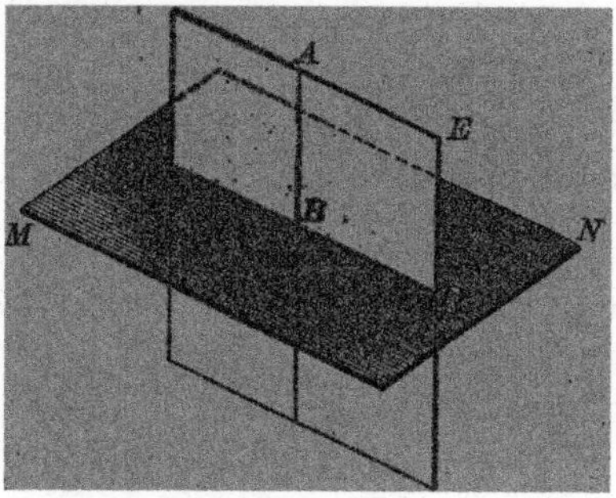

Hyp. AB is $\perp$ to plane MN.

To prove that plane EC containing AB is $\perp$ to MN.

Hint.—In plane MN at B, draw $BF \perp$ to CD, and prove that $\angle ABF$ is a right one.

PROPOSITION XVIII. THEOREM

503. *If two planes are perpendicular to each other, a straight line drawn in one of them, perpendicular to their intersection, is perpendicular to the other.*

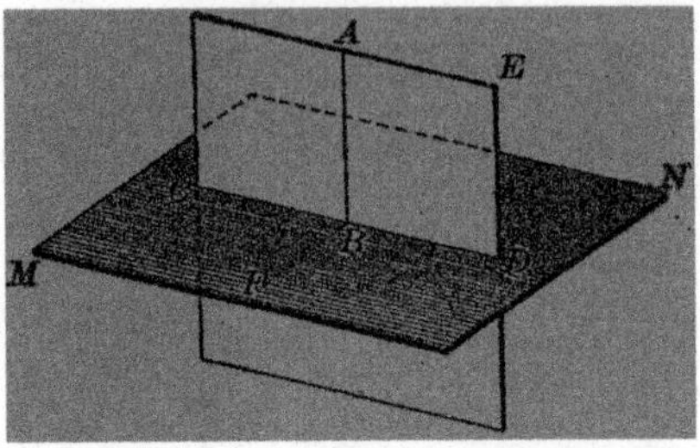

Hyp. Plane CE is $\perp$ to plane MN, and in CE line BA is $\perp$ to CD, the intersection of the two planes.

To prove AB is $\perp$ to plane MN.

HINT. — At B construct the plane angle and prove that AB is $\perp$ to two lines drawn through its foot.

504. COR. 1. If two planes are perpendicular to each other, a perpendicular to one of these, at any point of their intersection, lies in the other.

[Prove by indirect method.]

505. COR. 2. If two planes are perpendicular to each other, a straight line drawn from any point of one, perpendicular to the other, lies in the first.

[Prove by indirect method.]

Ex. 1050. If the opposite sides of a quadrilateral in space are equal, the opposite plane angles are equal.

Ex. 1051. Three points not in a straight line each equidistant from the ends of a given line determine a perpendicular bisecting plane of the given line.

PROPOSITION XIX. THEOREM

506. *If two intersecting planes are each perpendicular to a third plane, their intersection is perpendicular to that plane.*

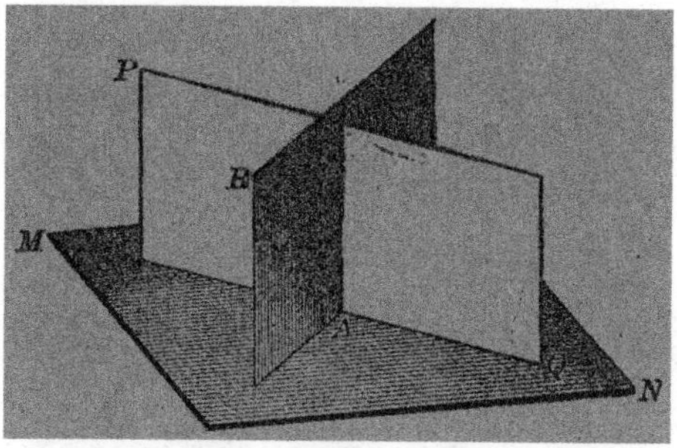

Hyp. The planes PQ and RA, intersecting in AB, are perpendicular to plane MN.

To prove AB is $\perp$ to MN.

Proof. At A draw a straight line $\perp$ to MN.
This line must lie in plane PQ and in plane RA. (504)
Hence it must coincide with AB.

Or AB is $\perp$ to plane MN. Q.E.D.

507. COR. If a plane is perpendicular to two other planes, perpendicular to each other, the intersection of any two planes is perpendicular to the third plane and perpendicular to the other intersections.

Ex. 1052. If a line is parallel to a plane, any other plane perpendicular to the line is perpendicular to the plane.

Ex. 1053. If two parallel planes are intersected by a third one, the interior angles on the same side of the third plane are supplementary.

PROPOSITION XX. THEOREM

508. *Every point in a plane bisecting a diedral angle is equidistant from the faces of the angle.*

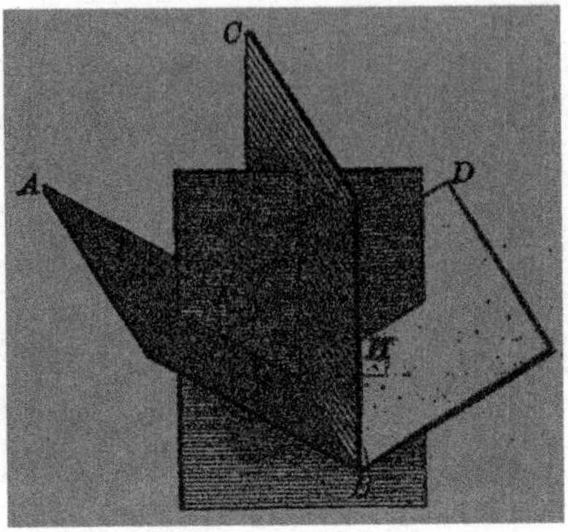

Hyp. Plane CB bisects diedral $\angle ABED$, and FG and FH are the respective distances of a point F in BC, from AB and BD.

To prove $FH = FG$.

Proof. Through FG and FH pass a plane intersecting the faces in EG and EH.

plane FGH is $\perp$ to AB and to DB. (502)

Therefore plane FGH is $\perp$ to BE. (506)

Hence EB is $\perp$ to EG, EF, and EH. (461)

$\therefore$ GEF and HEF are the plane angles of diedral angles $ABEC$ and $DBEC$.

[To be completed by the student. Compare Remark 73.]

Ex. 1054. The locus of a point equidistant from two intersecting planes consists of the two planes bisecting the diedral angles formed by the planes.

Ex. 1055. Find the locus of a point equidistant from two given intersecting lines (in space).

Proposition XXI. Theorem

509. *Through any given straight line not perpendicular to a plane, one plane can be passed perpendicular to that given plane, and only one.*

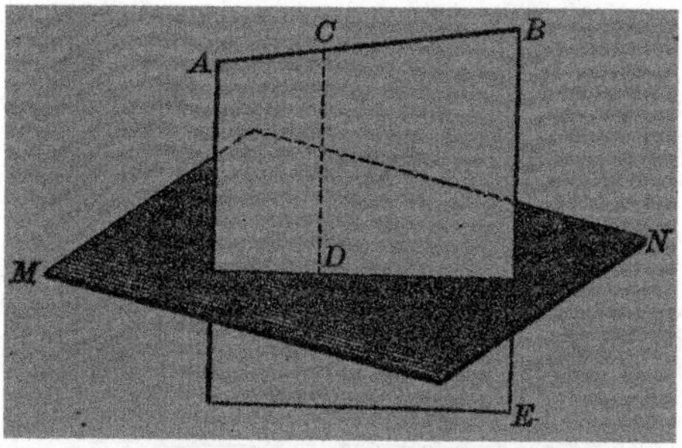

Hyp. *AB* is a line not ⊥ to plane *MN*.

To prove through *AB* one, and only one, plane can be passed ⊥ to *MN*.

Proof. From any point *C* in *AB*, draw *CD* ⊥ to *MN*.

Pass a plane *AE* through *AB* and *CD*.

Then plane *AD* is ⊥ to *MN*. (502)

If two planes could be passed through *AB* ⊥ to *MN*, *AB*, their intersection, would be perpendicular to *MN*. (506)

But this is impossible since it contradicts the hypothesis.

Hence, only one plane can be drawn perpendicular to *MN* and passing through *AB*. Q.E.D.

510. Cor. The projection of a line not perpendicular to a plane upon that plane is a straight line.

Ex. 1056. If the projection of a figure upon a plane is a straight line, the figure is a plane figure.

511. Def. The **inclination of a line to a plane,** or the **angle made by a line and a plane,** is the angle which the line forms **with its own projection upon that plane.**

PROPOSITION XXII. THEOREM

512. *The acute angle formed by a line and its projection upon a plane is the least angle which the line makes with any line in the plane.*

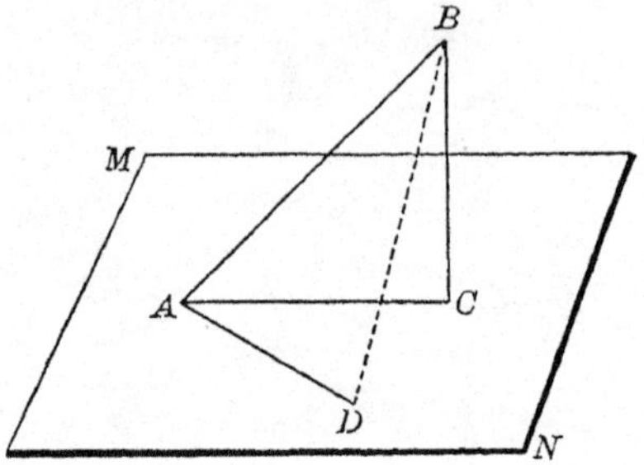

Hyp. The line AC is the projection of line AB upon plane MN, and AD is any other line drawn through A in MN.

To prove $\angle BAC < \angle BAD$.

Proof. Make $AD = AC$ and draw BD.

In $\triangle$ ABC and ABD,

AB is common,

$AC = AD$,

$BC < BD$. (Why?)

Hence $\angle BAC < \angle BAD$. (Why?) Q.E.D.

Ex. 1057. In the diagram for Prop. XXII, if, in plane MN, AF be drawn perpendicular to AC, prove that $\angle BAF$ is a right angle.

Ex. 1058. In the diagram for Prop. XXII, if $AC = 3$ inches, $\angle BAC = 60°$, and $\angle CAD = 90°$, find the length of BD.

Ex. 1059. The inclinations of two parallel lines to a plane are equal.

Proposition XXIII. Problem

513. *To draw a common perpendicular to two lines not in the same plane.*

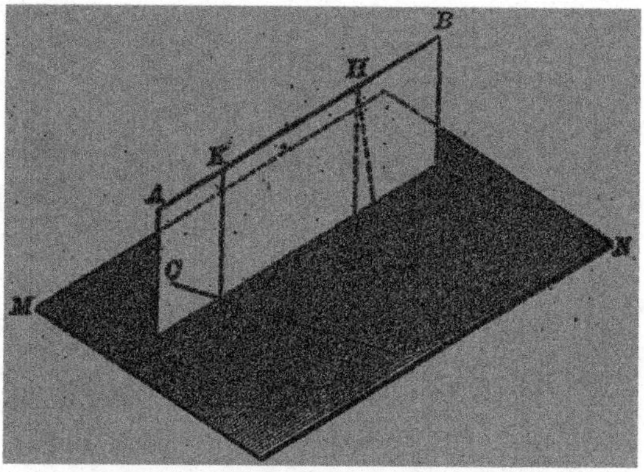

Given. Lines *AB* and *CD* not in the same plane.

Required. To draw a common perpendicular to *AB* and *CD*.

Construction. Through *CD* draw plane *MN* parallel to *AB*.

Through *AB* draw plane *AF* ⊥ to *MN*, intersecting *MN* in *EF*, and let *G* be the intersection of *EF* and *CD*.

At *G*, in plane *AF*, draw *GK* ⊥ to *EF*, meeting *AB* in *K*.

Then *GK* is the required perpendicular.

Proof. Since *EF* and *AB* are parallel, and *GK* is ⊥ to *EF*, *GK* must be ⊥ to *AB*. (88)

But *GK* is ⊥ to *MN*. (503)

∴ *GK* is ⊥ to *CD*. Q.E.F.

514. Cor. Only one common perpendicular can be drawn to two lines not in the same plane.

For, suppose there was another common perpendicular *HD*, it can easily be demonstrated that there would be two perpendiculars, *HD* and *HI*, from the point *H* to the plane *MN*, which is impossible.

POLYEDRAL ANGLES

515. Def. A **polyedral angle** or a **solid angle** is formed by three or more planes meeting in a point.

516. Def. The **vertex** of a polyedral angle is the common point in which the planes meet; the **edges** are the intersections of the planes; the **faces** are the planes bounded by the edges; and the **face angles** are the angles formed by the edges.

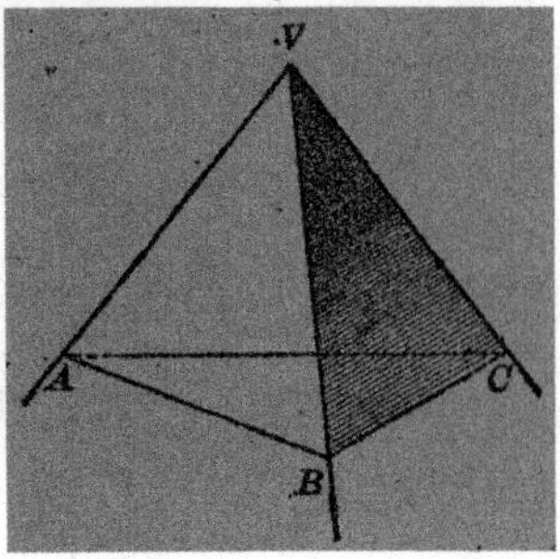

Thus, if three planes *ABV*, *BCV*, and *ACV* meet in *V*, *V* is the vertex, *VA*, *VB*, and *VC* are the edges, the planes *AVB*, *BVC*, *CVA* are the faces, and the angles *AVB*, *BVC*, *CVA* are the face angles of the polyedral angle *V–ABC*.

The magnitude of a polyedral angle depends upon the relative positions of its faces and not upon their extent.

In a polyedral angle each pair of adjacent faces forms a diedral angle, and each pair of adjacent edges forms a face angle. The face angles and the diedral angles are called the **parts** of the polyedral angle.

517. Def. **A polyedral angle is convex** if any section made by a plane intersecting all its faces is a convex polygon.

518. Def. A polyedral angle is called **triedral, tetraedral,** etc., according as it has three, four, etc., faces.

519. Def. A polyedral angle is called **rectangular, bi-rectangular,** or **tri-rectangular,** according as it has one, two, or three, right diedral angles.

520. Def. An **isosceles** triedral angle is one, two of whose face angles are equal.

521. Two *polyedral angles are equal* if the face and diedral angles of the one are respectively equal to the face and diedral angles of the other one, and all the parts are arranged in the same order (for evidently they can be made to coincide).

Two polyedral angles are symmetrical if the face and diedral angles of the one are respectively equal to the face and diedral angles of the other, and all parts are arranged in reverse order.

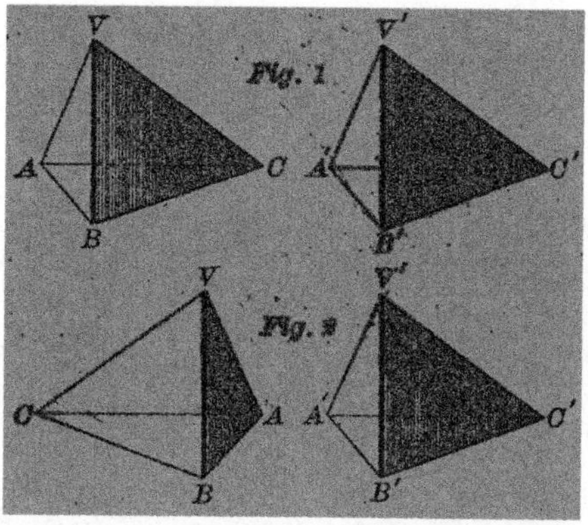

Thus, in Fig. 1 the triedral $\angle V\text{–}ABC =$ triedral $\angle V'\text{–}A'B'C'$ if $\angle AVB = \angle A'V'B'$, $\angle BVC = \angle B'V'C'$, $\angle CVA = \angle C'V'A'$, and diedral $\angle AV =$ diedral $\angle A'V'$, diedral $\angle BV =$ diedral $\angle B'V'$, and diedral $\angle CV =$ diedral $\angle C'V'$.

While in Fig. 2 $V\text{–}ABC$ and $V'\text{–}A'B'C'$ are symmetrical.

It is obvious that, in general, two symmetrical polyedral angles cannot be made to coincide.

Ex. 1060. If in diagram for 516, $VA = VB = VC = AB = BC = CA$, construct (by means of a plane construction) a diedral angle.

Ex. 1061. If in the same diagram $\angle AVB = \angle BVC$ and $\triangle AVB$ and AVC are given, find (by a plane construction) diedral angle VB.

Ex. 1062. Find (by a plane construction) the diedral angles of any triedral angle, if the face angles are given.

PROPOSITION XXIV. THEOREM

522. *The sum of any two face angles of a triedral angle is greater than the third face angle.*

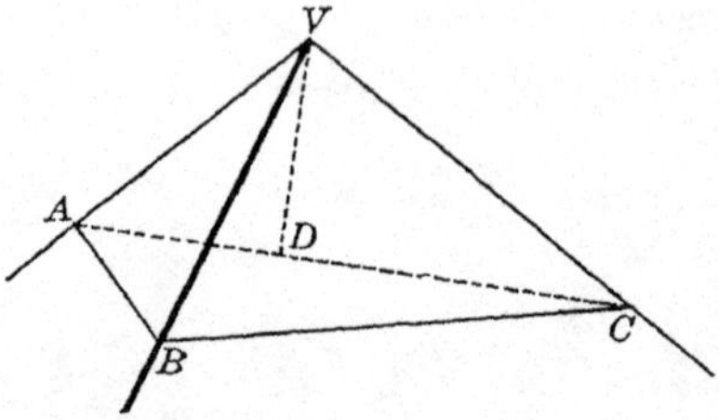

Hyp. $\angle AVC$ is the greatest face angle of triedral angle $V\text{-}ABC$.

To prove $\angle AVB + \angle BVC > \angle AVC$.

Proof. In the face AVC draw VD equal to VB, making

$$\angle DVA = \angle BVA.$$

In the face AVC, draw AC through D.
Draw AB and BC.

Then $\triangle AVB = \triangle AVD$. (Why ?)

But $AB + BC > AC$. (Why ?)

Subtracting the equals AB and AD,

$$BC > DC.$$

But in the $\triangle$s BVC and DVC,

VC is common,

$VD = VB$, (Why ?)

and $BC > DC$.

Hence $\angle BVC > \angle DVC$. (Why ?)

Adding the equal angles AVB and AVD,

$$\angle AVB + \angle BVC > \angle DVC + \angle AVD,$$

or

$$\angle AVB + \angle BVC > \angle AVC. \qquad \text{Q.E.D.}$$

Ex. 1063. In the diagram for Prop. XXIV,

$$\angle AVD + \angle DVB < \angle AVC + \angle CVB.$$

* **Ex. 1064.** In the same diagram, if E is a point in $\triangle ABC$,

$$\angle AVE + \angle BVE + \angle CVE < \angle AVB + \angle BVC + \angle CVA,$$

but

$$> \tfrac{1}{2}(\angle AVB + \angle BVC + \angle CVA).$$

Ex. 1065. In the diagram for Prop. XXII, if in plane MN, AF be drawn so as to make $\angle FAC$ obtuse, then $\angle FAB$ is obtuse.

Proposition XXV. Theorem

523. *The sum of the face angles of any convex polyedral angle is less than four right angles.*

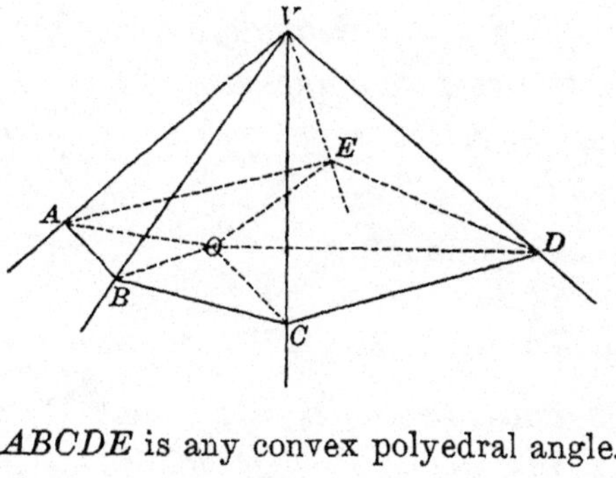

Hyp. V–$ABCDE$ is any convex polyedral angle.

To prove the sum of $\angle s$ AVB, BVC, etc., is less than four right angles.

Proof. Construct a plane intersecting the edges in A, B, C, etc., the faces in AB, BC, CD, etc. Then $ABCDE$ is a convex polygon.

Join any point O in plane ABC to A, B, C, etc.

Then $\angle VBA + \angle VBC > \angle ABC$,

and $\angle VCB + \angle VCD > \angle BCD$, etc. (522)

Therefore the sum of all the base angles of the triangles whose vertex is V is greater than the sum of all the base angles of the triangles whose vertex is O.

But the sum of all the angles of the triangles whose vertex is V is equal to the sum of all the angles of the triangles whose vertex is O. (Why?)

Subtracting the base angles from the equal sums, the sum of all the angles at V is less than the sum of all the angles at O.

But the sum of all the angles at O is equal to four right angles.

Whence, the sum of all the angles at V is less than four right angles. Q.E.D.

Proposition XXVI. Theorem

524. *Two triedral angles are equal if two face angles and the included diedral angle of the one are respectively equal to two face angles and the included diedral angle of the other, the equal parts being arranged in the same order.*

[Proof by superposition.]

Proposition XXVII. Theorem

525. *Two triedral angles are equal if two diedral angles and the included face angle of the one are respectively equal to two diedral angles and the included face angle of the other, all equal parts being arranged in the same order.*

[Proof by superposition.]

PROPOSITION XXVIII. THEOREM

526. *Two triedral angles are equal if the three face angles of the one are respectively equal to the three face angles of the other, and all are arranged in the same order.*

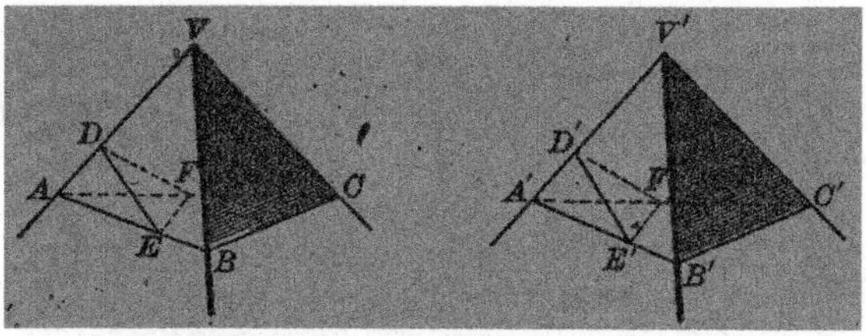

Hyp. In triedral angles $V\text{-}ABC$ and $V'\text{-}A'B'C'$,

$$\angle AVB = \angle A'V'B',\ \angle BVC = \angle B'V'C',$$

and $$\angle CVA = \angle C'V'A'.$$

To prove $$V\text{-}ABC = V'\text{-}A'B'C'.$$

Proof. On the six edges lay off

$$VA = VB = VC = V'A' = V'B' = V'C'.$$

Draw AB, BC, CA, $A'B'$, $B'C'$, and $C'A'$.

Then $$\triangle AVB = \triangle A'V'B',$$ (Why?)

and $$AB = A'B'.$$

In like manner, $$BC = B'C',$$

and $$CA = C'A'.$$

On AV and $A'V'$, respectively, lay off AD equal to $A'D'$. Draw DE in face AVB, and DF in face AVC, perpendicular to VA. These lines meet AB and AC, respectively, since triangles AVB and AVC are isosceles.

In like manner, draw $D'E'$ and $D'F'$, and join EF and $E'F'$.
[To be completed by the student.]

HINT. — Prove $\triangle ABC = \triangle A'B'C'$.

$\triangle ADE = \triangle A'D'E'$.

$\triangle ADF = \triangle A'D'F'$.

$\triangle AEF = \triangle A'E'F'$.

$\triangle DEF = \triangle D'E'F'$.

527. DEF. **Vertical polyedral angles** are polyedral angles which have a common vertex, and the edges of one are prolongations of the edges of the other.

PROPOSITION XXIX

528. *Two triedral angles are symmetrical:*

(1) *If two face angles and the included diedral angle of the one are respectively equal to two face angles and the included diedral angle of the other,*

or, (2) *If two diedral angles and the included face angle of the one are respectively equal to two diedral angles and the included face angle of the other,*

or, (3) *Three face angles of the one are equal respectively to three face angles of the other,*

provided all equal parts are arranged in reverse order.

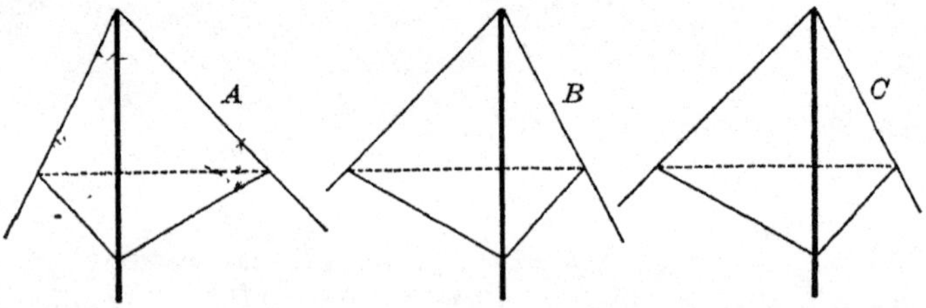

Proof. If A and B are the given triedral angles, construct triedral angle C symmetrical to A.

Since C and A have all parts arranged in reverse order, C and B have all parts arranged in the same order.

Hence $$C = B.$$

But since C is symmetrical to A, its equal B is symmetrical to A. Q E D

529. COR. Vertical triedral angles are symmetrical.

Ex. 1066. Vertical polyedral angles are symmetrical.

EXERCISES

Ex. 1067. If two face angles of a triedral angle are equal, the opposite diedral angles are equal.

Ex. 1068. Find the locus of a point, having a given distance from a given plane.

Ex. 1069. Planes bisecting supplementary adjacent diedral angles are perpendicular to each other.

Ex. 1070. The projections of parallel lines upon the same plane are parallel.

Ex. 1071. The bisectors of the three diedral angles of a triedral angle meet in the same straight line.

Ex. 1072. A line is twice as long as its projection upon a plane. What is the inclination of the line to the plane?

Ex. 1073. If the three face angles of a triedral angle are equal, the three diedral angles are equal.

Ex. 1074. If the opposite face angles of a tetraedral angle are equal, the opposite diedral angles are equal.

Ex. 1075. If the three face angles of a triedral angle are right angles, the three diedral angles are right ones (a tri-rectangular triedral angle).

Ex. 1076. The bisecting planes of alternate interior diedral angles made by parallel planes are parallel to each other.

Ex. 1077. If a straight line intersects two parallel planes, it makes equal angles with both planes.

Ex. 1078. In a triedral angle, the greater face angle is opposite the greater diedral angle.

Ex. 1079. If the sum of the four angles of a quadrilateral is equal to four right angles, its vertices lie in a plane.

Ex. 1080. The common perpendicular is the shortest distance that can be drawn between two straight lines not in the same plane.

BOOK VII

POLYEDRONS, CYLINDERS, AND CONES

POLYEDRONS

530. Def. A **polyedron** is a solid bounded by planes. The **faces** of a polyedron are the bounding planes; the **edges** are the intersections of the faces; and the **vertices** are the intersections of the edges.

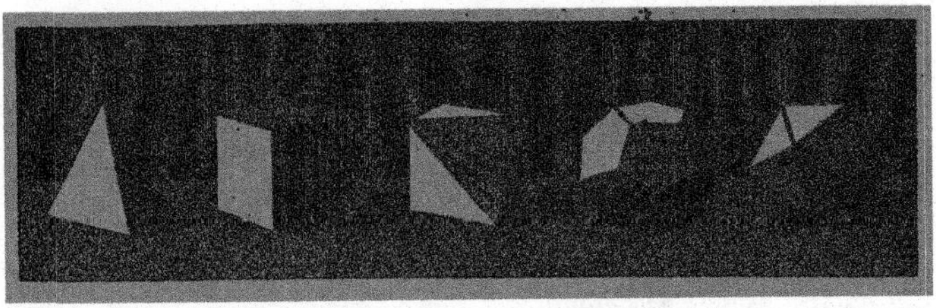

531. Def. A **tetraedron** is a polyedron of four faces; a **hexaedron,** one of six faces; an **octaedron,** one of eight faces; a **dodecaedron,** one of twelve faces; an **icosaedron,** one of twenty faces.

Note. — The least number of faces that a polyedron can have is *four;* for *three* planes intersecting in a common point form a *triedral angle*, and, therefore, one more plane is needed to form a solid.

532. Def. A **diagonal** of a polyedron is the straight line joining any two vertices not in the same face.

533. Def. A **convex** polyedron is one, every section of which is a convex polygon.

Note. — All the polyedrons treated of in this book are convex.

PRISMS AND PARALLELOPIPEDS

534. DEF. A **prism** is a polyedron, two of whose faces are equal polygons in parallel planes, and whose other faces are parallelograms. The **bases** of a prism are the equal polygons; the **lateral faces** are the parallelograms; the **lateral edges** are the intersections of the lateral faces; and the **lateral area** is the sum of the areas of the lateral faces.

From definition we have *the lateral edges of a prism are equal and parallel.*

535. DEF. The **altitude** of a prism is the perpendicular distance between the planes of the bases.

536. DEF. A **right prism** is a prism whose lateral edges are perpendicular to the planes of the bases. In a right prism each lateral edge is equal to the altitude.

537. DEF. A **regular prism** is a right prism whose bases are *regular polygons*

538. DEF. An **oblique prism** is a prism whose lateral edges are not perpendicular to the planes of the bases.

A prism is **triangular, quadrangular,** etc., according as its bases are triangles, quadrilaterals, etc.

539. DEF. A **truncated prism** is the part of a prism included between a base and a section made by a plane not parallel to the base and cutting all of the lateral edges.

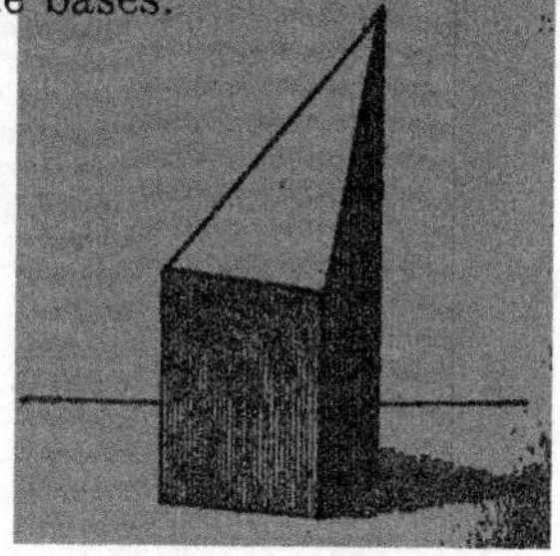

540. Def. A **right section** of a prism is a section made by a plane perpendicular to all the lateral edges of the prism.

541. Def. A **parallelopiped** is a prism whose bases are parallelograms; hence one, all of whose faces are parallelograms.

542. Def. A **right parallelopiped** is a parallelopiped whose lateral edges are perpendicular to the bases.

543. Def. A **rectangular parallelopiped** is a right parallelopiped whose bases are rectangles; hence all of whose faces are rectangles.

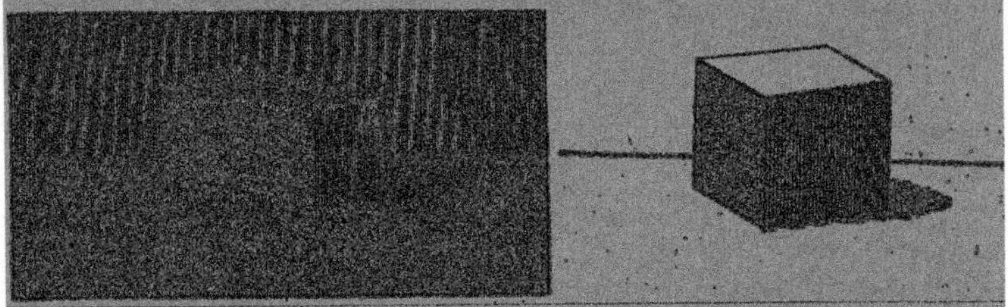

544. Def. A **cube** is a parallelopiped all of whose faces are squares.

Proposition I. Theorem

545. *The sections of a prism made by parallel planes cutting all the lateral edges are equal polygons.*

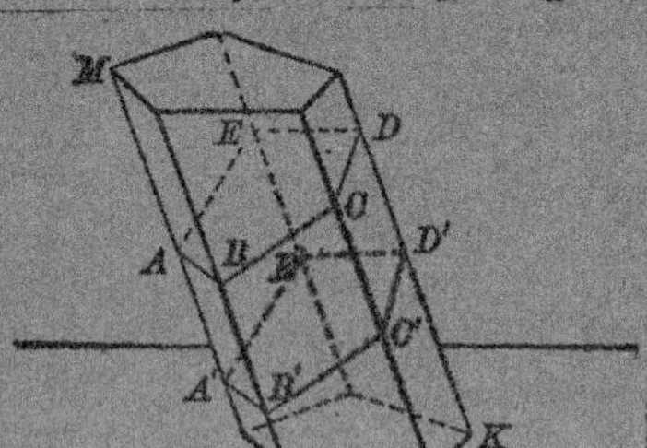

Hyp. *KM* is a prism intersected by parallel planes *AD* and *A'D'*, cutting all the lateral edges.

To prove section $AD =$ section $A'D'$.

Proof. *AB, BC, CD*, etc., are parallel to *A'B', B'C', C'D*, etc., respectively. (Why?)

$\therefore$ ∡ *ABC, BCD*, etc., are equal to ∡ *A'B'C', B'C'D'*, etc., respectively. (Why?)

AB, BC, CD, etc., are equal respectively to *A'B', B'C', C'D'*, etc. (Why?)

$\therefore$ section $AB =$ section $A'B'$. (149) Q E D.

546. Cor. Every section of a prism made by a plane parallel to the base is equal to the base.

Proposition II. Theorem

547. *The lateral area of a prism is equal to the product of the perimeter of a right section and a lateral edge.*

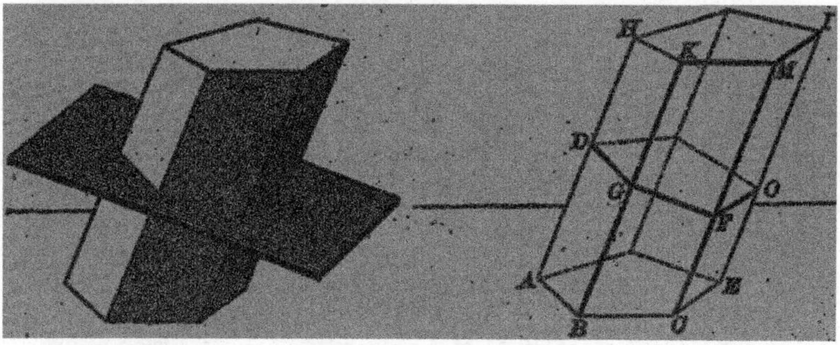

Hyp. *AI* is a prism, *S* its lateral area, *E* a lateral edge, and *P* the perimeter of right section *DO*.

To prove $S = P \times E$.

T

Proof. $HA = KB = MC =$ etc. (Why?)

$DG \perp$ to BK, $GF \perp CM$, etc. (Why?)

$\therefore$ area $BH = DG \times BK$. (Why?)

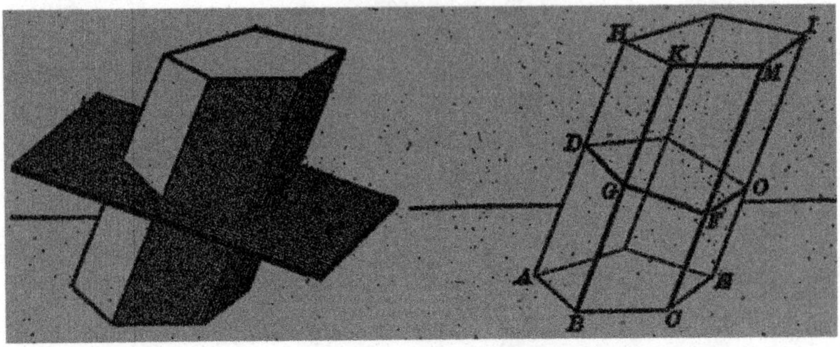

Similarly area $CK = GF \times CM = GF \times BK$, and so on.

But $S =$ sum of areas of lateral faces, BH, CK, etc.

$$\therefore S = DG \times BK + GF \times BK + FO \times BK + \text{etc.}$$
$$= (DG + GF + FO + \text{etc.}) \times BK$$
$$= P \times E.$$

Q.E.D.

548. Cor. The lateral area of a right prism equals the perimeter of the base multiplied by the altitude.

Ex. 1081. Find the lateral area of a right prism if its altitude is 15 inches, and its base is a triangle the sides of which are 8 inches, 10 inches, and 11 inches, respectively.

Ex. 1082. Find the lateral area of a right prism if its altitude is 18 inches, and its base is a quadrilateral the sides of which are 7 inches, 8 inches, 11 inches, and 12 inches, respectively.

Ex. 1083. Find the altitude of a right prism if its base is an equilateral triangle inscribed in a circle of radius 5 inches, and its lateral area is 135 square inches.

Ex. 1084. Find the radius of the circle circumscribing an equilateral triangle which forms the base of a right prism whose altitude is 9, and whose lateral area is $540\sqrt{3}$.

PROPOSITION III. THEOREM

549. *If two prisms have the three faces including a triedral angle of one equal respectively to the three faces including a triedral angle of the other, and the faces similarly placed, they are equal.*

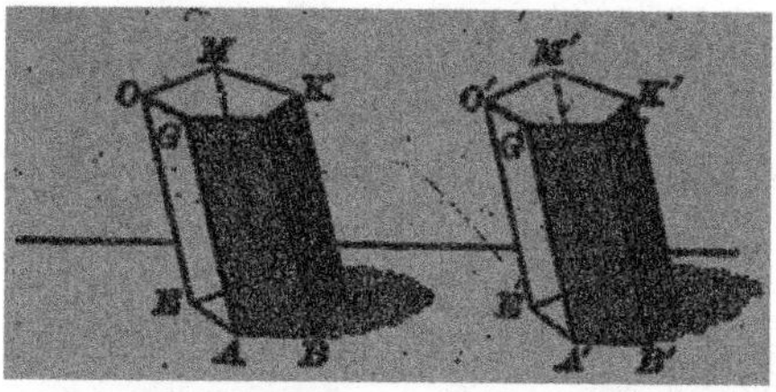

Hyp. The faces AD, AH, AO, of prism $AM =$ the faces $A'D'$, $A'H'$, $A'O'$ of prism $A'M'$ respectively, and are similarly placed.

To prove $AM = A'M'$.

Proof. $\angle BAE = \angle B'A'E'$, $\angle BAG = \angle B'A'G'$,

$\angle EAG = \angle E'A'G'$. (Why?)

$\therefore$ triedral angle $A =$ triedral angle A'. (526)

Place triedral $\angle A$ on its equal $\angle A'$.

Then face AD coincides with face $A'D'$, face AH with face $A'H'$, and face AO with face $A'O'$.

The point C falls at C', and D at D'. (Why?)

CK will fall on $C'K'$, and DM on $D'M'$.

(Since the lateral edges of a prism are parallel.)

The points G, H, and O coincide respectively with G', H', and O'. (Why?)

$\therefore$ planes GM and $G'M'$ coincide. (454, 2)

Hence point K coincides with K', and M with M'. (Why?)

$\therefore$ prisms coincide throughout and are equal. Q.E.D.

550. Cor. 1. Two right prisms having equal bases and equal altitudes are equal.

551. Cor. 2. Two truncated prisms are equal when they fulfil the hypothesis of the above theorem.

552. Def. The **volume** of any solid is the number of times it contains the unit of volume. The **unit of volume** is a cube whose edges are equal to the linear unit.

553. Def. **Equivalent solids** are solids whose volumes are equal.

Proposition IV. Theorem

554. *An oblique prism is equivalent to a right prism whose base is a right section of the oblique prism, and whose altitude is equal to a lateral edge of the oblique prism.*

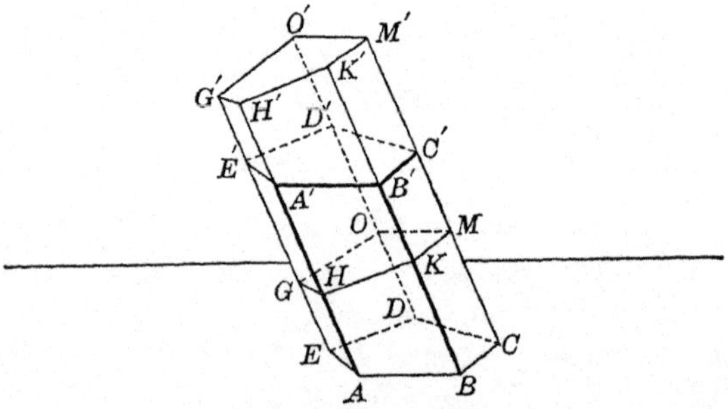

Hyp. GM is a right section of oblique prism AD', and GM' a right prism whose altitude is equal to a lateral edge of AD'.

To prove $AD' \approx GM'$.

Proof. The lateral edges of GM' = lateral edges of AD'. (Why?)

$\therefore AH = A'H'$, $BK = B'K'$, $CM = C'M'$, etc. (Why?)

Face AK = face $A'K'$, face BM = face $B'M'$, and face BE = face $B'E'$, (Why ?)

$\therefore$ truncated prism AM = truncated prism $A'M'$. (551)

Hence $AM' - AM = AM' - A'M'$, (Ax. 3)

or $AD' \approx GM'$. Q.E.D

Proposition V. Theorem

555. *The opposite lateral faces of a parallelopiped are equal and parallel.*

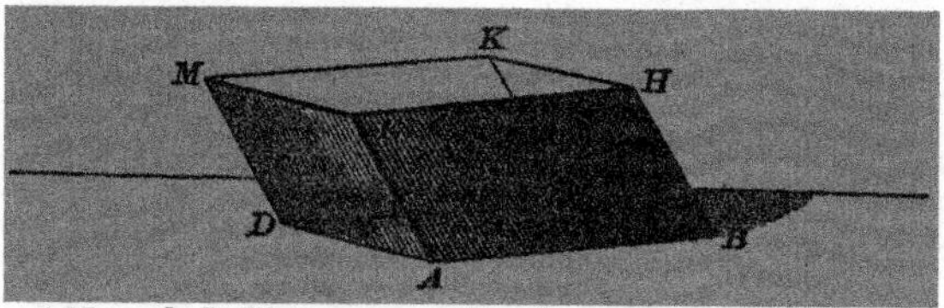

Hyp. AH and DK are opposite faces of parallelopiped AK.

To prove $AH = DK$ and $AH \parallel$ to DK.

Proof. $AB =$ and $\parallel$ to DC. (Why ?)

$AE =$ and $\parallel$ to DM.

$\therefore \angle EAB = \angle MDC$. (468)

$\therefore AH = DK$, (142)

and $AH \parallel DK$. (486)

Similarly faces AM and BK are equal and parallel.

556. Cor. Any two opposite faces of a parallelopiped may be taken as bases.

Ex. 1085. Find the total area of a right prism if its altitude is 12 inches, and its base is a triangle the sides of which are 9 inches, 6 inches, and 5 inches respectively.

Ex. 1086. Find the total area of a right prism if its altitude is 9 inches, and its base is a rhombus of side 4 inches and having an acute angle of 60°.

Proposition VI. Theorem

557. *The plane passed through two diagonally opposite edges of a parallelopiped divides it into two equivalent triangular prisms.*

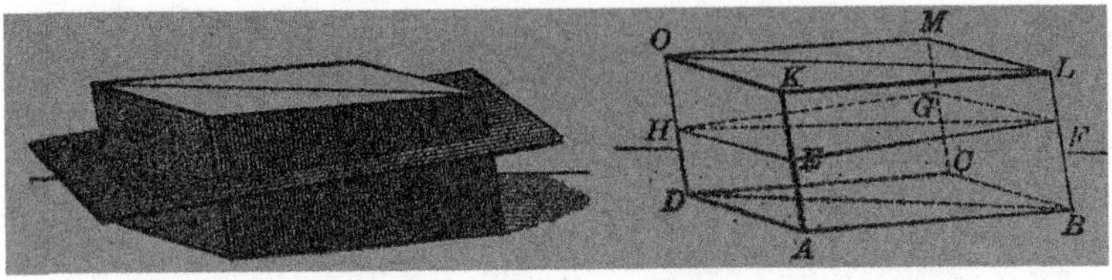

Hyp. $DOLB$ is a plane passing through edges DO and BL.

To prove prism $ABD\text{-}K \approx$ prism $BCD\text{-}M$.

Proof. Draw right section $EFGH$, intersecting $DOLB$ in HF.

$AL \parallel DM,\ AO \parallel BM.$ (Why?)

$\therefore\ EF \parallel HG.$ (478)

And EH is $\parallel$ to FG.

$\therefore\ EFGH$ is parallelogram.

$\triangle EFH = \triangle FGH.$ (Why?)

The prism $ABD\text{-}K$ is equivalent to a right prism whose base is $\triangle EFH$ and altitude AK, and prism $BDC\text{-}M$ is equivalent to a right prism whose base is $\triangle FGH$ and altitude AK. (554)

$\therefore$ prism $ABD\text{-}K \approx$ prism $BDC\text{-}M$. Q.E.D

Ex. 1087. The diagonals of a parallelopiped bisect each other.

Ex. 1088. Find the edge of a cube having a surface equal to the sum of the surfaces of two cubes whose edges are 120 inches and 209 inches.

Ex. 1089. The dimensions of a rectangular solid are proportional to 3, 4, and 5. Find them if the entire surface contains 2350 square inches.

PROPOSITION VII. THEOREM

558. *Two rectangular parallelopipeds having equal bases are to each other as their altitudes.*

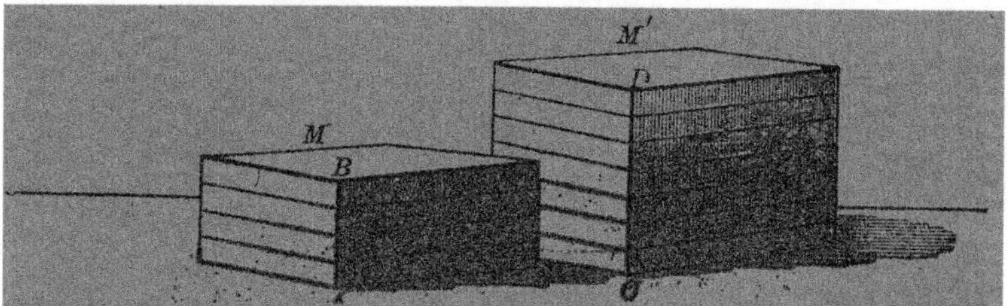

Hyp. AB and CD are the altitudes of parallelopipeds M and M' which have equal bases.

To prove $M : M' :: AB : CD.$

CASE I. When the altitudes are commensurable.

HINT. — Apply common measure to altitudes AB and CD, and through the points of division pass planes ∥ to bases. Use method of Prop. I, Bk. IV.

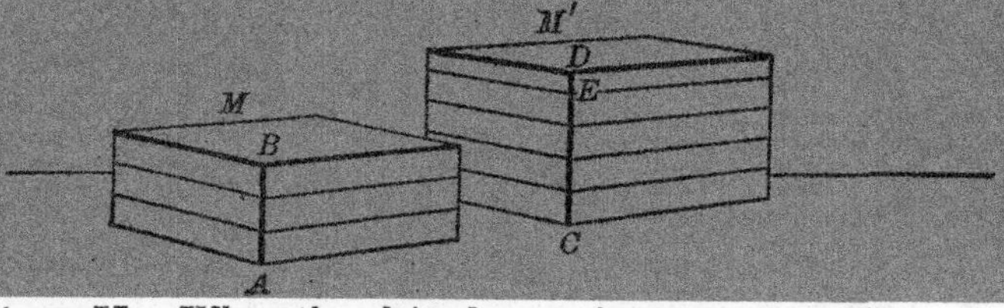

CASE II. When the altitudes are incommensurable.

HINT. — Divide AB into any number of equal parts, apply one of these to CD, leaving a remainder ED. Through E pass a plane. Use method of Prop. I, Bk. IV.

559. DEF. The **dimensions** of a rectangular parallelopiped are the three edges that meet at the same vertex.

560. SCHOLIUM. The preceding theorem may be stated: Two rectangular parallelopipeds which have two dimensions in common are to each other as the third dimension.

Proposition VIII. Theorem

561. *Two rectangular parallelopipeds having equal altitudes are to each other as their bases.*

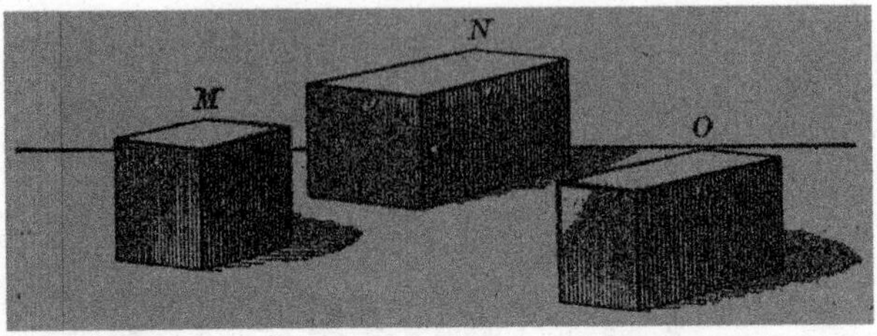

Hyp. a, b, c and a, b', c' are the three dimensions of rectangular parallelopipeds M and N, respectively; a being the equal altitude.

To prove $M : N = b \times c : b' \times c'$.

Proof. Construct O a rectangular parallelopiped with dimensions a, b', and c.

Then $$\frac{M}{O} = \frac{b}{b'},$$

and $$\frac{O}{N} = \frac{c}{c'}.$$ (Why?)

Multiplying these equations together,

we have $$\frac{M}{N} = \frac{b \times c}{b' \times c'}.$$

562. Scholium. This may be stated: two rectangular parallelopipeds that have one dimension in common are to each other as the products of their other two dimensions.

Ex. 1090. Two rectangular parallelopipeds have equal altitudes and bases whose dimensions are 4 and 7, and 5 and 9 respectively. Find the ratio of their volumes.

Ex. 1091. What is the ratio of the volumes of two rectangular parallelopipeds having equal bases and altitudes a and b respectively?

Proposition IX. Theorem

563. *Two rectangular parallelopipeds are to each other as the products of their three dimensions.*

Hyp. a, b, c and a', b', c' are the three dimensions of the rectangular parallelopipeds M and N respectively.

To prove $M : N = a \times b \times c : a' \times b' \times c'$.

Hint. — Construct O a rectangular parallelopiped with dimensions a, b', and c. Find the ratios $\frac{M}{O}$ and $\frac{O}{N}$ (Props. VII and VIII), and multiply them together.

Proposition X. Theorem

564. *The volume of a rectangular parallelopiped is equal to the product of its three dimensions.*

Hyp. a, b, c are the dimensions of a rectangular parallelopiped M.

To prove volume of $M = a \times b \times c$.

Hint. — Construct $V =$ unit of volume, and apply Prop. IX.

565. Cor. 1. The volume of a cube equals the cube of its edge.

566. Cor. 2. The volume of a rectangular parallelopiped equals the product of its altitude by its base.

EXERCISES

Ex. 1092. What is the ratio of the volumes of two rectangular parallelopipeds whose dimensions are 6, 8, and 9 and 6, 7, and 9 respectively?

Ex. 1093. Find the area of the base of a rectangular parallelopiped whose altitude is 6 and one edge of the base is 4, if the solid is equivalent to a rectangular parallelopiped whose dimensions are 8, 12, and 15.

Ex. 1094. Find volume of a rectangular parallelopiped the dimensions of whose base are 12 and 20, and the area of whose entire surface is 800.

Ex. 1095. Find the volume of a rectangular parallelopiped the dimensions of whose base are 12 in. and a in., and the area of whose entire surface is $(120 + 34a)$ sq. in.

Proposition XI. Theorem

567. *The volume of any parallelopiped is equal to the product of its base and altitude.*

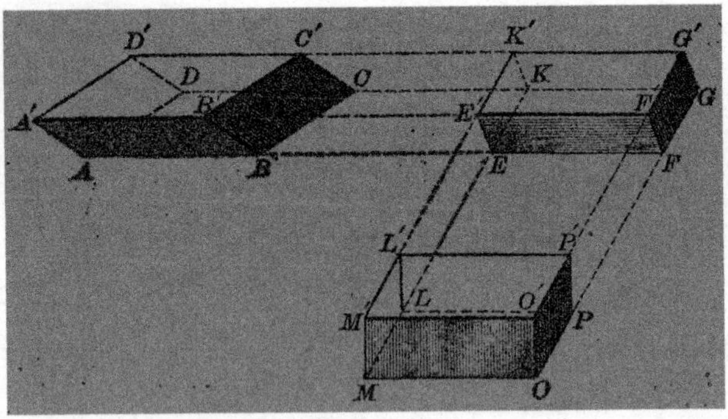

Hyp. AC' is any parallelopiped whose base is $ABCD$ and altitude H.

To prove volume $AC' = ABCD \times H$.

Proof. Produce the edge AB and the edges $\parallel$ to it. Take EF equal to AB, and through E and F pass planes EK' and $FG' \perp EF$, forming the right parallelopiped EG'.

Then $EG' \approx AC'$. (Why?)

Produce the edge KE and the edges $\parallel$ to it. Take LM equal to KE, and through L and M pass planes LP' and $MO' \perp LM$, forming the parallelopiped LO'.

Then $LO' \approx EG'$.

$\therefore LO' \approx AC'$. (1) (Why?)

Now $LMOP \approx KEFG$, (342)

and $KEFG \approx ABCD$. (342)

$\therefore LMOP \approx ABCD$. (2)

Also the planes of the upper and lower bases are parallel, and hence the three solids have a common altitude H.

By construction the planes MK' and OG' are perpendicular to plane MG, and the planes LP' and MO' are also perpendicular to plane MG, hence LO' is a rectangular parallelopiped.

Now volume $LO' = LMOP \times H$. (564)

Hence, from (1) volume $AC' = LMOP \times H$,

and by substitution from (2)

volume $AC' = ABCD \times H$. Q.E.D.

Ex. 1096. Find the volume of a parallelopiped whose base is 50 and whose lateral edge is 20, if the inclination of the lateral edge to the base is 30°.

Ex. 1097. Find the volume of a parallelopiped whose base is 20 and whose lateral edge is 10, if the inclination of the lateral edge to the base is 45°.

Ex. 1098. In the diagram for Prop. XI, find the volume of AC', if $AB = 4$, $BC = 5$, $AA' = 6$, $\angle ABC = 30°$, and the inclination of AA' to the base AC is 60°.

PROPOSITION XII. THEOREM

568. *The volume of a triangular prism is equal to the product of its base by its altitude.*

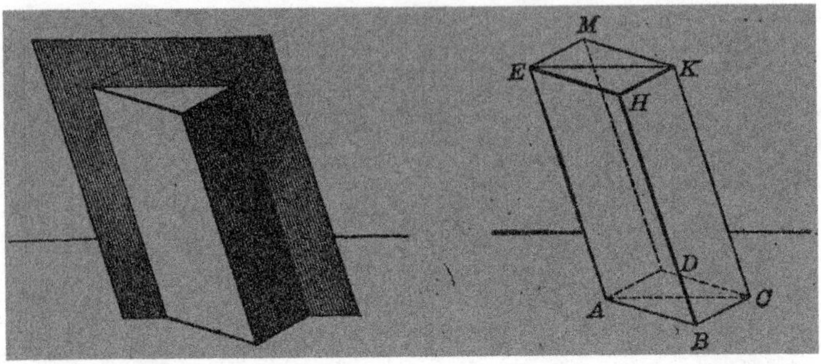

Hyp. V denotes the volume, B the base, and a the altitude of the triangular prism $ABC\text{–}H$.

To prove $V = B \times a$.

Proof. Upon the edges AB, BC, BH, construct the parallelopiped $ABCD\text{–}H$.

$ABC\text{–}H \approx \frac{1}{2} ABCD\text{–}H$. (557)

The volume of $ABCD\text{–}H =$ base $ABCD \times a$. (Why ?)

The base $ABCD = 2\,B$. (Why ?)

$\therefore V = \frac{1}{2}(2\,B \times a)$.

$\therefore V = B \times a$. Q.E.D

Ex. 1099. In the diagram for Prop. XII, find the volume of $ABC\text{–}H$, if $AB = 2$, $BC = 3$, $BH = 5$, $\angle ABC = 30°$, and the projection of BH upon base ABC equals 4.

Ex. 1100. In the same diagram find the area of $\triangle ABC$, if the volume of $ABC\text{–}H$ is 200 and the altitude is 20.

PROPOSITION XIII. THEOREM

569. *The volume of any prism is equal to the product of its base by its altitude.*

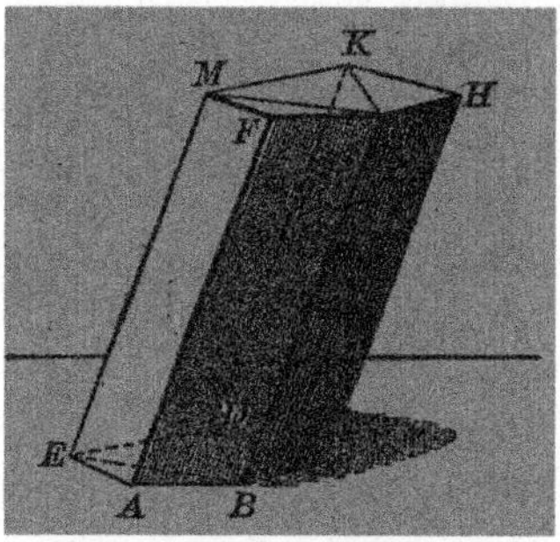

Hyp. V denotes the volume, B the base, and a the altitude of any prism.

To prove $V = B \times a$.

Proof. Pass planes through one lateral edge BG, and the diagonals BE, BD, etc., of the base.

The prism will be divided into triangular prisms, all of which will have the altitude a.

The volume of the given prism is equal to the sum of the volumes of the triangular prisms.

The volume of each triangular prism equals its base multiplied by a. (Why ?)

$\therefore$ the volume of the given prism equals the sum of the bases of the triangular prisms multiplied by the common altitude a.

But the sum of the bases of the triangular prism equals B.

$$\therefore V = B \times a.$$

---

570. COR. 1. Prisms are to each other as the products of their bases by their altitudes.

571. COR. 2. Prisms that have equivalent bases are to each other as their altitudes.

572. Cor. 3. Prisms that have equal altitudes are to each other as their bases.

573. Cor. 4. Prisms that have equal altitudes and equivalent bases are equivalent.

Ex. 1101. In the diagram for Prop. XIII, find the volume V, if $ABCDE = 40$ sq. in., $AF = 13$ in., and the projection of AF upon the base equals 5 in.

Ex. 1102. In the diagram for Prop. XIII, find AF, if $V = 200$, $ABCDE = 25$, and the projection of AF upon the base is 6.

Ex. 1103. In the same diagram find the inclination of AF to the base, if $AF = 6$, $ABCDE = 20$, and $V = 60$.

PYRAMIDS

574. Def. A **pyramid** is a polyedron, one of whose faces is a polygon, and whose other faces are triangles having a common vertex.

575. Def. The **base** of the pyramid is the polygon, the **lateral faces** are the triangles, the **vertex of the pyramid** is the common vertex of the triangles, the **lateral edges** are the intersections of the lateral faces, and the **lateral area** is the sum of the areas of the lateral faces.

576. Def. A pyramid is **triangular, quadrangular**, etc., according as its base is a triangle, a quadrilateral, etc.

Note. — A tetraedron is a triangular pyramid.

577. Def. The **altitude** of a pyramid is the length of the perpendicular from the vertex of the pyramid to the plane of the base.

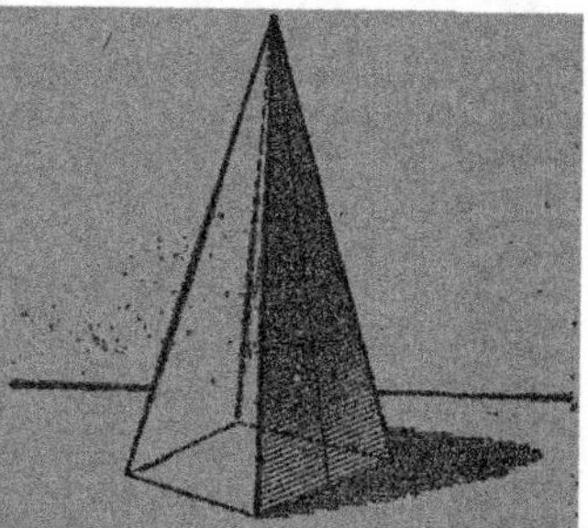

578. Def. A **regular pyramid** is one whose base is a regular polygon, the center of which coincides with the foot of the altitude.

579. Def. The **axis** of a regular pyramid is its altitude.

Note. — The lateral edges of a regular pyramid are equal since they cut off equal distances from the foot of the altitude. (473)

Hence the lateral faces of a regular pyramid are equal isosceles triangles.

580. Def. The **slant height of a regular pyramid** is the altitude of any one of the lateral faces.

581. Def. A **truncated pyramid** is the portion of a pyramid included between the base and a section formed by a plane cutting all the lateral edges.

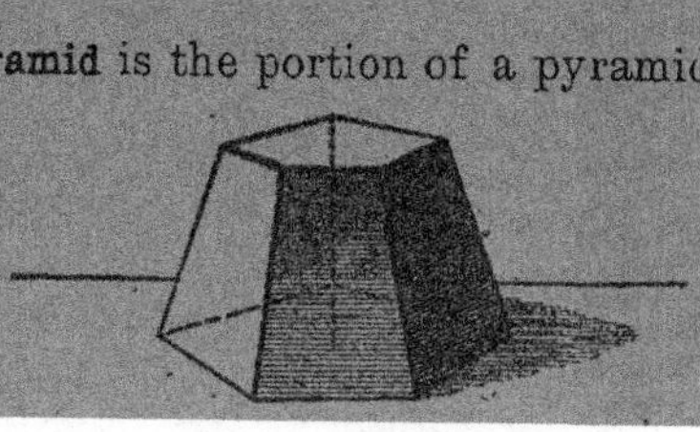

582. Def. A **frustum of a pyramid** is a truncated pyramid in which the plane of the section is parallel to the base.

583. Def. The **altitude of a frustum** is the perpendicular distance between its bases.

584. Def. The **lateral area of the frustum** is the sum of the areas of its lateral faces; the **lateral faces of a frustum** of a regular pyramid are equal isosceles trapezoids.

585. Def. The **slant height of a frustum** of a regular pyramid is the altitude of one of these trapezoids which form the lateral faces.

Proposition XIV. Theorem

586. *If a pyramid be cut by a plane parallel to its base:*

(1) *The lateral edges and the altitude are divided proportionally.*

(2) *The section is a polygon similar to the base.*

Hyp. Pyramid $O\text{–}ABCD$ is cut by plane ∥ to the base, E, F, G, H are the points of intersection with edges, forming section $EFGH$, and N the point of intersection with altitude OM.

1st. *To prove* $\frac{OA}{OE} = \frac{OB}{OF} = \cdots = \frac{OM}{ON}$.

This follows directly from 488

2d. *To prove* $EFGH$ similar to $ABCD$.

Proof. $EF \parallel AB$, $FG \parallel BC$, $GH \parallel CD$, etc., (478)

$$\therefore \frac{EF}{AB} = \left(\frac{OF}{OB}\right) = \frac{FG}{BC} = \left(\frac{OG}{OC}\right) = \frac{HG}{CD}, \text{ etc.} = \frac{ON}{OM}. \quad \text{(Why?)}$$

and $\angle EFG = \angle ABC$, $\angle FGH = \angle BCD$, etc. (468)

$\therefore$ polygon $EFGH$ is similar to $ABCD$. (289)

587. Cor. 1. A section of a pyramid parallel to the base is to the base as the square of its distance from the vertex is to the square of the altitude of the pyramid.

$$EFGH : ABCD = \overline{EF}^2 : \overline{AB}^2 \quad (371)$$

But $\overline{EF}^2 : \overline{AB}^2 = \overline{OE}^2 : \overline{OA}^2 = \overline{ON}^2 : \overline{OM}^2$. (Why?)

$$\therefore EFGH : ABCD = \overline{ON}^2 : \overline{OM}^2.$$

588. Cor. 2. If two pyramids that have equal altitudes are cut by planes parallel to the bases, and at equal distances from the vertices, the sections have the same ratio as the bases.

Hint. — Use Cor. 1.

589. Cor. 3. If two pyramids have equal altitudes and equivalent bases, sections made by planes parallel to the bases, and at equal distances from the vertices, are equivalent.

Ex. 1104. The altitude of a regular triangular pyramid, each side of whose base is 4 ft., is 12 ft. Find the area of a section made by a plane parallel to the base and 4 ft. from the vertex.

Ex. 1105. Find the area of a section of a regular quadrangular pyramid, each side of whose base is 3 ft. and whose altitude is 8 ft., made by a plane 3 ft. from the vertex and parallel to the base.

Ex. 1106. The base of a regular hexagonal pyramid measures 10 in. on a side. If the altitude be 5 in., how far from the vertex must a plane be passed to form a section the area of which is $3\sqrt{3}$?

PROPOSITION XV. THEOREM

590. *The lateral area of a regular pyramid is equal to half the product of the slant height by the perimeter of the base.*

Hyp. *A* is the lateral area, *Sl. H.* the slant height, and *P* the perimeter of the base of the regular pyramid *O–ABCDE*.

To prove $A = \frac{1}{2}$ *Sl. H.* $\times P$.

Proof. $\triangle OAB = \triangle OBC = \triangle OCD$, etc. (Why?)

The altitude of each $\triangle$, *OK* is *Sl. H.*

Area of each $\triangle = \frac{1}{2}$ *Sl. H.* $\times$ its base.

Sum of areas of all the $\triangle$s $= \frac{1}{2}$ *Sl. H.* $\times$ sum of bases.

$= \frac{1}{2}$ *Sl. H.* $\times P$.

$\therefore A = \frac{1}{2}$ *Sl. H.* $\times P$. Q.E.D.

591. COR. The lateral area of a frustum of a regular pyramid is equal to one-half the sum of the perimeters of the bases multiplied by the slant height.

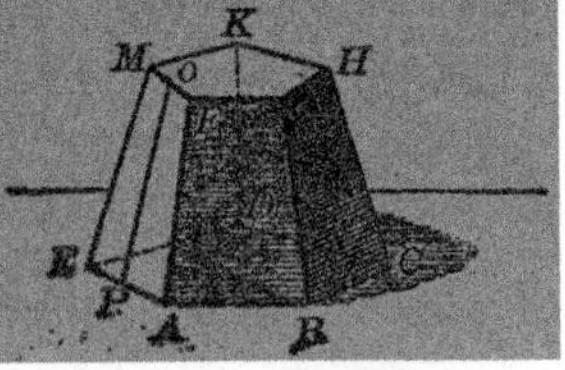

HINT. — *What is the shape of the lateral faces?*

Ex. 1107. Find the lateral area of a regular quadrangular pyramid whose altitude is 12, and whose base has sides equal to 10.

Ex. 1108. Find the lateral area of a frustum of a regular pyramid of five sides whose slant height is 5, and the sides of whose bases are 7 and 3, respectively.

PROPOSITION XVI. THEOREM

592. *Two triangular pyramids which have equivalent bases and equal altitudes are equivalent.*

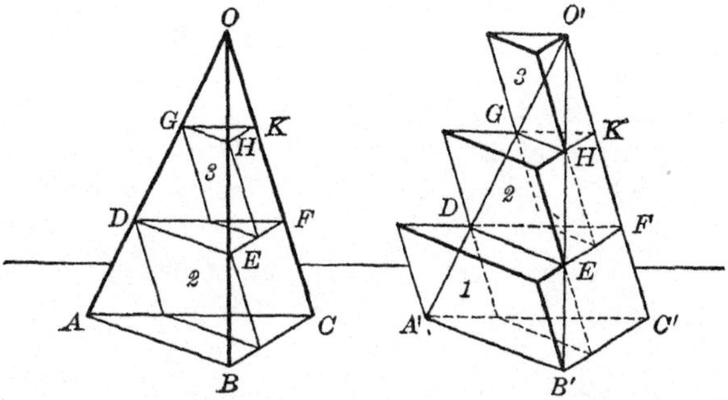

Hyp. Two triangular pyramids $O\text{–}ABC$ and $O'\text{–}A'B'C'$ have equal altitudes, equivalent bases, and the volumes V and V' respectively.

To prove $V = V'$.

Proof. If they are not equivalent, assume $V' > V$.

Place the pyramids so that their bases are in the same plane.

Divide the altitudes into n equal parts, calling the length of each part h.

Through the points of division pass planes parallel to the bases.

The corresponding sections thus formed are equivalent. (589)

On the base $A'B'C'$, and on each parallel section of $O'\text{–}A'B'C'$ as a lower base, construct a prism with lateral edges parallel to $O'C'$ and altitude equal to h. On each section of $O\text{–}ABC$

as an upper base construct a prism with lateral edges parallel to OC and altitude equal to h.

Each prism of $O–ABC$ is equivalent to the prism next above it in $O'–A'B'C'$, hence the *difference between the two sets of prisms* is the lowest prism of the first set. By increasing the number of parts into which each altitude is divided, each part h, and hence the volume of the lowest prism in $O'–A'B'C'$, can be diminished indefinitely. Or if P' and P denote the volumes of the two sets of prisms, the limit of $P' - P = 0$.

But $P' > V'$ and $V > P$. (Ax. 9.)

$$\therefore P' - P > V' - V.$$

If, however, V' was greater than V, $V' - V$ would be a positive quantity, or a positive quantity would be less than a quantity which approaches zero as a limit, which is impossible.

$\therefore V'$ is not greater than V. In the same manner we may prove that pyramid V is not greater than pyramid V'.

$$\therefore V = V'.$$

Proposition XVII. Theorem

593. *The volume of a triangular pyramid is equal to one third of the product of its base by its altitude.*

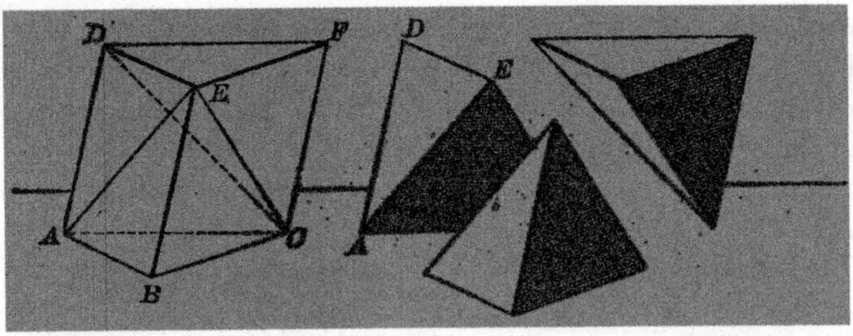

Hyp. V is the volume, B the base, and a the altitude of the triangular pyramid $E–ABC$.

To prove. $V = \frac{1}{3} B \times a.$

Proof. On ABC construct the prism $ABC\text{–}DEF$, having its lateral edges equal and parallel to EB.

The prism is composed of the pyramid $E\text{–}ABC$ and the quadrangular pyramid $E\text{–}ACFD$.

Through ED and EC pass a plane cutting $ACFD$ in DC, forming two triangular pyramids, $E\text{–}DCF$ and $E\text{–}ADC$.

$$E\text{–}DCF \approx E\text{–}ADC. \qquad (1)$$

(*Same altitude and equivalent bases.*)

Pyramid $E\text{–}ADC$ is the pyramid $C\text{–}ADE$, since any face of a tetraedron may be taken as base,

and $\triangle ADE = \triangle EAB.$ (Why?)

$C\text{–}ADE \approx C\text{–}ABE.$ (2) (Why?)

$\therefore$ $C\text{–}ABE$ or $E\text{–}ABC = \frac{1}{3} ABC\text{–}EDF.$ (From (1) and (2))

$\therefore$ $V = \frac{1}{3} B \times a.$ (568)

594. Cor. 1. The volume of any pyramid is equal to one-third the product of its base by its altitude.

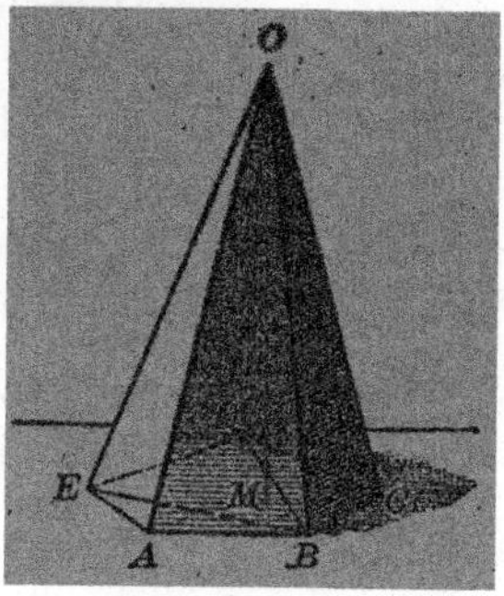

For the pyramid may be divided into triangular pyramids by passing planes through any edge and the corresponding diagonals of the base.

The bases of the triangular pyramids together form the base of the given pyramid, and the sum of the volumes of the triangular pyramids equals the volume of the given pyramid.

Therefore the volume of any pyramid equals one-third the product of its base by its altitude.

595. Cor. 2. The volumes of two pyramids are to each other as the products of their bases and altitudes.

596. Cor. 3. Pyramids having equivalent bases are to each other as their altitudes, and those having equal altitudes are to each other as their bases.

597. Cor. 4. Pyramids having equal altitudes and equivalent bases are equivalent.

598. Scholium. The volume of any polyedron may be obtained by dividing it into pyramids and finding their volumes separately.

Proposition XVIII. Theorem

599. *The frustum of a triangular pyramid is equivalent to the sum of three pyramids each having the same altitude as the frustum and whose bases are the upper and lower bases of the frustum, and the mean proportional between these bases.*

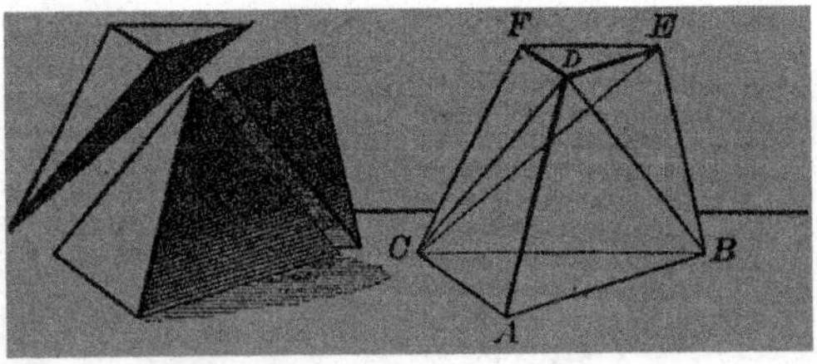

Hyp. *ABC–DEF* is a frustum of any triangular pyramid, with bases *ABC* [*B*], and *DEF* [*b*], and altitude *H*.

To prove *ABC–DEF* equivalent to the sum of three pyramids whose common altitude is *H*, and whose bases are *ABC* and *DEF*, and a mean proportional between them.

Proof. Through the vertices C, D, B, and C, D, E, pass planes, dividing the frustum into three pyramids.

The pyramids $D\text{–}CAB$ and $C\text{–}FDE$ have the common altitude H, and for bases respectively ABC and DEF, the lower and upper bases of the frustum.

It remains to show that $C\text{–}DEB$ is equivalent to a pyramid whose altitude is H, and whose base is a mean proportional between ABC and DEF $[\sqrt{B \times b}]$.

Pyramids $C\text{–}ABD$ and $C\text{–}BDE$ have their bases in the plane $ABED$.

$$\therefore\ C\text{–}ABD : C\text{–}BDE = \triangle ABD : \triangle BDE. \qquad \text{(Why?)}$$

$$\triangle ABD : \triangle BDE = AB : DE. \qquad \text{(Why?)}$$

Hence, $C\text{–}ABD : C\text{–}BDE = AB : DE.$ (1) (Why?)

The pyramid $C\text{–}DBE$ may be read $D\text{–}BCE$, and we have the pyramids $D\text{–}CEF$ and $D\text{–}BCE$ with bases in the plane $FEBC$ and vertices at D.

$$\therefore\ D\text{–}BCE : D\text{–}CEF :: \triangle BCE : \triangle CEF.$$

But $\triangle BCE : \triangle CEF :: BC : EF.$

Hence, $D\text{–}BCE : D\text{–}CEF :: BC : EF.$ (2)

The $\triangle$s ABC and DEF are similar,

hence, $AB : DE = BC : EF.$

$$\therefore\ C\text{–}ABD : C\text{–}BDE = D\text{–}BCE : D\text{–}CEF.$$

But $C\text{–}ABD = \frac{1}{3} H \times B$, and $D\text{–}CEF = \frac{1}{3} H \times b.$ (Why?)

$$\therefore\ C\text{–}BED = \sqrt{\tfrac{1}{3}H \times B \times \tfrac{1}{3} H \times b} = \tfrac{1}{3} H \sqrt{B \times b}.$$

Hence the frustum of the pyramid is equivalent to the sum of three pyramids, whose bases are the upper and lower bases of the frustum, and the mean proportional between these bases, and having the altitude, H.

600. Cor. 1. Since the volume of a pyramid is equal to the base multiplied by one-third of its altitude, we have

Vol. of frustum of a triangular pyramid $= \frac{1}{3} H(B + b + \sqrt{B \times b})$.

601. Cor. 2. The volume of the frustum of any pyramid is equal to one-third of its altitude multiplied by the sum of its bases and a mean proportional between them.

Ex. 1109. Find the volume of a pyramid whose base is 60, and whose altitude is 5.

Ex. 1110. Find the altitude of a pyramid whose volume is 200, and whose base is a rectangle having the sides 5 and 12 respectively.

Ex. 1111. The three sides of the base of a pyramid are respectively 10, 17, and 21. Find the volume if the altitude is 5.

Ex. 1112. The three sides of the base of a pyramid are respectively 9, 10, 17. Find the volume if a lateral edge is 20, and its projection upon the base equals 12.

Ex. 1113. A lateral edge of a pyramid equals 10, and its inclination to the base is 30°. Find the area of the base if the volume of the pyramid is 100.

Ex. 1114. The base of a pyramid is a rhombus whose diagonals are respectively 10 and 12. Find the volume if the altitude is 6.

Ex. 1115. The diagonals of a parallelopiped divide the figure into six equivalent pyramids.

Ex. 1116. If any point within a parallelopiped be joined to the 8 vertices, 6 pyramids are formed, of which the sum of any opposite two is equal to the sum of any other opposite two.

Ex. 1117. Each edge of a triangular pyramid is equal to 10. Find the volume.

Ex. 1118. The perimeter of the triangular base of a regular pyramid is 40. Find the volume if the altitude is 12.

Ex. 1119. The base of a pyramid is a parallelogram of base 10 and altitude 8. Find the volume if a lateral edge is equal to 6, and forms with the base an angle of 45°.

Ex. 1120. The base of a pyramid is a rectangle having sides respectively equal to a and b. A lateral edge is equal to c, and is inclined to the base 30°. Find the volume.

Ex. 1121. Find the altitude of a pyramid of base b, equivalent to another pyramid of base a and altitude h.

Ex. 1122. Find the volume of the frustum of a triangular pyramid, if the lower base is 9, the upper 4, and the altitude 5.

Ex. 1123. Find the volume of the frustum of a regular quadrangular pyramid, if the edge of the lower base is 7, of the upper 6, and if the altitude is 8.

Ex. 1124. The upper base of a frustum of a pyramid is 2, the lower 18. Find the altitude if the volume is 260.

Proposition XIX. Theorem

602. *A truncated triangular prism is equivalent to the sum of three pyramids whose common base is the base of the prism and whose vertices are the three vertices of the inclined section.*

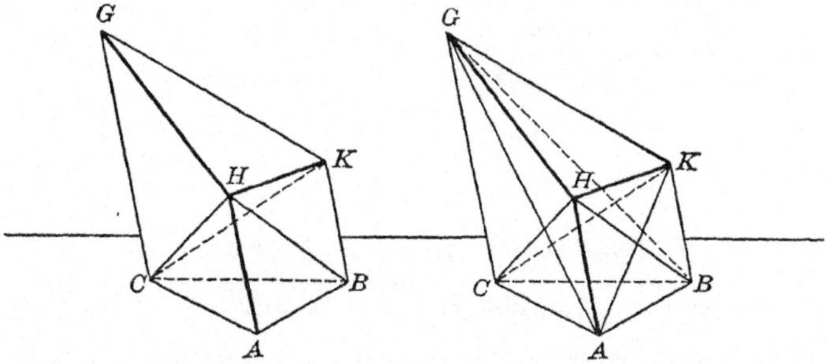

Hyp. *ABC–HKG* is a truncated triangular prism of which *ABC* is the base.

To prove *ABC–HGK* ⇔ *H–ABC* + *K–ABC* + *G–ABC*.

Proof. Pass planes through *H, B, C*, and through *H, K, C*, forming the pyramids *H–ABC*, *H–BCK*, and *H–CKG*.

H–ABC is evidently one of the required pyramids.

H–BCK may be read *C–HBK*.

C–HBK ⇔ *C–AKB*.

(*Equivalent bases and the same altitude.*)

C–AKB may be read *K–ABC*.

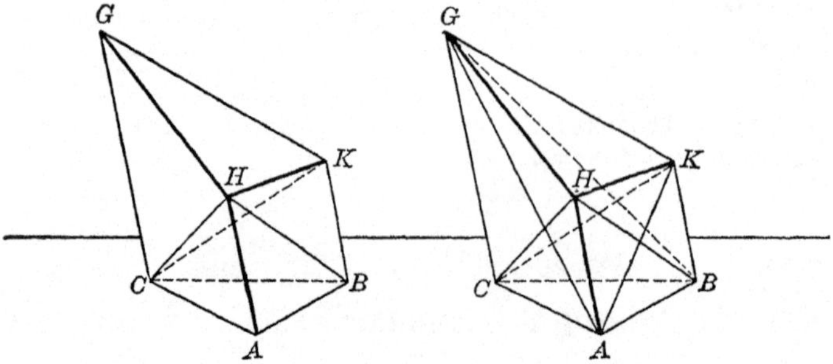

$\therefore$ *H–BCK* ≎ *K–ABC*.

H–CKG may be read *K–CHG*.

K–CHG ≎ *K–ACG*.

(*Equivalent bases and the same altitude.*)

K–ACG ≎ *B–ACG*.

(*BK* ∥ *plane AG*.)

B–ACG may be read *G–ABC*.

$\therefore$ *H–CKG* ≎ *G–ABC*.

$\therefore$ *ABC–HKG* ≎ *H–ABC* + *K–ABC* + *G–ABC*.

Q.E.D.

603. Cor. 1. The volume of a truncated right triangular prism is equal to the product of its base by one-third the sum of its lateral edges.

Hint. — The lateral edges are the altitudes of the three pyramids which form the truncated prism.

604. Cor. 2. The volume of any truncated triangular prism is equal to the product of its right section by one-third the sum of its lateral edges.

Hint. — The right section divides the truncated prism so that two truncated right prisms are formed.

PROPOSITION XX. THEOREM

605. *The volumes of two triangular pyramids, that have a triedral angle of one equal to a triedral angle of the other, are to each other as the product of the three edges of these angles.*

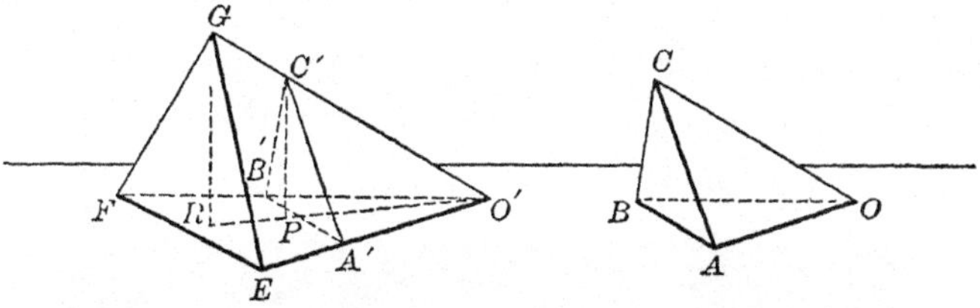

Hyp. V and V' are volumes of the triangular pyramids $O\text{–}ABC$ and $O'\text{–}EFG$, which have the equal triedral angles O and O'.

To prove $$\frac{V}{V'} = \frac{OA \times OB \times OC}{O'E \times O'F \times O'G}.$$

Proof. Place the pyramids so that triedral $\angle\ O$ will coincide with triedral $\angle\ O'$.

Draw $C'P$ and GR perpendicular to face $O'FE$, and let the plane of $C'P$ and GR intersect face $O'EF$ in line $O'PR$.

Since $O'A'B'$ and $O'EF$ are bases, and $C'P$ and GR the altitudes of pyramids $C'\text{–}O'A'B'$ and $G\text{–}O'EF$,

$$\frac{V}{V'} = \frac{\triangle\ O'A'B' \times C'P}{\triangle\ O'EF \times GR} = \frac{\triangle\ O'A'B'}{\triangle\ O'EF} \times \frac{C'P}{GR}. \quad (595)$$

$$\frac{\triangle\ O'A'B'}{\triangle\ O'EF} = \frac{O'A' \times O'B'}{O'E \times O'F}. \quad (369)$$

$$\frac{C'P}{GR} = \frac{O'C'}{O'G}. \quad \text{(Why ?)}$$

$$\therefore\ \frac{V}{V'} = \frac{O'A' \times O'B'}{O'E \times O'F} \times \frac{O'C'}{O'G} = \frac{OA \times OB \times OC}{O'E \times O'F \times O'G}. \quad \text{Q.E.D.}$$

SIMILAR POLYEDRONS

606. Def. **Similar polyedrons** are polyedrons that have the same number of faces similar each to each and similarly placed, and have the homologous polyedral angles equal.

Proposition XXI. Theorem

607. *Two similar polyedrons can be decomposed into the same number of tetraedrons similar each to each and similarly placed.*

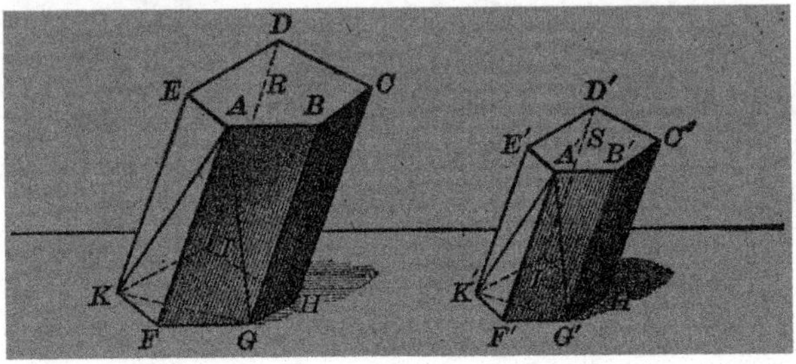

Hyp. R and S are two similar polyedrons; A and A' homologous vertices.

To prove R and S can be decomposed into the same number of tetraedrons, similar each to each, and similarly placed.

Proof. Divide all the faces of R and S except those having A and A' as vertices, *i.e.* the faces $FGHIK$, GC, etc., and $F'G'H'I'K'$, $G'C'$, etc., into ⧍.

Pass planes through A and the vertices of the ⧍ in the faces of R, and through A' and the vertices of ⧍ in the faces of S.

Let $AKFG$ and $A'K'F'G'$ be two homologous tetraedrons formed by these planes.

$A\text{-}KFG$ is similar to $A'\text{-}K'F'G'$.

For

$$\left.\begin{array}{l}\triangle AKF \text{ is similar to } \triangle A'K'F' \\ \triangle AFG \text{ is similar to } \triangle A'F'G' \\ \triangle KFG \text{ is similar to } \triangle K'F'G'\end{array}\right\} \qquad (303)$$

$$\frac{AK}{A'K'}=\frac{KF}{K'F'}=\frac{KG}{K'G'}=\frac{FG}{F'G'}=\frac{AG}{G'A'}. \qquad \text{(Why?)}$$

$$\therefore \triangle AKG \sim A'K'G'. \qquad \text{(Why?)}$$

The tetraedrons $A-KFG$ and $A'-K'F'G'$ have all homologous triedral angles equal.

$\therefore$ tetraedron $A-KFG$ is similar to tetraedron $A'-K'F'G'$ (606)

When $A-KFG$ and $A'-K'F'G'$ are removed, the polyedrons are similar, for their faces remain similar, and the polyedral angles are equal.

In like manner any other two homologous tetraedrons can be proved similar.

Since the homologous faces of R and S will be divided into the same number of similar $\triangle$, there will be the same number of planes passed through A and A', and consequently the polyedrons will be decomposed into the same number of tetraedrons similar each to each and similarly placed. Q.E.D.

608. Cor. 1. The homologous edges of similar polyedrons are proportional.

609. Cor. 2. Any two homologous lines in two similar polyedrons have the same ratio as any two homologous edges.

610. Cor. 3. Two homologous faces of similar polyedrons are to each other as the squares of any two homologous edges.

611. Cor. 4. The total areas of two similar polyedrons are to each other as the squares of any two homologous edges.

PROPOSITION XXII. THEOREM

612. *The volumes of two similar tetraedrons are to each other as the cubes of their homologous edges.*

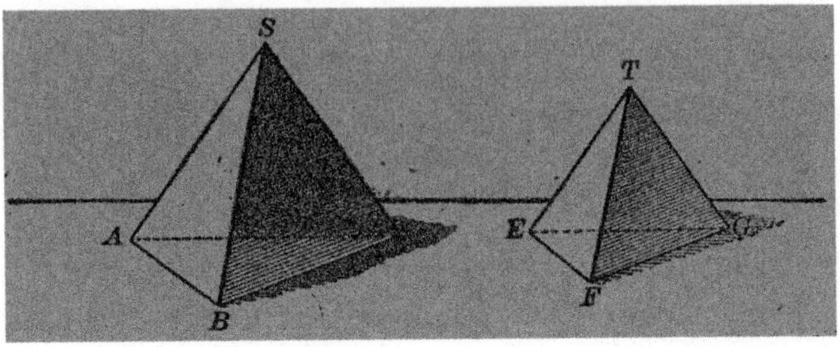

Hyp. V and V' are the volumes of the similar tetraedrons $S\text{–}ABC$ and $T\text{–}EFG$.

To prove $$\frac{V}{V'} = \frac{\overline{SB}^3}{\overline{TF}^3} = \frac{\overline{SA}^3}{\overline{TE}^3}, \text{ etc.}$$

Proof. Triedral $\angle S =$ triedral $\angle T$. (606)

$$\frac{V}{V'} = \frac{SB \times SC \times SA}{TF \times TG \times TE} \text{ or } \frac{SB}{TF} \times \frac{SC}{TG} \times \frac{SA}{TE}. \quad (605)$$

But $$\frac{SB}{TF} = \frac{SC}{TG} = \frac{SA}{TE}. \quad \text{(Why ?)}$$

$$\therefore \frac{V}{V'} = \frac{\overline{SB}^3}{\overline{TF}^3}.$$ Q E D

Ex. 1125. In the diagram for Prop. XXII, if $SA = 3$ and $TE = 2$, find the ratio of V to V'.

Ex. 1126. In the same diagram, find TF, if $SB = 2$, and $V : V' = 1 : 2$.

613. COR. The volumes of two similar polyedrons are to each other as the cubes of any two homologous edges.

Similar polyedrons can be decomposed into the same number of similar tetraedrons similarly placed which are to each other as the cubes of homologous edges.

And in a series of equal ratios, the sum of antecedents is to sum of consequents as any antecedent is to its consequent.

REGULAR POLYEDRONS

614. DEF. A **regular polyedron** is a polyedron whose faces are equal regular polygons, and whose polyedral angles are equal.

PROPOSITION XXIII. PROBLEM

615. *To determine the number of regular convex polyedrons possible:*

(a) A convex polyedral angle must have at least three faces.

(b) The sum of its face angles must be less than 360°. (523)

1. A convex polyedral angle may be formed by combining **three, four,** or **five equilateral triangles.** Since each angle of an equilateral △ is 60°, the sum of six such angles is 360°, a sum greater than that of the face angles of a convex polyedral angle (*b*).

Hence three regular convex polyedrons are possible with equilateral ⃤ as faces.

2. A convex polyedral angle may be formed by combining **three squares.**

(*Why cannot four squares be used?*)

Hence one regular convex polyedron is possible with squares as faces.

3. A convex polyedral angle may be formed by combining **three regular pentagons.**

(*Why cannot four be used?*)

Hence one regular convex polyedron can be formed having regular pentagons as faces.

4. (*Why cannot a convex polyedral angle be formed by combining regular hexagons, heptagons, or octagons?*)

Hence only five regular convex polyedrons are possible, **tetraedrons, octaedrons,** and **icosaedrons,** having *equilateral triangles* as faces; **hexaedrons** having *squares* as faces; and **dodecaedrons,** having regular **pentagons** as faces.

616. To construct the regular polyedrons, draw on stiff paper or cardboard the following diagrams. Cut partly through the paper along the dotted lines. Fold over and hold the edges in contact by pasting strips of paper along them. (Compare diagrams on page 270.)

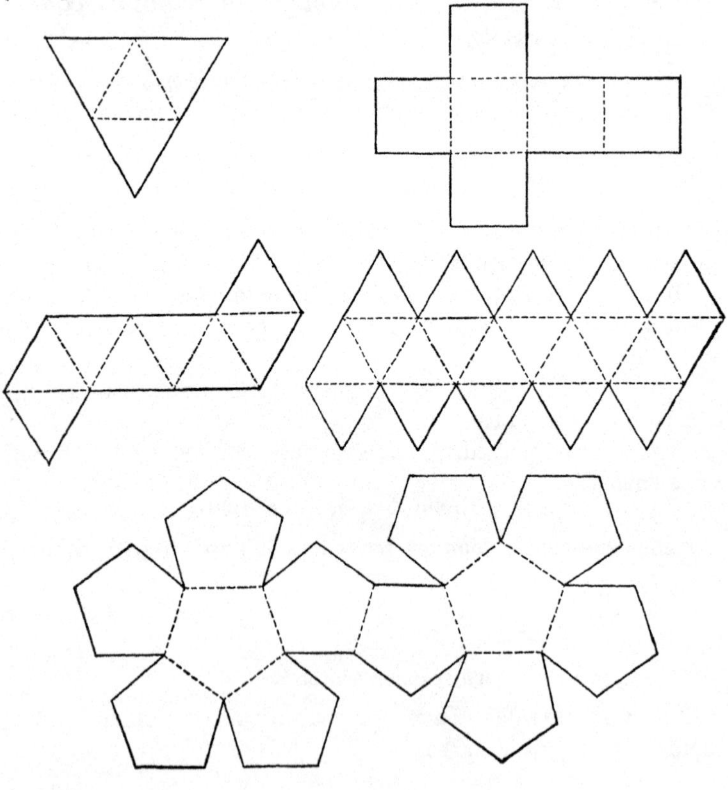

CYLINDERS

617. DEF. A **cylindrical surface** is a curved surface generated by a moving straight line which continually intersects a given

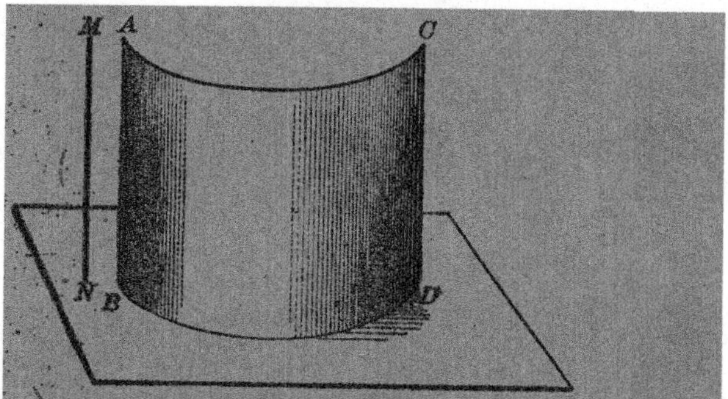

fixed curve and is always parallel to a given straight line not in the same plane with the curve.

618. DEF. The **generatrix** of the surface is the moving straight line; the **directrix** is the given curve; and an **element** of the surface is the moving line in any of its positions.

619. DEF. A **cylinder** is a solid bounded by a cylindrical surface and two parallel planes; the **bases** of a cylinder are the parallel planes; and the **lateral surface** is the cylindrical surface.

The elements of a cylinder are equal since they are ∥ lines included between ∥ planes.

620. DEF. A **circular cylinder** is a cylinder whose bases are circles.

621. DEF. A **right cylinder** is a cylinder whose elements are perpendicular to the bases.

622. Def. An **oblique cylinder** is one whose elements are oblique to the bases

623. Def. The **altitude** of a cylinder is the perpendicular distance between the bases.

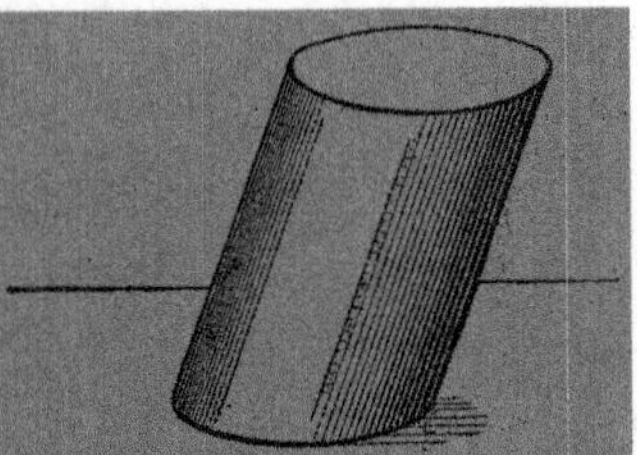

624. Def. A **cylinder of revolution** is a right circular cylinder because it may be generated by a rectangle revolving about one of its sides as an **axis.**

625. Def. **Similar cylinders of revolution** are cylinders generated by similar rectangles revolving about homologous sides as axes.

626. Def. A **tangent line to a cylinder** is a straight line, which touches the lateral surface in one point but does not intersect it. A **tangent plane** to a cylinder is a plane which contains one element of the cylinder and but one, and does not intersect the cylinder.

627. Def. A prism is **inscribed** in a cylinder when its lateral edges are elements of the cylinder and its bases are inscribed in the bases of the cylinder.

628. DEF. A prism is **circumscribed** about a cylinder when its lateral edges are parallel to the elements of the cylinder and its bases are circumscribed about the bases of the cylinder.

629. DEF. A **section of a cylinder** is the figure formed when the cylinder is intersected by a plane; a **right section** is a section formed by a plane perpendicular to the elements.

PROPOSITION XXIV. THEOREM

630. *Every section of a cylinder made by a plane passing through an element is a parallelogram.*

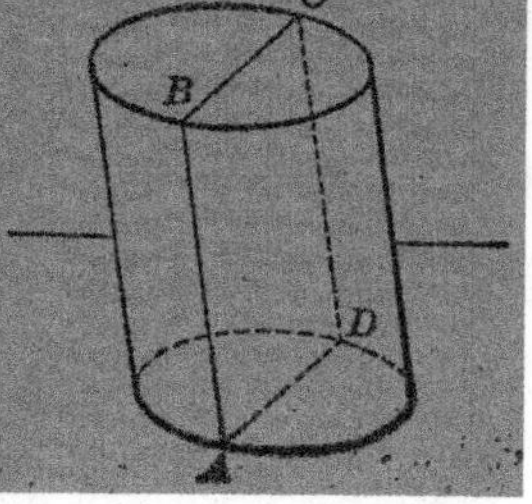

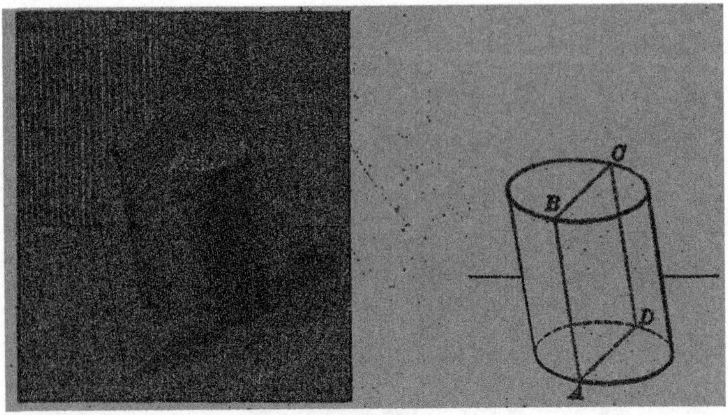

Hyp. *ABCD* is a section of cylinder *AC*, made by plane through element *AB*.

To prove *ABCD* is a parallelogram.

Proof. Any straight line through *D* in plane *AC* ∥ to *AB* is an element of the cylindrical surface. (Why?)

Since this line is in the plane *AC* and is an element of the cylindrical surface, it must be their intersection, and therefore coincides with *DC*.

∴ *DC* is a straight line ∥ to *AB*.

Also *AD* is a straight line ∥ to *BC*. (Why?)

∴ *ABCD* is a parallelogram. Q.E.D.

631. Cor. Every section of a right cylinder made by a plane passing through an element is a rectangle.

Ex. 1127. The altitude of a right cylinder is 12 inches and the radius of the base is 6 inches. Find the area of a section made by a plane passing through an element and perpendicular to a radius at a point whose distance from the center measured on this radius is

(*a*) two inches. (*b*) five inches.

Ex. 1128. What is the area of the figure formed in the preceding exercise when the plane passes through the center of the base?

PROPOSITION XXV. THEOREM

632. *The bases of a cylinder are equal.*

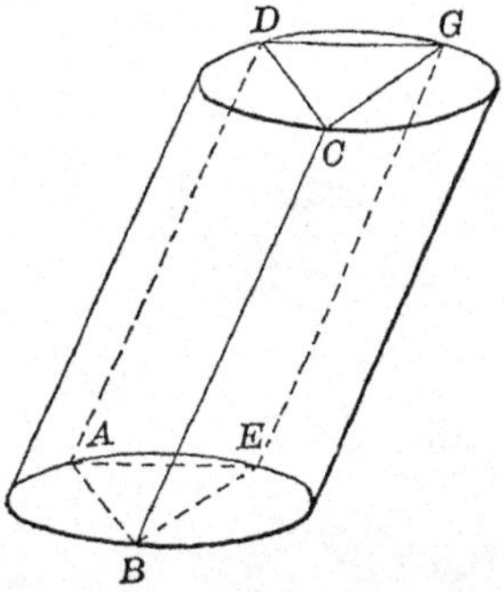

Hyp. ABE and DCG are the bases of the cylinder BG.

To prove Base ABE = base DCG.

Proof. A, B, E, are any three points in lower base of cylinder, AD, BC, and EG are elements.

Draw AE, AB, BE, DG, DC, and CG.

The figures AC, BG, and ED are parallelograms. (Why?)

Hence $AB = DC$, $BE = CG$, $AE = DG$.

$$\therefore \triangle ABE = \triangle DGC.$$

Place the lower base on the upper base so that the equal △ coincide. Then the bases will coincide, because the points A, B, and E are *any* three points in the perimeter of the lower base, and therefore every point in the perimeter of the lower base will fall in the perimeter of the upper base.

∴ the bases are equal. Q.E.D.

633. COR. 1. Any two parallel sections cutting all the elements of a cylinder are equal.

HINT. — (*What is the solid included between these sections?*)

634. COR. 2. Any section of a cylinder parallel to the base is equal to the base.

PROPOSITION XXVI. THEOREM

635. *The lateral area of a cylinder is equal to the product of the perimeter of a right section of the cylinder by an element.*

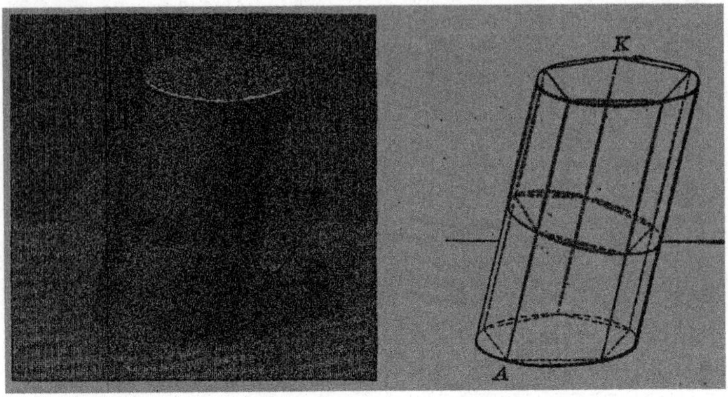

Hyp. S is the lateral area, P the perimeter of a right section, and E an element of the cylinder AK; S' is the lateral area, P' the perimeter of a section of a prism with a regular polygon as base, inscribed in cylinder AK.

To prove $S = P \times E$.

Proof. The edge of the inscribed prism coincides with an element of the cylinder. (Why ?)

$\therefore S' = P' \times E.$ (547)

If the number of faces of the inscribed prism be increased, S' will approach S as a limit, and P' will approach P as a limit.

But $S' = P' \times E.$

$\therefore S = P \times E.$ (213)

636. COR. 1. The lateral area of a cylinder of revolution is the product of the circumference of its base by its altitude.

637. Cor. 2. If S denote the lateral area, T the total area, H the altitude, and R the radius of a cylinder of revolution,

$$S = 2\,\pi RH, \text{ and}$$
$$T = 2\,\pi R\,(H + R).$$

Proposition XXVII. Theorem

638. *The volume of a circular cylinder is equal to the product of its base by its altitude.*

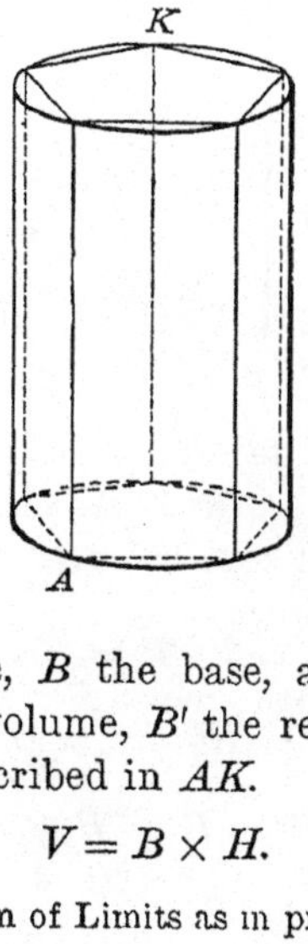

Hyp. V the volume, B the base, and H the altitude, of cylinder AK; V' the volume, B' the regular polygon forming the base of a prism inscribed in AK.

To prove $V = B \times H$.

Hint.—Use the Theorem of Limits as in preceding proposition.

639. Cor. For a cylinder of revolution with radius of base R,

$$V = \pi R^2 \times H.$$

Ex. 1129. Two cylinders of revolution have equal altitudes and their radii are respectively 3 and 4. Find a third cylinder of revolution of the same altitude and equivalent to the sum of the two given cylinders.

Proposition XXVIII. Theorem

640. *The lateral areas, or the total areas, of similar cylinders of revolution are to each other as the squares of their radii, or as the squares of their altitudes; and their volumes are to each other as the cubes of their radii, or as the cubes of their altitudes.*

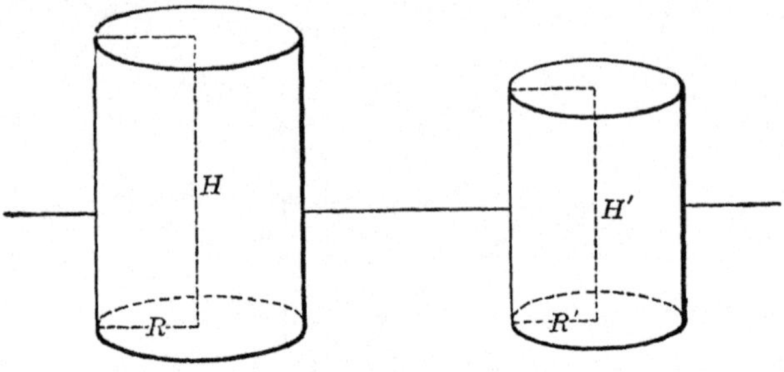

Hyp. S, S', are the lateral areas; T, T', the total areas, V, V', the volumes, R, R', the radii, and H, H', the altitudes of two similar cylinders of revolution.

To prove $S : S' = T : T' = R^2 : R'^2 = H^2 : H'^2$,

and $V : V' = R^3 : R'^3 = H^3 : H'^3$.

Proof. $$\frac{H}{H'} = \frac{R}{R'} = \frac{H+R}{H'+R'}, \qquad \text{(Why?)}$$

$$\frac{S}{S'} = \frac{2\pi RH}{2\pi R'H'} = \frac{R}{R'} \times \frac{H}{H'} = \frac{R^2}{R'^2} = \frac{H^2}{H'^2}, \qquad \text{(Why?)}$$

and $$\frac{T}{T'} = \frac{2\pi R(H+R)}{2\pi R'(H'+R')} = \frac{R}{R'} \times \frac{H+R}{H'+R'} = \frac{R^2}{R'^2} = \frac{H^2}{H'^2}, \qquad \text{(Why?)}$$

and $$\frac{V}{V'} = \frac{\pi R^2 H}{\pi R'^2 H'} = \frac{R^2}{R'^2} \times \frac{H}{H'} = \frac{R^3}{R'^3} = \frac{H^3}{H'^3}. \qquad \text{(Why?)}$$ Q.E.D.

CONES

641. Def. A **conical surface** is a surface generated by a moving straight line which continually intersects a given fixed curve and constantly passes through a fixed point not in the same plane with the curve.

642. Def. The **generatrix** of the surface is the moving straight line; the **directrix** is the given curve; and the **vertex** is the fixed point.

643. Def. An **element** of a conical surface is the generatrix in any position.

644. Def. The **upper and lower nappes** are the portions of the conical surface formed above and below the vertex when the length of the generatrix is unlimited.

645. Def. A **cone** is a solid bounded by a conical surface and a plane cutting all its elements.

646. Def. The **lateral area of the cone** is the conical surface; the **base of the cone** is the plane surface; the **vertex of the cone** is the vertex of the conical surface; and the **elements of the cone** are the elements of the conical surface.

647. Def. A **circular cone** is a cone whose base is a circle; the **axis** of a circular cone is the straight line joining the vertex and the center of the base.

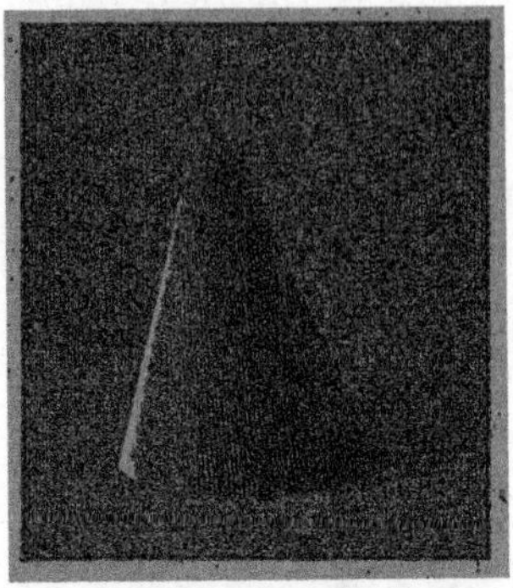

648. Def. A **right circular cone** is one whose axis is perpendicular to the base; an **oblique circular cone** is one whose axis is not perpendicular to the base.

Note. — In this book only circular cones are treated of.

649. Def. A **cone of revolution** is a right circular cone, for this latter may be generated by a right triangle revolving about one of its arms as an axis.

650. Def. The **altitude** of a cone is the perpendicular distance from the vertex to the plane of the base.

651. Def. **Similar cones of revolution** are cones generated by the revolution of similar right triangles about homologous arms.

652. Def. A **tangent line to a cone** is a line which touches the cone in one point only and does not intersect it.

653. Def. A **tangent plane to a cone** is a plane which contains one element of the cone and but one, and does not intersect the cone.

654. Def. A pyramid is **inscribed in a cone** when its base is inscribed in the base of the cone and its vertex coincides with the vertex of the cone.

655. A pyramid is circumscribed about a cone when its base is circumscribed about the base of the cone and its vertex coincides with the vertex of the cone.

656. Def. A **frustum** of a cone is the portion included between its base and a plane parallel to the base. The **lower base** of the frustum is the base of the cone, and the **upper base** is the section made by the plane.

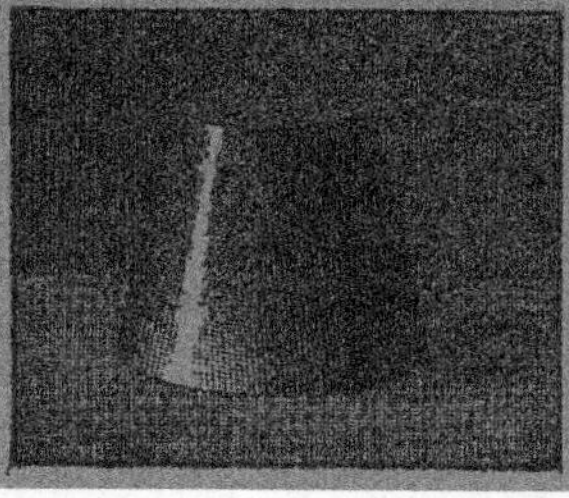

PROPOSITION XXIX. THEOREM

657. *Every section of a cone made by a plane passing through its vertex is a triangle.*

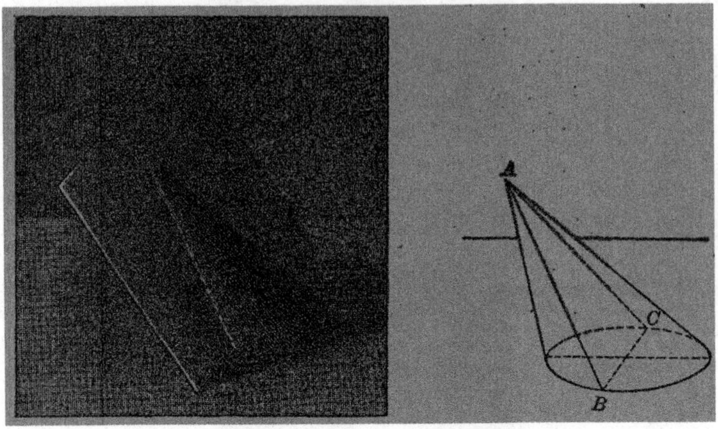

Hyp. ABC is a section of the cone made by plane passing through vertex A.

To prove ABC a triangle.

Proof. BC is a straight line. (Why ?)

AB and AC are elements of the conical surface.

$\therefore$ AB and AC are straight lines.

AB and AC are in the cutting plane. (Why ?)

$\therefore$ they are intersections of the plane and the conical surface. (Why ?)

$\therefore$ ABC, the section, is a triangle. Q.E.D.

658. COR. Every section of a right cone made by a plane passing through its vertex is an isosceles triangle.

Proposition XXX. Theorem

659. *Every section of a circular cone made by a plane parallel to the base is a circle.*

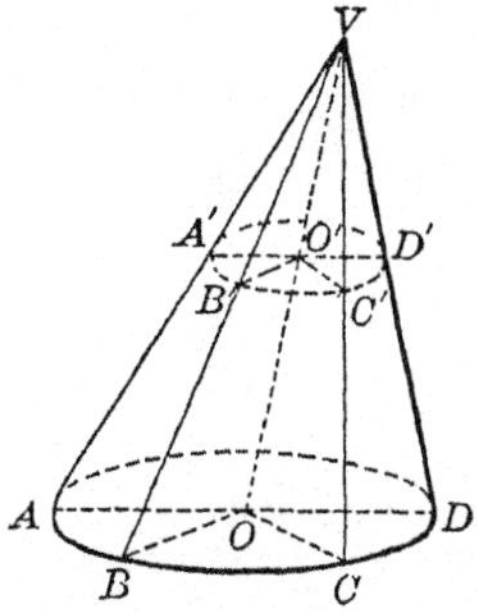

Hyp. $A'B'C'D'$ is a section of cone $V\text{-}ABCD$ made by a plane ∥ to $ABCD$, O the center of the base, O' the point where axis VO intersects the section. OA, OD, and $O'A'$, $O'D'$ are intersections of a plane through V and any element VA with $ABCD$ and $A'B'C'D'$ respectively.

To prove $A'B'C'D'$ a circle.

Proof. $O'A'$ and $O'D'$ are ∥ to OA and OD respectively. (Why?)

$\therefore \triangle VO'A'$ and $\triangle VOA$ are similar. (Why?)

$$\therefore \frac{O'A'}{OA} = \frac{VO'}{VO} = \frac{O'D'}{OD}.$$

But $OA = OD$. (Why?)

$\therefore O'A' = O'D'$.

$\therefore A'B'C'D'$ is a circle. Q.E.D.

660. Cor. 1. The axis of a circular cone passes through the center of every section which is parallel to the base, or

The locus of the centers of the sections of a circular cone made by planes parallel to the base is the axis of the cone.

661. Cor. 2. Sections made by planes parallel to the bases of a circular cone are to each other as the squares of their radii, or as the squares of their distances from the vertex of the cone.

Proposition XXXI. Theorem

662. *The lateral area of a cone of revolution is equal to half the product of the slant height by the circumference of the base.*

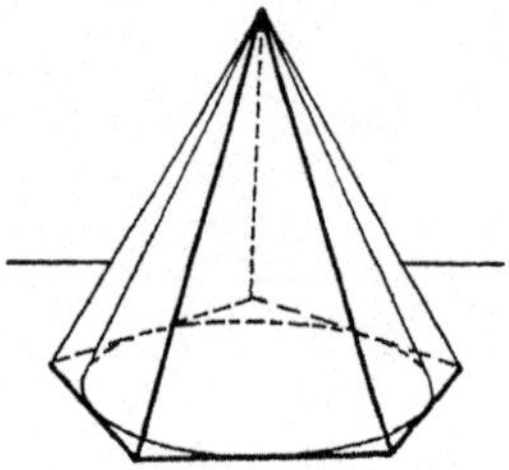

Hyp. S is lateral area, C the circumference of the base, and L the slant height of the cone; S' the lateral area, P the perimeter of the regular polygon forming the base of a circumscribed pyramid.

To prove $S = \frac{1}{2} C \times L$.

Hint. — Circumscribe a pyramid; its slant height is L. Use Theorem of Limits.

663. Cor. If S is the lateral area, T the total area, H the altitude, L the slant height, R the radius of the base, of a cone of revolution,

$$S = \pi RL.$$

$$T = \pi R(L + R).$$

Ex. 1130. Find the lateral area of a cone of revolution if the hypotenuse of the generating triangle be 10 inches and the acute angles be 45° each.

PROPOSITION XXXII. THEOREM

664. *The volume of a circular cone is equal to one-third the product of its base by its altitude.*

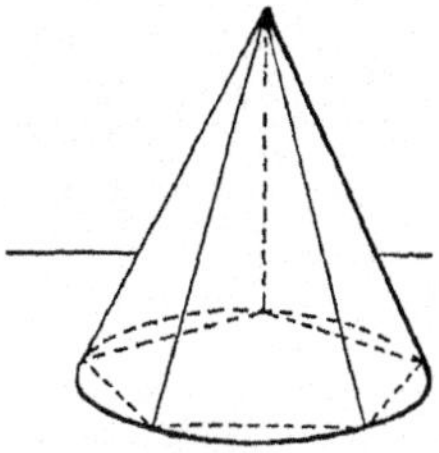

Hyp. V is the volume, B the base, and H the altitude of the cone; V' the volume, B' the regular polygon forming the base of an inscribed pyramid.

To prove $V = \frac{1}{3} B \times H.$

$V' = \frac{1}{3} B' \times H.$ (594)

HINT. — Use Theorem of Limits.

665. COR. If the cone is a cone of revolution, with R as radius of B, then

$$V = \frac{1}{3} \pi R^2 H.$$

Ex. 1131. Find the lateral area of a cone of revolution whose radius is 4 and whose altitude is 3.

Ex. 1132. Find the volume of a cone of revolution with radius 6 and altitude 2.

Ex. 1133. Find the volume of a cone of revolution whose radius is 5 and whose slant height is 13.

Ex. 1134. Find the lateral surface of a cone of revolution if its volume is 314 and its altitude is 2.

PROPOSITION XXXIII. THEOREM

666. *The lateral areas, or the total areas, of two similar cones of revolution are to each other as the squares of their altitudes, as the squares of their radii, or as the squares of their slant heights; and their volumes are to each other as the cubes of their altitudes, as the cubes of their radii, or as the cubes of their slant heights.*

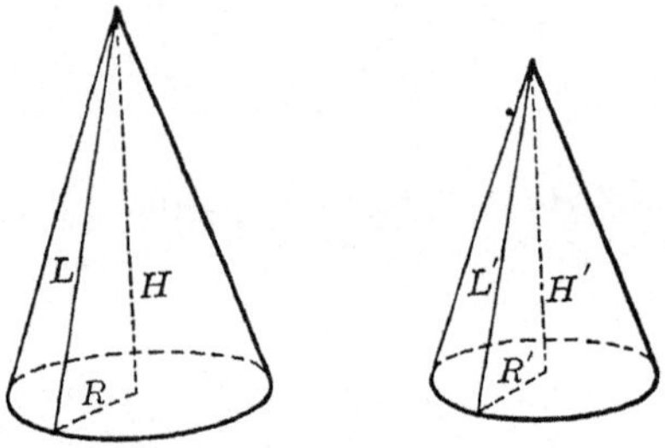

Hyp. S, S' are the lateral areas, T, T' the total areas, V, V' the volumes, H, H' the altitudes, R, R' the radii, L, L' the slant heights of two similar cones of revolution.

To prove $S : S' = T : T' :: H^2 : H'^2 = R^2 : R'^2 = L^2 : L'^2$

and $V : V' = H^3 : H'^3 = R^3 : R'^3 = L^3 : L'^3$

The proof is similar to that of Prop. XXVIII.

Ex. 1135. Find the ratio of the total areas of two similar cones of revolution whose altitudes are

(*a*) 12 inches and 24 inches respectively.
(*b*) 5 inches and *d* inches respectively.

Ex. 1136. The volume of a cone of revolution is 343 cubic inches and its altitude is 5 inches. Find volume of a similar cone of altitude

(*a*) 7 inches.
(*b*) 15 inches.

PROPOSITION XXXIV. THEOREM

667. *The lateral area of a frustum of a cone of revolution is equal to one-half the sum of the circumferences of its bases multiplied by its slant height.*

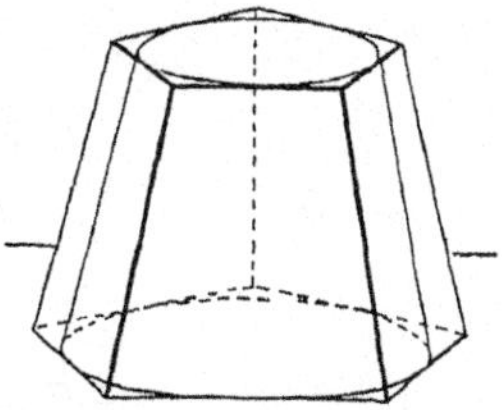

Hyp. S is the lateral area, C and C' the circumferences of the bases, R and R' their radii, and L the slant height of the frustum; S' the lateral area, P and P' the perimeters of the bases, of a frustum of a regular pyramid circumscribed about the cone.

To prove $S = \frac{1}{2}(C + C') \times L.$

Proof. The slant height of the frustum of the circumscribed pyramid $= L$, (Why ?)

then $S' = \frac{1}{2}(P + P') \times L.$ (Why ?)

[To be completed by the student. HINT.—Use Theorem of Limits.]

NOTE.—It can be shown that the limit of the sum of a finite number of variables is equal to the sum of their respective limits.

668. COR. The lateral area of a frustum of a cone of revolution is equal to the circumference of a section equidistant from its bases multiplied by its slant height. For

$$C = 2\pi R, \text{ and } C' = 2\pi R', \text{ and } 2\pi\left(\frac{R + R'}{2}\right)$$

is the circumference of a section equidistant from the bases.

PROPOSITION XXXV. THEOREM

669. *The volume of a frustum of a circular cone is equivalent to the sum of its bases, and the mean proportional between its bases, multiplied by one-third its altitude.*

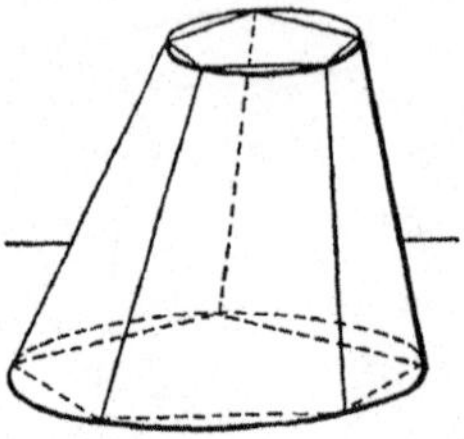

Hyp. V is the volume, B the lower base, b the upper base, H the altitude of the frustum, V' the volume, B' the lower base, b' the upper base, of an inscribed frustum of a pyramid with regular polygons as bases.

To prove $V = \frac{1}{3} H (B + b + \sqrt{B \times b})$.

Proof. The altitude of the frustum of the pyramid $= H$.

Then $V' = \frac{1}{3} H (B' + b' + \sqrt{B' \times b'})$. (601)

By increasing the number of the lateral faces of the inscribed frustum indefinitely,

V' becomes a variable with V as a limit,

B' approaches B as a limit,

b' approaches b as a limit, and $B' \times b'$, $B \times b$ as a limit.

$\therefore B' + b' + \sqrt{B' \times b'}$ approaches

$B + b + \sqrt{B \times b}$ as a limit.

$\therefore V = \frac{1}{3} H (B + b + \sqrt{B \times b})$. Q.E.D.

NOTE. — See note, p. 49. It can also be shown that the limit of the product of two or more variables is the product of their respective limits when no one of these limits is zero.

670. Cor. If the frustum is that of a cone of revolution, with R and R' as radii of its bases,

$$V = \frac{1}{3}\pi H(R^2 + R'^2 + RR').$$

Note. — The preceding proposition may be stated: —

The volume of a frustum of a circular cone is equivalent to the sum of the volumes of three cones whose common altitude is the altitude of the frustum, and whose bases are the lower base, the upper base, and the mean proportional between the bases of the frustum. For, the value of V may be written,

$$\frac{1}{3}H \times B + \frac{1}{3}H \times b + \frac{1}{3}H \times \sqrt{B \times b}.$$

MISCELLANEOUS EXERCISES

Ex. Find the lateral edge, lateral area, and volume of a regular triangular pyramid each side of whose base is 9 and whose altitude is 3.

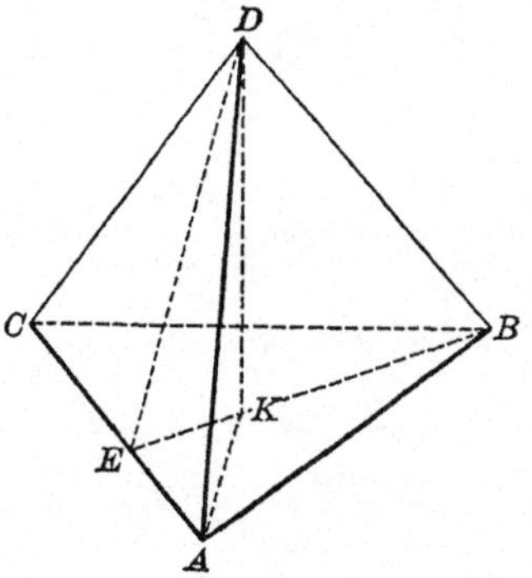

Let K be the center of the base ABC of the regular triangular pyramid D-ABC. Draw the median BE and the altitude DK. Join A and K, and D and E.

$$AK = \frac{2}{3}\left(\text{alt. of equil. } \triangle \text{ of side } 9\right) = \frac{2}{3}\left(\frac{9\sqrt{3}}{2}\right) = 3\sqrt{3}.$$

$$DA = \sqrt{\overline{AK}^2 + \overline{DK}^2} = \sqrt{(3\sqrt{3})^2 + 9} = 6.$$

$$DE = \sqrt{\overline{DA}^2 - \overline{EA}^2} = \sqrt{36 - \tfrac{81}{4}} = \tfrac{3}{2}\sqrt{7}.$$

$$\text{Lat. Area} = 3 \times 9 \times \frac{DE}{2} = \frac{81\sqrt{7}}{4}.$$

Volume = area of base × alt. ÷ 3

$$= \frac{81}{4}\sqrt{3} \times 3 - 3 = \frac{81\sqrt{3}}{4}.$$

Ex. Find the lateral edge, lateral area, and volume of a frustum of a regular hexagonal pyramid, the sides of whose bases are 14 and 6, respectively, and whose altitude is 8.

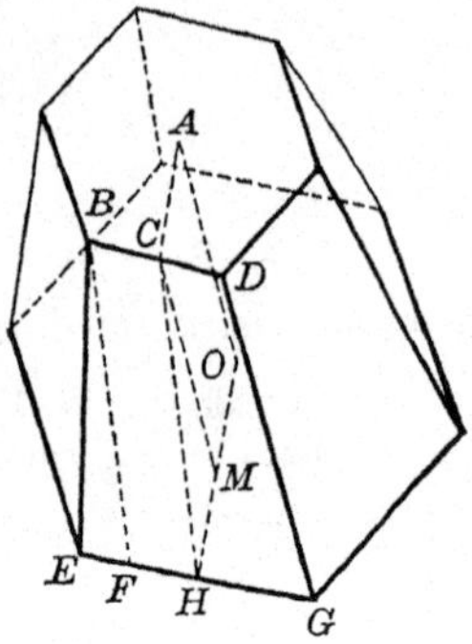

Let A and O be the centers of the upper and lower bases, respectively, of the frustum of a regular hexagonal pyramid. Draw AC and OH perpendicular to BD and EG, respectively. Draw BF perpendicular to EG.

$$HM = OH - AC = 7\sqrt{3} - 3\sqrt{3} = 4\sqrt{3}.$$

$$CH = \sqrt{\overline{CM}^2 + \overline{HM}^2} = \sqrt{64 + 48} = 4\sqrt{7}.$$

$$BE = \sqrt{\overline{BF}^2 + \overline{FE}^2} = \sqrt{\left(4\sqrt{7}\right)^2 + \left(\frac{EG - BD}{2}\right)^2} = 8\sqrt{2}.$$

$$\text{Lat. area} = 6\left[(14 + 6)\frac{4\sqrt{7}}{2}\right] = 240\sqrt{7}.$$

$$\text{Volume} = (B + b + \sqrt{B \times b})\frac{\text{alt.}}{3}$$

$$= (294\sqrt{3} + 54\sqrt{3} + 126\sqrt{3})\tfrac{8}{3} = 1264\sqrt{3}.$$

Find lateral edge, area, and volume

Ex. 1137. Of a regular triangular pyramid, each side of whose base is 10 and whose altitude is 18.

Ex. 1138. Of a regular quadrangular pyramid, each side of whose base is 6 and whose altitude is 12.

Ex. 1139. Of a regular hexagonal pyramid, each side of whose base is 6 and whose altitude is 16.

Ex. 1140. Of a frustum of a regular triangular pyramid, the sides of whose bases are 12 and 4, and whose altitude is 15.

Ex. 1141. Of a frustum of a regular quadrangular pyramid, the sides of whose bases are 12 and 8 and whose altitude is 14.

Ex. 1142. Of a frustum of a regular hexagonal pyramid, the sides of whose bases are 10 and 5, and whose altitude is 18.

Ex. 1143. Find the volume of a truncated right triangular prism, the sides of whose base are 6, 8, and 10, and whose lateral edges are 5, 8, and 7, respectively.

Ex. 1144. The total surface of a cube is 336 sq. in. Find its volume.

Ex. 1145. The lateral area of a right prism is 140, and its base is a triangle whose sides are 5, 7, and 8. Find its volume.

Ex. 1146. Find the volume of an oblique prism whose base is a regular hexagon inscribed in a circle of radius 5 in., if a lateral edge be 8 in. and its projection on the plane of the base be 3 in

Ex. 1147. The base of a pyramid whose altitude is 24 is a square of side 10. Find the area of a section 4 in. from the vertex and parallel to the base.

Ex. 1148. How far from the vertex, on the lateral edge a, must a plane ∥ to the base be passed to divide a pyramid into two equivalent parts?

Ex. 1149. The altitude of a pyramid, having a regular hexagon, of side 8 in., for its base, is 28 in. Find the distance from the vertex of a section whose area is $\frac{3\sqrt{3}}{2}$ sq. in.

Ex. 1150. Find the surface and volume of a cube in which the diagonal of each face is 15 in.

Ex. 1151. Find the surface and volume of a cube whose diagonal is 30 in.

Ex. 1152. The right-section of a right prism is a quadrilateral $ABCD$, in which $AB = 7$ in., $BC = 20$ in., $CD = 15$ in., $DA = 24$ in., and the angles at A and C are right angles. If the height of the prism is 12 in., find its entire surface and volume.

Ex. 1153. Find the lateral area of a right pyramid having the same base and height as a cube whose edge is 10 in.

Ex. 1154. The diagonals of a rectangular parallelopiped are equal.

Ex. 1155. Find the lateral area of the frustum of a pyramid the altitude of which is 15 in., and the bases are squares of sides 40 in. and 24 in., respectively.

Ex. 1156. A cone 12 in. in height and 16 in. in diameter at the base is cut by a plane parallel to the base and 9 in. from it. Find the lateral area and volume of the frustum so formed.

Ex. 1157. The midpoints of two pairs of opposite edges of a tetraedron determine a parallelogram.

Ex. 1158. The length of a right prism is 10 in., and the base is a regular hexagon of side 8 in. Find the surface and volume of the greatest possible cylinder, of the same axis, that can be cut from the prism.

Ex. 1159. The base of a right pyramid is a regular hexagon of side 20 in., and the lateral faces are inclined to the base at an angle of 60°. Find the volume.

Ex. 1160. Lines joining the midpoints of opposite edges of a tetraedron meet in a point and bisect each other.

Ex. 1161. The altitude of a cone of revolution is 27 in., and its curved surface is 7 times the area of its base. Find the radius of the base.

Ex. 1162. The sum of the squares of the four diagonals of a parallelopiped is equal to the sum of the squares of the twelve edges. (Ex. 663.)

Ex. 1163. In any tetraedron the straight lines which join the intersections of the medians of any face with the opposite vertex meet in a point which divides each line in the ratio 1 : 3.

Ex. 1164. Find the edge of the greatest cube that can be cut from a right pyramid a inches high, and having a square base of side b inches, one face of the cube being in the plane of the base of the pyramid.

Ex. 1165. Find the total surface of a regular tetraedron, if the perpendicular from one vertex to the opposite face is 5 inches.

Ex. 1166. The ends of a frustum of a cone are respectively 8 inches and 2 inches in diameter. If the lateral area is equal to the area of a circle whose radius is 5 inches, find the altitude of the frustum.

Ex. 1167. The corners of a cube are cut off by planes which pass through the middle points of each set of edges meeting at a common vertex. If the edge of the cube is 2 feet, find the volume of the solid formed.

BOOK VIII

THE SPHERE

671. Def. A **sphere** is a solid bounded by a surface, all the points of which are equally distant from a point within called the *center*.

672. Def. The **radius** of a sphere is a straight line drawn from the center to any point in the surface.

673. Def. The **diameter** of a sphere is a straight line passing through the center and terminated at either end by the surface.

674. From the definitions it follows that

(1) *All the radii of a sphere* are equal, and all diameters are equal.

(2) A semicircle rotating about its diameter generates a sphere.

(3) Two spheres are equal if their radii are equal, and conversely.

(4) A point is without a sphere if its distance from the center is greater than the radius.

Ex. 1168. The radii of two spheres are respectively 10 in. and 4 in., their lines of centers (*i.e.* the line joining their centers) is 7 in. Is every point of the smaller sphere lying within the larger one?

PROPOSITION I. THEOREM

675. *Every section of a sphere made by a plane is a circle.*

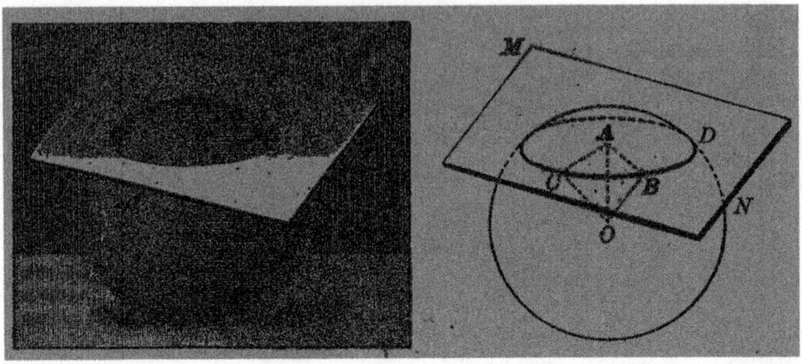

Hyp. *CBD* is the intersection of plane *MN*, and a sphere whose center is *O*.

To prove *CBD* is a circle.

Proof. From *O* draw $OA \perp$ to *MN*.
Draw *CA*, *AB*, *CO*, and *OB*.

Since $OC = OB$,

$CA = AB$. (474)

Since any two points *B* and *C* in *CBD* are equidistant from *A*, all points must be equidistant from *A*.
Or *CBD* is a circle. Q.E.D.

676. COR. 1. A circle nearer to the center of a sphere is greater than one more remote.
For since $\overline{AC}^2 = \overline{OC}^2 - \overline{AO}^2$, *AC* is the smaller, the greater *AO*.

677. DEF. **A great circle of a sphere** is a section made by a plane passing through the center.

678. Def. **A small circle of a sphere** is a section made by a plane not passing through the center.

679. Def. The **axis of a circle of a sphere** is the diameter perpendicular to the plane of the circle; its ends are the **poles** of the circle.

680. Cor. 2. The axis of a circle passes through the center of the circle, and conversely.

681. Cor. 3. All great circles of a sphere are equal.

682. Cor. 4. Any two great circles of a sphere bisect each other.

For since the plane of each contains the center of the sphere, their intersection is a diameter and bisects both circles.

683. Cor. 5. Every great circle bisects the sphere.

684. Cor. 6. One and only one circle may be drawn through any three points in the surface of a sphere. (A plane is determined by three points.)

685. Cor. 7. A great circle may be drawn through two points B and C in the surface of a sphere. (Three points, B, C, and the center O, determine a plane.)

Generally there is only one great circle which passes through two given points, but if the given points are ends of a diameter, any number of great circles can be passed through these points.

686. Def. The **distance between two points on the surface of a sphere** is the length of the minor arc of a great circle between them.

Ex. 1169. What is the radius of a small circle, if the distance of its plane from the center of the sphere is 9 in., and the radius of the sphere is 15 in.

PROPOSITION II. THEOREM

687. *All points in the circumference of a circle of a sphere are equidistant from a pole of the circle.*

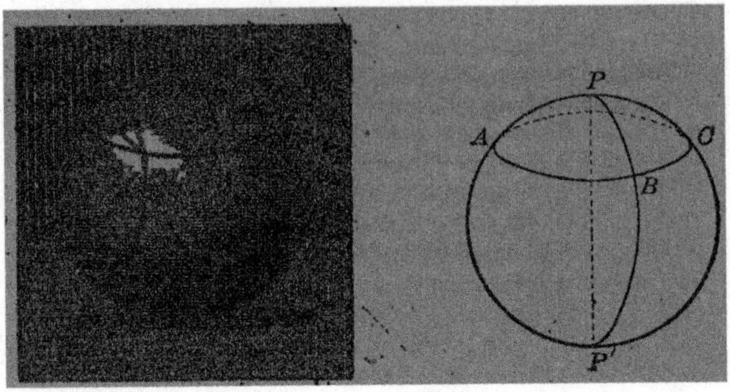

Hyp. P and P' are the poles of circle ABC of a sphere.

To prove arc $PA =$ arc $PB =$ arc PC,

and arc $P'A =$ arc $P'B =$ arc $P'C$.

HINT. — Prove by means of (473) the equality of straight lines PA, PB, and PC.

688. DEF. The **polar distance of a circle of a sphere** is the distance of a point in the circumference from the nearer pole.

689. SCHOLIUM. A **quadrant** in Spherical Geometry is the fourth part of the circumference of a great circle.

690. COR. The polar distance of a great circle is a quadrant.

Ex. 1170. The polar distance of a circle of a sphere is 60°, and the radius of the sphere is 13 in. Find

(*a*) the distance of its plane from the center.
(*b*) the radius of the circle.

PROPOSITION III. THEOREM

691. *On the surface of a sphere, a point at a quadrant's distance from two other points, not the extremities of a diameter, is the pole of a great circle passing through these two points.*

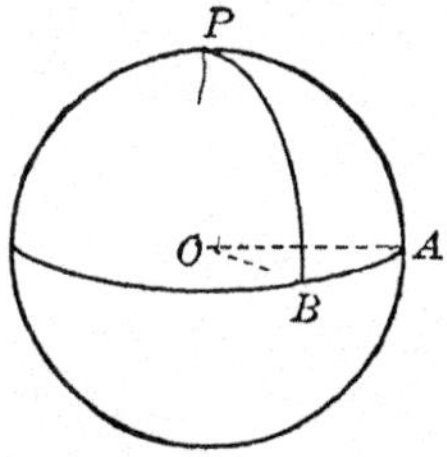

Hyp. P, A, and B are three points on the surface of a sphere, and PA and PB are quadrants.

To prove P is the pole of a great circle AB.

HINT. — Prove that a line OP is perpendicular to plane ABO by means of (461).

692. SCHOLIUM. Theorem III enables us to construct a great circle through two points on the surface of a material sphere by means of the compasses.

From the given points A and B as centers draw arcs with a radius equal to the chord of a quadrant, intersecting in P. From P, with the same radius, draw a circle which is the required one.

693. DEF. **A plane is tangent to a sphere** when it has one and only one point common with the surface of a sphere.

694. DEF. **A line is tangent to a sphere** when it has one and only one point common with the surface of a sphere.

695. Def. **Two spheres are tangent** if their surfaces have one and only one point common.

696. Def. A **sphere is inscribed in a polyedron** if all the faces of the polyedron are tangent to the sphere.

697. Def. A **sphere is circumscribed about a polyedron** if all the vertices of the polyedron lie in the surface of the sphere.

Proposition IV. Theorem

698. *A plane perpendicular to a radius of a sphere at its extremity is tangent to the sphere.*

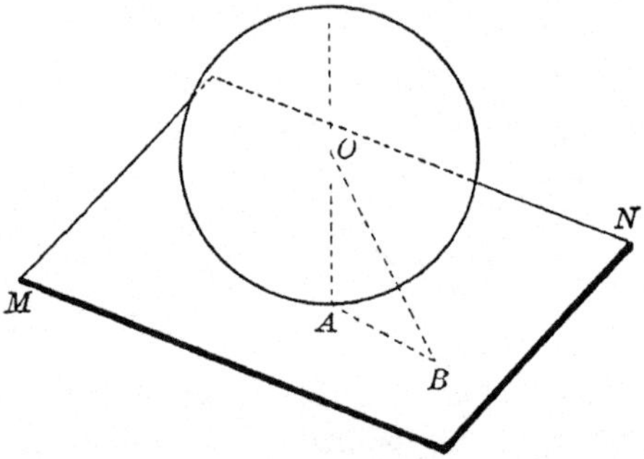

Hyp. Plane MN is $\perp$ to radius OA of sphere O at its extremity A.

To prove MN is tangent to the sphere.

Proof. Join any other point B in MN to A and O.

Then $OB > AO$. (476)

Hence B lies without the sphere.

Therefore, every point of MN, except A, lies without the sphere, and MN is tangent to the sphere. Q.E.D.

699. Cor. 1. Any straight line perpendicular to the radius of a sphere at its extremity is a tangent to the sphere.

700. Cor. 2 Any line or plane tangent to a sphere is perpendicular to the radius drawn to the point of contact.

Proposition V. Theorem

701. *A sphere may be circumscribed about any tetraedron.*

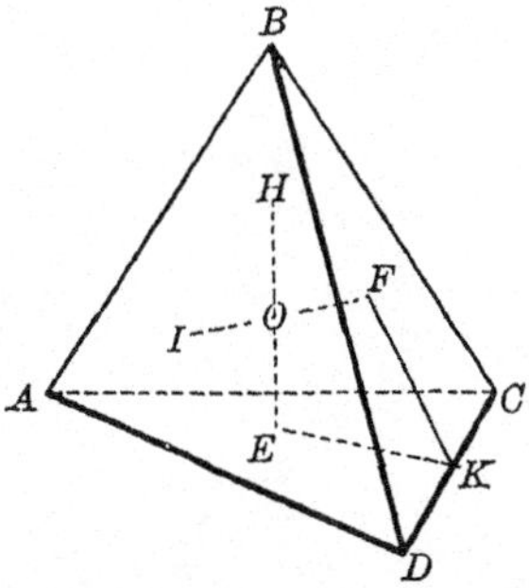

Hyp. *ABCD* is a tetraedron.

To prove a sphere may be circumscribed about *ABCD*.

Proof. Let *E* and *F* be respectively the centers of the circumscribed circles of ⧍ *ADC* and *BCD*.

Draw *EH* ⊥ to plane *ADC*, and *FI* ⊥ to plane *BCD*. Join *K*, the midpoint of *DC*, to *F* and *E*.

Then *KF* and *KE* are ⊥ to *DC*. (111)

Hence plane *FKE* is ⊥ to *DC*. (461)

∴ plane *FKE* is ⊥ to *ACD*. (502)

∴ *HE* lies in plane *FKE*, (504)

and in like manner *IF* lies in this plane.

Hence HE and IF must meet in some point, as O.

Since O is in the perpendicular EH, it must be equidistant from A, D, and C. (473)

And since it is in FI it is equidistant from B, C, and D. (473)

Hence O is equidistant from A, B, C, and D, and a sphere described from O as a center with a radius equal to AO will be circumscribed about the tetraedron.

702. Cor. Four points not in the same plane determine a sphere.

PROPOSITION VI. THEOREM

703. *A sphere may be inscribed in any given tetraedron.*

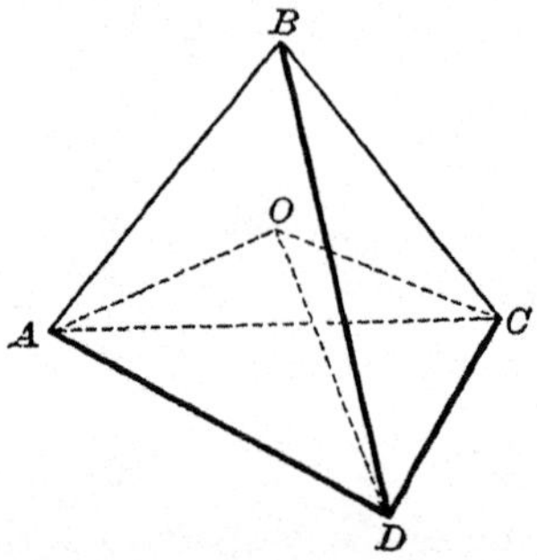

Hyp. $ABCD$ is a tetraedron.

To prove a sphere may be inscribed in $ABCD$.

Proof. Bisect any three diedral angles as AD, DC, and AC by the planes OAD, ODC, and OAC.

Then O is equidistant from the four faces of the tetraedron.

[To be completed by the student.]

704. Cor. The six planes bisecting the diedral angles of a tetraedron meet in a point.

PROPOSITION VII. PROBLEM

705. *To construct the diameter of a material sphere.*

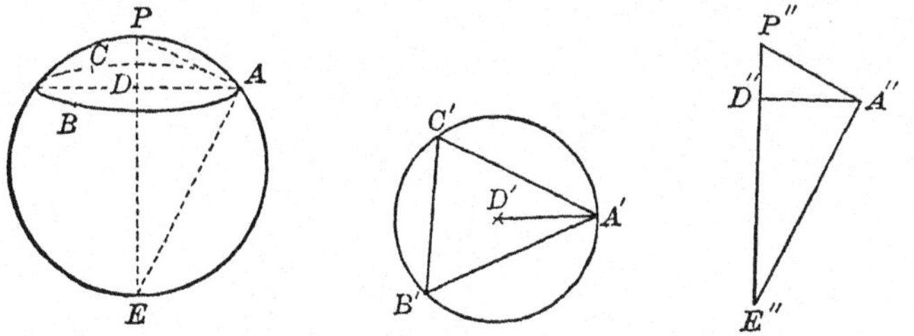

Given. A material sphere $ABCE$.

Required. The diameter of the sphere.

Construction. From any point P as center, with any [opening of compasses as] radius, draw circumference ABC.

Measure [with compasses] the three chords AB, BC, and CA, and in any plane construct $\triangle A'B'C'$, having its sides respectively equal to AB, BC, and CA.

Construct $D'A'$ the radius of the circumscribed circle $A'B'C'$.

Draw right triangle $P''A''D''$, having the hypotenuse $P''A'' = PA$, and one arm $D''A'' = D'A'$.

At A'' draw $A''E'' \perp$ to $P''A''$, meeting $P''D''$ produced in E''.

Then $P''E''$ is the required diameter.

[The proof is left to the student.]

706. DEF. The **angle between two intersecting curves** is the angle formed by the tangents at the point of contact.

707. DEF. A **spherical angle** is the angle between two intersecting great circles.

Ex. 1171. The volume of any tetraedron is equivalent to its surface multiplied by one-third the radius of the inscribed sphere.

Ex. 1172. If in the diagram for Prop. VII, $A'B\ C'$ is an equilateral triangle, find the angles between chord $A'B'$ and arc $A\ B'$.

Proposition VIII. Theorem

708. *A spherical angle is measured by the arc of a great circle described from its vertex as a pole, and included between its sides, produced if necessary.*

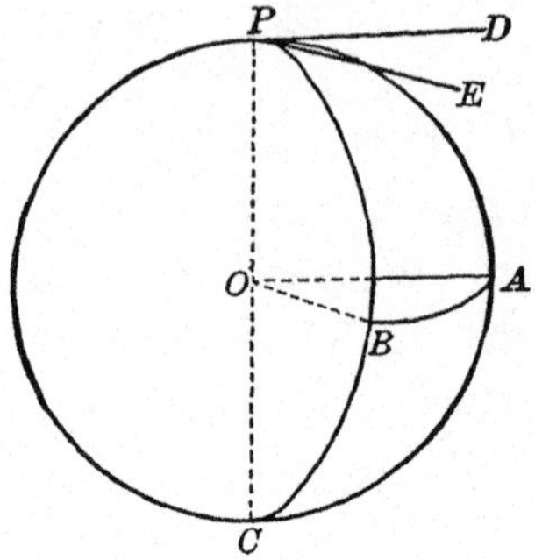

Hyp. *DPE* is the spherical angle formed by the great circles *PAC* and *PBC*. *AB* is an arc of a great circle having *P* for its pole.

To prove $\angle DPE$ is measured by arc *AB*.

Proof. Draw radii *OA* and *OB*.

$\angle POA$ is measured by quadrant *PA*.

Hence AO is $\perp$ to OP.

But DP is $\perp$ to OP, (192)

and AO and DP are in the plane of $\odot PAO$.

Hence AO is $\parallel$ to DP. (84)

Similarly BO is $\parallel$ to EP.

$\therefore \angle DPE = \angle AOB$. (468)

But $\angle AOB$ is measured by arc *AB*.

Hence $\angle DPE$ is measured by arc *AB*. Q.E.D.

709. Cor. An angle formed by two great circles is equal to the plane angle of the diedral angle formed by their planes.

SPHERICAL POLYGONS

710. Def. **A spherical polygon** is a portion of the surface of a sphere bounded by three or more arcs of great circles.

The arcs are the **sides**, their points of intersection are the **vertices**, and the spherical angles formed by the sides are the **angles** of the polygon.

711. Def. **A diagonal of a spherical polygon** is an arc joining any two non-adjacent vertices.

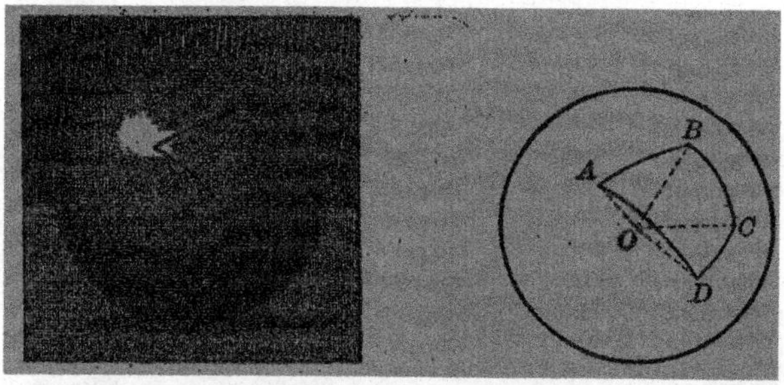

Thus, *ABCD* is a spherical polygon, *AB*, *BC*, etc., its sides, *A*, *B*, *C*, etc., its vertices, and ∠s *ABC*, *BCD*, etc., its angles.

712. Def. **A spherical triangle** is a spherical polygon of three sides. It is called **isosceles, equilateral,** etc., in the same cases in which a plane triangle would be so called.

The planes of the sides of a spherical polygon form at the center a polyedral angle $(O - ABCD)$ which is said to *correspond* with the spherical polygon.

The sides of the spherical polygon are measured by the sides of the corresponding polyedral angle, its angles are equal to the diedral angles of the corresponding polyedral angle.

713. Remark. — **By means of the relations between the parts of a spherical polygon and those of its corresponding polyedral angle, we can deduce from any theorem of polyedral angles an analogous one of spherical polygons.**

714. A spherical *polygon* is *convex* if its corresponding polyedral angle is convex. All spherical polygons are supposed to be convex polygons unless stated otherwise.

715. Spherical *polygons* are *symmetrical* if their corresponding polyedral angles are symmetrical. Evidently, their parts must be respectively equal, but follow in reverse order.

In general, two symmetrical spherical polygons cannot be made to coincide.

The sides of a spherical polygon are usually measured in degrees.

PROPOSITION IX. THEOREM

716. *The sum of two sides of a spherical triangle is greater than the third side.*

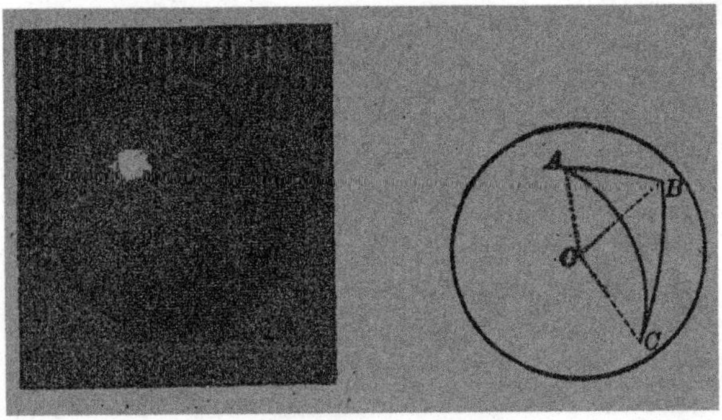

Hyp. ABC is a spherical triangle.

To prove $AB + BC > AC$.

Proof. Draw radii OA, OB, and OC.

Then $\angle AOB + \angle BOC > \angle AOC$. (522)

But the central angle is measured by the intercepted arc.

$\therefore AB + BC > AC$. Q.E.D.

PROPOSITION X. THEOREM

717. *The sum of the sides of any spherical polygon is less than four right angles.*

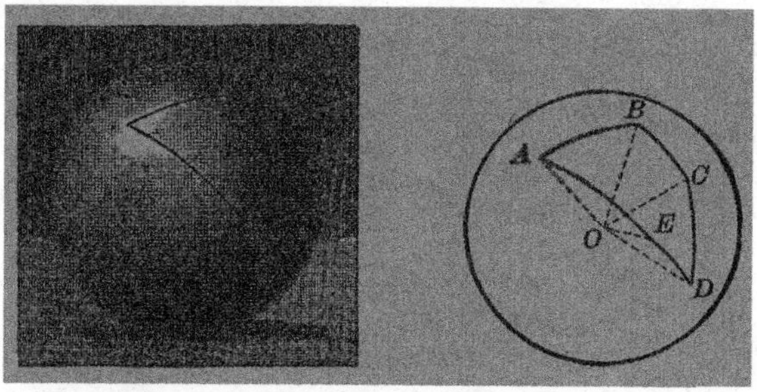

Hyp. $ABCDE$ is a spherical polygon.

To prove $AB + BC + CD + DE + EA < 360°$.

HINT. — Construct the corresponding polyedral angle and compare Remark (713).

PROPOSITION XI. THEOREM

718. *Two triangles on the same sphere are equal:*

(1) *If two angles and the included side of the one are respectively equal to two angles and the included side of the other,*

(2) *If two sides and the included angle of the one are respectively equal to two sides and the included angle of the other,*

(3) *If three sides of the one are respectively equal to three sides of the other,*

Provided the equal parts are arranged in the same order.

HINT. — Prove the equality of the corresponding polyedral angles.

719. COR. Two symmetrical isosceles triangles are equal.

PROPOSITION XII. THEOREM

720. *Two triangles on the same sphere are symmetrical:*

(1) *If two angles and the included side of the one are respectively equal to two angles and the included side of the other,*

(2) *If two sides and the included angle of the one are respectively equal to two sides and the included angle of the other,*

(3) *If three sides of the one are respectively equal to three sides of the other,*

Provided the equal parts are arranged in the reverse order.

HINT.—Prove that the corresponding polyedral angles are symmetrical.

721. REMARK.—The equality of spherical angles and arcs is usually proven by means of equal or symmetrical triangles.

PROPOSITION XIII. THEOREM

722. *The base angles of an isosceles spherical triangle are equal.*

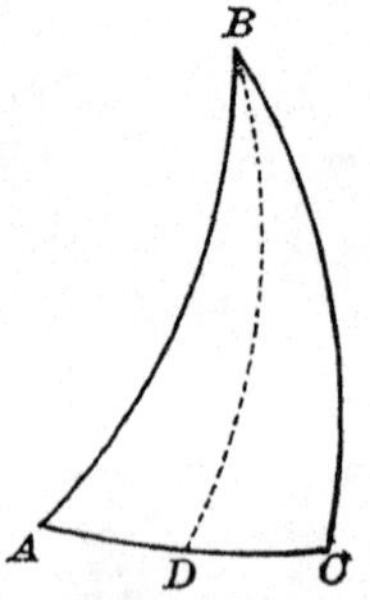

HINT.—Bisect the vertical angle and prove that two symmetrical triangles are formed.

723. Cor. An equilateral spherical triangle is also equiangular.

724. Remark. — **Many theorems of Spherical Geometry may be proved by methods analogous to those of Plane Geometry.**

725. Note. — In propositions relating to spherical figures, the words, "lines," "bisectors," "perpendiculars," etc., are often used for arcs of great circles, arcs of great circles bisecting spherical angles, etc.

Ex. 1173. Every point in a perpendicular bisector of an arc of a great circle is equidistant from the ends of the arc.

Ex. 1174. Two points equidistant from the ends of an arc of a great circle determine the perpendicular bisector of the arc.

Ex. 1175. To bisect a spherical angle.

Ex. 1176. To bisect an arc of a great circle.

Ex. 1177. At a point in a given arc of a great circle, to draw a perpendicular to the arc.

Ex. 1178. From a point without, to draw a perpendicular to a given great arc.

Ex. 1179. If the opposite sides of a spherical quadrilateral are equal, the opposite angles are equal.

Ex. 1180. Vertical spherical angles are equal.

Ex. 1181. If two semicircumferences have common ends, they include equal angles.

Ex. 1182. If the opposite sides of a spherical quadrilateral are equal, the diagonals bisect each other.

Ex. 1183. To circumscribe a circle about a spherical triangle.

Ex. 1184. At a given point in a great circle, to draw an angle equal to a given angle.

Ex. 1185. To construct a spherical triangle having given two sides and the included angle.

Ex. 1186. To construct a spherical triangle having given three sides.

Ex. 1187. To construct a spherical triangle having given the base, the altitude, and the median corresponding with the base.

Ex. 1188. The three bisectors of a spherical triangle meet in a point.

Ex. 1189. The bisectors of the base angles of an isosceles spherical triangle are equal.

Ex. 1190. A central angle of a circle on a sphere is measured by the intercepted arc.

Ex. 1191. If two small circles on a sphere intersect, their line of centres bisects the common chord at right angles.

Ex. 1192. If a radius bisects a chord of a small circle on a sphere, it is perpendicular to the chord.

Ex. 1193. In a circle on a sphere, equal chords are equidistant from the pole.

726. Def. Two spherical *polygons* are **vertical** if their corresponding polyedral angles are vertical.

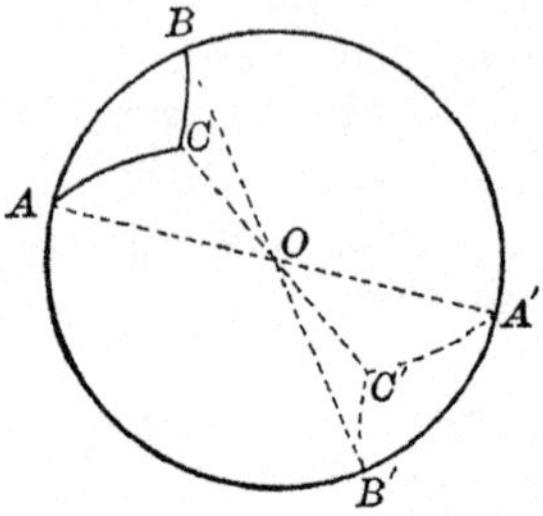

Ex. 1194. Two vertical spherical triangles are symmetrical.

POLAR TRIANGLES

727. Def. If from the vertices of any spherical triangle as poles arcs of great circles are described, another triangle will be formed which is called the **polar triangle** of the first.

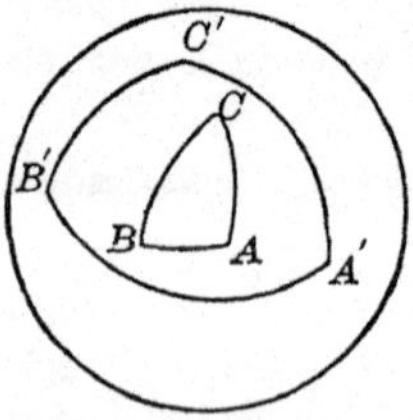

Thus, if A is the pole of great circle $B'C'$, B the pole of great circle $A'C'$, and C the pole of great circle $A'B'$, $\triangle A'B'C'$ is the polar triangle of ABC.

If from the poles A, B, and C entire circles should be described, eight triangles would be formed.

The polar triangle is selected by the following method: Denote the vertex formed by the intersection of the arcs from B and C by A', then of the four points of intersection that one is A' whose distance from A is less than a quadrant. Similarly for B' and C'.

PROPOSITION XIV. THEOREM

728. *If one spherical triangle is the polar triangle of another, then the second spherical triangle is the polar triangle of the first.*

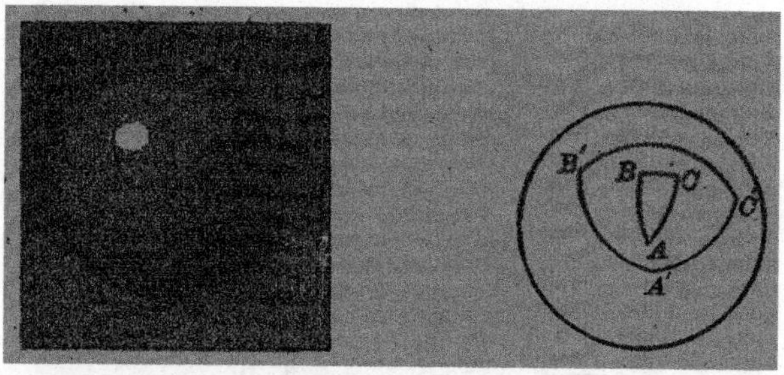

Hyp. $A'B'C'$ is the polar triangle of ABC.

To prove ABC is the polar triangle of $A'B'C'$.

Proof. Since A is the pole of $B'C'$, the distance AB' is a quadrant, and since C is a pole of $A'B'$, the distance $B'C$ is a quadrant.

Therefore, B' is the pole of AC. (691)

Similarly, A' is the pole of BC,

and C' is the pole of AB.

But the distances AA', BB' CC' are less than a quadrant.

Hence ABC is the polar triangle of $A'B'C'$. Q.E.D.

PROPOSITION XV. THEOREM

729. *In two polar triangles, each angle of the one is measured by the supplement of the opposite side of the other.*

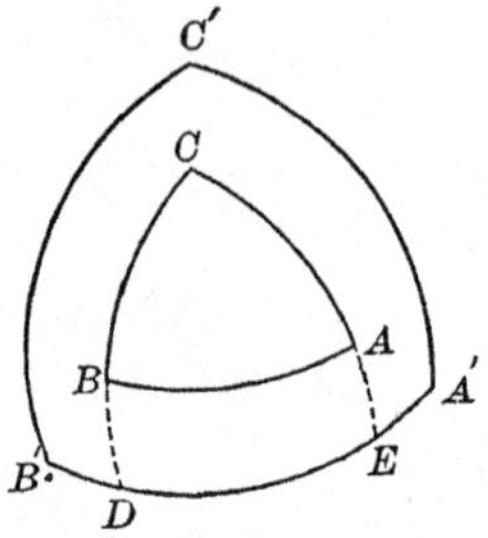

Hyp. ABC and $A'B'C'$ are polar triangles.

To prove $\angle C$ is the supplement of $B'A'$.

Proof. Produce the sides of C till they meet $B'A'$ in D and E respectively.

Then $\angle C$ is measured by arc DE, (708)

and $B'E$ and DA' are each 90°.

But $DE + B'A' = B'E + DA'$ or 2 rt. $\angle$s.

Hence DE and $B'A'$ are supplementary.

$\therefore \angle C$ and $B'A'$ are supplementary. Q.E.D.

Ex. 1195. The sides of a spherical triangle are 60°, 50°, and 100°. Find the angles of the polar triangle.

Ex. 1196. The angles of a spherical triangle are 20°, 90°, 80°. Find the sides of the polar triangle.

730. REMARK.—**Polar spherical triangles are the means of proving propositions relating to angles by the corresponding propositions relating to sides, and vice versa.**

PROPOSITION XVI. THEOREM

731. *The sum of the angles of a spherical triangle is greater than two and less than six right angles.*

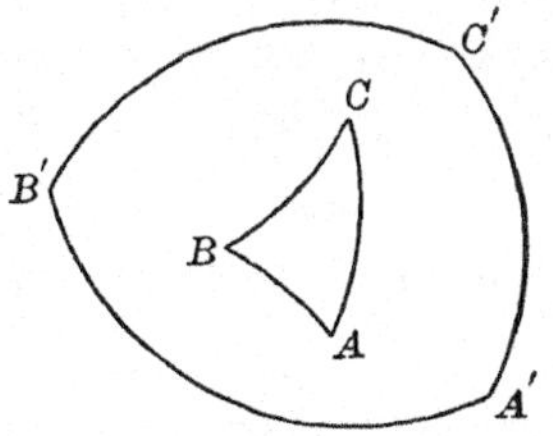

Hyp. ABC is a spherical triangle.

To prove $\angle A + \angle B + \angle C > 180°$,

and $\angle A + \angle B + \angle C < 540°$.

Proof. Construct the polar triangle $A'B'C'$, and denote the number of degrees in $B'C'$, $C'A'$, and $A'B'$, respectively, by a, b, and c.

Then $A = 180° - a$. (729)

$B = 180° - b$.

$C = 180° - c$.

Adding the equations, we obtain

$$A + B + C = 540° - (a + b + c).$$

Hence $A + B + C < 540°$.

But $(a + b + c) < 360°$. (717)

Hence $A + B + C > 180°$. Q.E.D.

732. COR. A spherical triangle may have one, two, or three right angles, also one, two, or three obtuse angles.

733. DEF. A **bi-rectangular spherical triangle** is one which contains two right angles.

734. Def. A **tri-rectangular spherical triangle** is one which contains three right angles.

735. Cor. 1. If three planes be passed through the center of a sphere, each perpendicular to the other two, the surface is divided into eight equal tri-rectangular triangles having all their sides equal to quadrants.

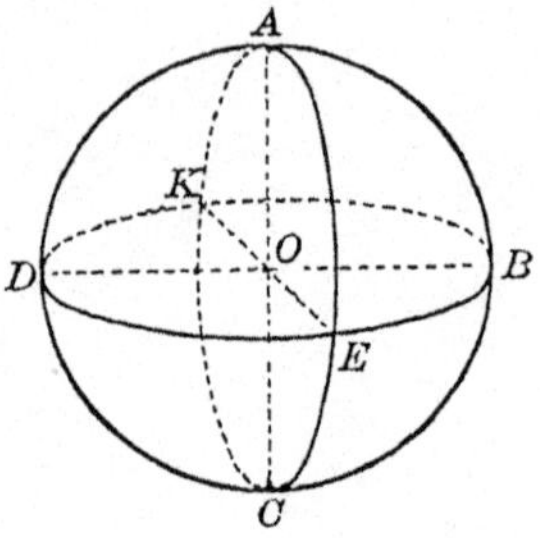

736. Cor. 2. The surface of a sphere is equal to eight times the surface of a tri-rectangular triangle.

Ex. 1197. The exterior angle of a spherical triangle is less than the sum of the two remote interior angles.

Ex. 1198. The sum of the angles of a spherical quadrilateral is greater than four right angles.

Ex. 1199. A tri-rectangular triangle is equal to its polar triangle.

Proposition XVII. Theorem

737. *If two angles of a spherical triangle are equal, the sides opposite are equal.*

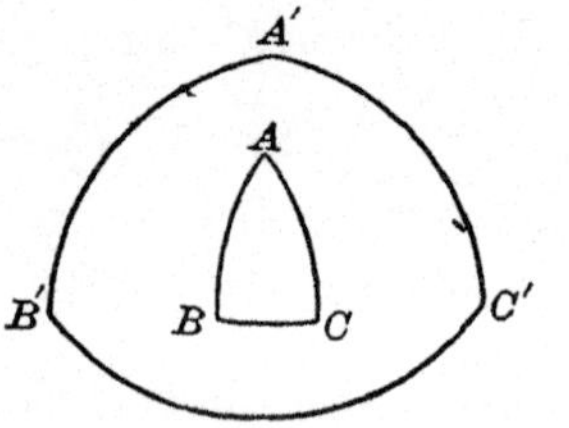

Hyp. In spherical triangle ABC, $\angle B = \angle C$.

To prove $AB = AC$.

Proof. Construct polar $\triangle A'B'C'$.

Then $A'B'$ is the supplement of $\angle C$, and $A'C'$ the supplement of $\angle B$.

Hence $$A'B' = A'C'.$$

$$\therefore \angle B' = \angle C'. \qquad (722)$$

But AB and AC are respectively the supplements of C' and B'.

Whence $$AB = AC. \qquad \text{Q.E.D.}$$

Ex. 1200. An equiangular spherical triangle is also equilateral.

Ex. 1201. If on the three sides AB, BC, and CA of the equilateral spherical triangle ABC, equal distances AA', BB', and CC' be laid off, then $A'B'C'$ is also an equilateral triangle.

Proposition XVIII. Theorem

738. *In any spherical triangle, the greater side is opposite the greater angle.*

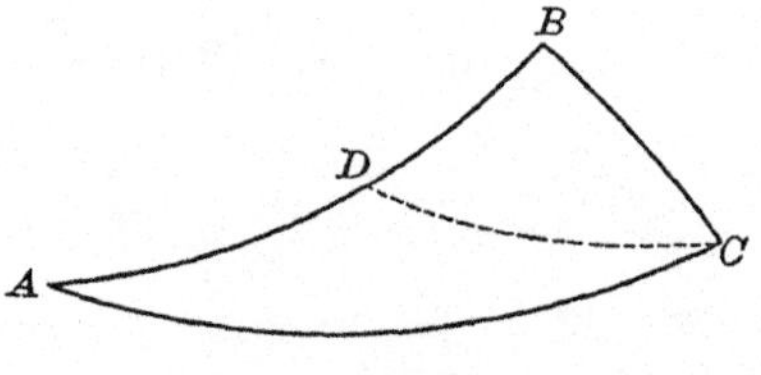

Hyp. In spherical triangle ABC, $\angle BCA > \angle BAC$.

To prove $$AB > BC.$$

Proof. Draw an arc of a great circle CD so that $\angle DCA = \angle A$, and let D be the intersection of AB and DC.

Then $$AD = DC. \qquad (737)$$

But $$DB + DC > BC. \qquad (716)$$

$$\therefore DB + AD > BC.$$

Or $$AB > BC. \qquad \text{Q.E.D.}$$

Ex. 1202. If BC is the base of an isosceles spherical triangle ABC, and D any point in AC, then $BD > CD$.

PROPOSITION XIX. THEOREM

739. *In any spherical triangle, the greater angle is opposite the greater side.*

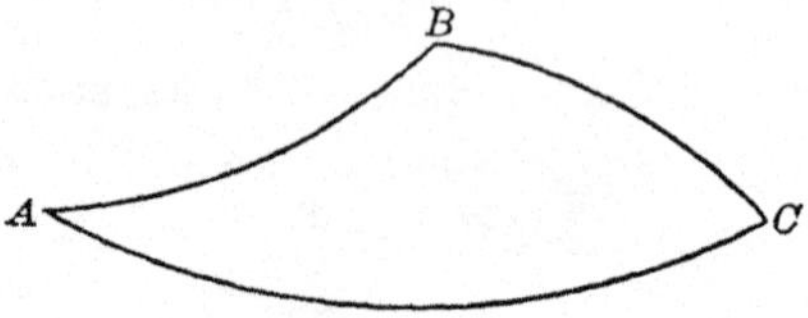

Hyp. In spherical $\triangle ABC$, $AB > BC$.

To prove $\angle C > \angle A$.

HINT. — Prove by the indirect method.

Ex. 1203. Prove Prop. XIX by means of polar triangles.

PROPOSITION XX. THEOREM

740. *If two triangles on the same sphere are mutually equiangular, they are mutually equilateral, and either equal or symmetrical.*

Hyp. Spherical $\triangle$s A and A' are mutually equiangular.

To prove A and A' are mutually equilateral and either equal or symmetrical.

Proof. Construct P the polar triangle of A, and P' the polar triangle of A'.

Since A and A' are mutually equiangular, (Hyp.)

P and P' are mutually equilateral. (729)

Hence P and P' are mutually equiangular. (718, 720)
But A and A' are the polar triangles of P and P'.
Hence A and A' are mutually equilateral. (729)
Hence A and A' are either equal or symmetrical. Q.E.D.

Ex. 1204. If the opposite angles of a spherical quadrilateral are equal, the opposite sides are equal.

HINT. — Produce two opposite sides in both directions until they meet, and find two equal triangles.

Ex. 1205. If two angles including a side of a spherical quadrilateral are equal, and the other two angles are equal, then the sides included by the unequal angles are equal.

Ex. 1206. If two great arcs are intersected by a transversal so as to make the alternate interior angles equal, the segments of the two great arcs are respectively equal.

PROPOSITION XXI. THEOREM.

741. *The shortest line that can be drawn between two points on the surface of a sphere is the arc of a great circle, not greater than a semicircumference, joining these points.*

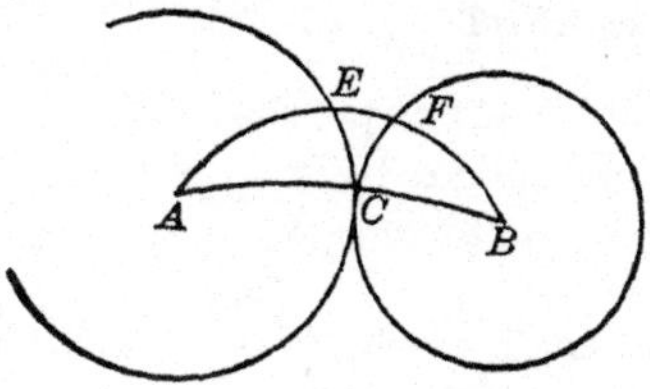

Hyp. AB is an arc of a great circle, smaller than a semicircumference.

To prove AB is shorter than any other line on the sphere joining A and B.

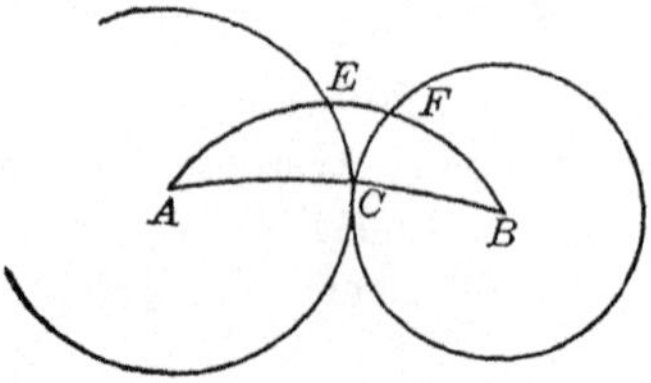

Proof. Take any point C on AB, and from A and B as centers, with radii respectively equal to AC and BC, draw two circles. These two circles cannot meet in any other point, for if they should meet in some point D, the sum of two sides AD and DB of $\triangle ADB$ would not be greater than the third side AB.

Any other line $AEFB$ between A and B must, therefore, intersect the circumferences in two points E and F, respectively.

But $AEFB$ cannot be the shortest line between A and B, for by revolving AE about A, and BF about F till E and F coincide with C, we would obtain a line joining A and B which is shorter than $AEFB$.

Hence, the shortest line between A and B must pass through C.

But since C is any point in AB, the shortest line between A and B must coincide with AB.

Or AB is the shortest line joining A and B. Q.E.D.

MEASUREMENT OF SPHERICAL FIGURES

742. Def. A **lune** is a portion of a spherical surface bounded by two semi-circumferences of great circles, as $ABCD$.

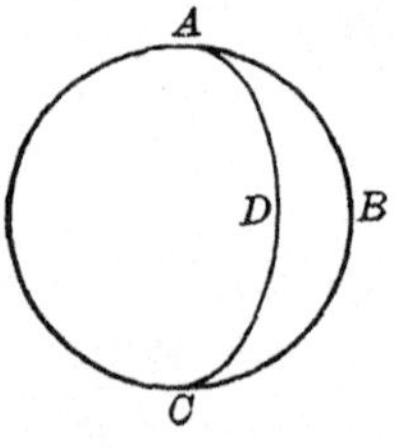

743. Def. The **angle of the lune** is the spherical angle included by the arcs.

Evidently, lunes on the same sphere are equal if their angles are equal.

PROPOSITION XXII. THEOREM

744. *Two lunes on the same sphere or equal spheres are to each other as their angles.*

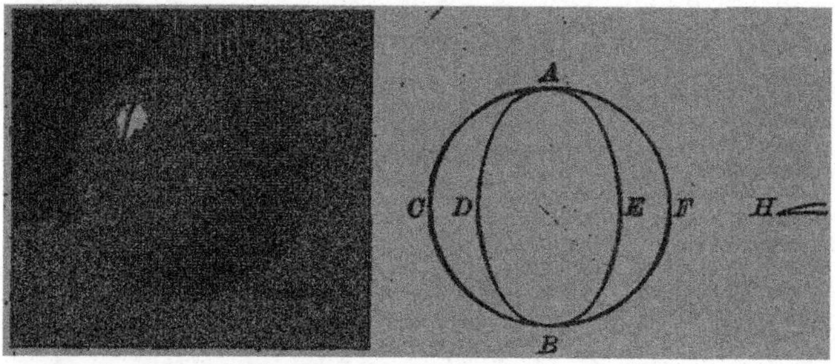

Hyp. $ACBD$ and $AEBF$ are lunes on the same sphere.

To prove $$\frac{ACBD}{AEBF}=\frac{\angle CAD}{\angle EAF}.$$

Proof. CASE I. The angles are commensurable.

Let a common measure H be contained in $\angle CAD$ m times, and in $\angle EAF$ n times.

Then $$\frac{\angle CAD}{\angle EAF}=\frac{m}{n}.$$

Producing the lines of division, lune $ACBD$ will be divided into m, and lune $AEBF$ into n equal parts.

Hence $$\frac{ACBD}{AEBF}=\frac{m}{n}.$$

Thence $$\frac{ACBD}{AEBF}=\frac{\angle CAD}{\angle EAF}.$$

CASE II. The angles are incommensurable.

[Proof by student who may compare (214).]

745. Cor. A lune is to the surface of the sphere on which it lies as its angle is to four right angles.

746. Scholium. The area of a tri-rectangular triangle is often taken as the unit of surface in Spherical Geometry. The area of the sphere is then measured by 8, the area of a hemisphere by 4, and the area of a lune by twice its angle.

For if A be the angle of the lune expressed in right angles, then lune : $8 = A : 4$, or lune $= 2A$.

Proposition XXIII. Theorem

747. *Symmetrical triangles on the same sphere are equivalent.*

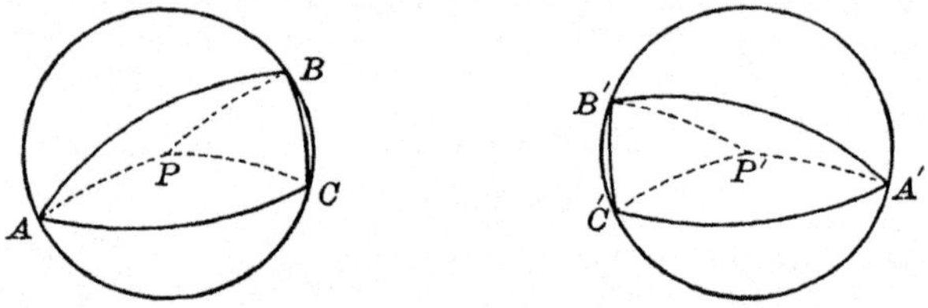

Hyp. ABC and $A'B'C'$ are symmetrical, spherical triangles.

To prove $\triangle ABC \approx \triangle A'B'C'$.

Proof. Let P and P' be the poles of small circles passing through A, B, C and A', B', C', respectively.

Since arcs AB, BC, CA are equal respectively to arcs $A'B'$, $B'C'$, $C'A'$,

Chords AB, BC, CA are equal respectively to chords $A'B'$, $B'C'$, $C'A'$.

Hence plane $\triangle ABC$ and $A'B'C'$ are equal. (106)

Hence circle $ABC =$ circle $A'B'C'$. (181)

Join PA, PB, PC, $P'A'$, $P'B'$, and $P'C'$.

These arcs are equal. (687)

Hence $\triangle ABP$ and $A'B'P'$ are isosceles and symmetrical.

Hence $\triangle ABP = \triangle A'B'P'$. (719)

Similarly, $\triangle ACP = \triangle A'C'P'$,

$\triangle BCP = \triangle B'C'P'$.

Hence by addition

$$\text{area} \triangle ABC = \text{area} \triangle A'B'C'. \qquad \text{Q.E.D.}$$

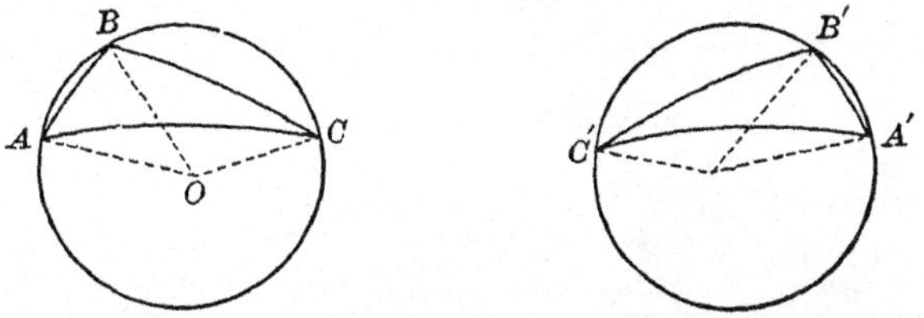

748. SCHOLIUM. If the poles of $\triangle ABC$ and of $\triangle A'B'C'$ lie without the triangles, the proof has to be slightly modified.

HINT. $\triangle ABC \approx ABO + BCO - ACO$.

749. COR. If two arcs of great circles ABC and DBE intersect on the surface of a hemisphere, the sum of the areas of two opposite spherical triangles ABE and DBC is equivalent to a lune whose angle is equal to the angle ABE, included by the two arcs.

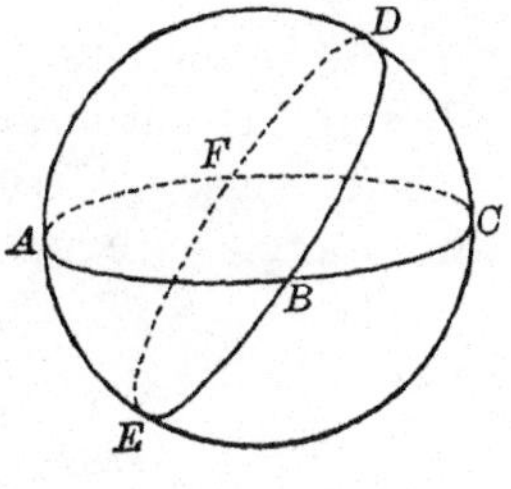

HINT. $\triangle ABE \approx \triangle DCF$.

750. DEF. The **spherical excess** of a spherical triangle is the excess of the sum of its three angles over two right angles.

Thus, if A, B, and C are the three angles of a triangle expressed in right angles, and the spherical excess is denoted by E, then

$$E = A + B + C - 2.$$

Proposition XXIV. Theorem

751. *If the right angle be taken as angular unit, and the tri-rectangular triangle as unit of surface, the area of a spherical triangle is measured by the spherical excess.*

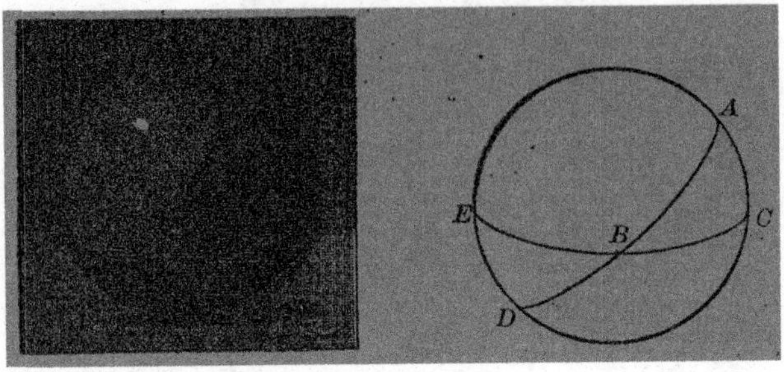

Hyp. ABC is a spherical triangle.

To prove. Area $ABC = A + B + C - 2$, if the angular unit is the right angle, and the unit of surface the tri-rectangular triangle.

Proof. Complete the circumference $ACDE$, and produce AB and CB until they meet ACD in D and E respectively.

Then, since $\triangle ABC + \triangle BED$ is equivalent to the lune, whose angle is equal to $\angle ABC$, (749)

$$ABC + BED \approx 2B.$$

But $ABC + ABE \approx 2C$, (746)

$ABC + BCD \approx 2A$. (746)

By adding these equations, we obtain

$$3\,ABC + BED + ABE + BCD = 2(A + B + C).$$

But $ABC + BED + ABE + BCD$ equals a hemisphere or 4.

Hence $2\,ABC + 4 = 2(A + B + C)$,

or $ABC + 2 = A + B + C$.

Whence area $ABC = A + B + C - 2$. Q.E.D.

752. COR. The area of a spherical triangle is to the surface of the entire sphere as its spherical excess is to 8.

Ex. 1207. If the surface of a sphere is 40 sq. ft., find the area of a spherical triangle whose angles are respectively 40°, 60°, and 100°.

Ex. 1208. What fraction of the surface of a sphere is covered by a spherical triangle whose angles are 90°, 100°, and 110°, respectively?

Ex. 1209. How large is the angle of an equiangular spherical triangle whose area is one-fourth of the surface of the sphere?

Ex. 1210. What is the ratio of a lune whose angle is equal to 80°, and the area of an equilateral spherical triangle whose angle is equal to 80°?

Ex. 1211. The arms of an isosceles spherical triangle whose base angles are 80° are produced so as to form a lune. Find the vertical angle if the triangle is the third part of the lune.

Ex. 1212. The angles of a spherical triangle are 80°, 90°, and 100°. Find the angle of an equivalent lune.

PROPOSITION XXV. THEOREM

753. *If the right angle be taken as the angular unit, and the tri-rectangular triangle as unit of surface, the area of any spherical polygon is equal to the sum of its angles diminished by twice the number of its sides less two.*

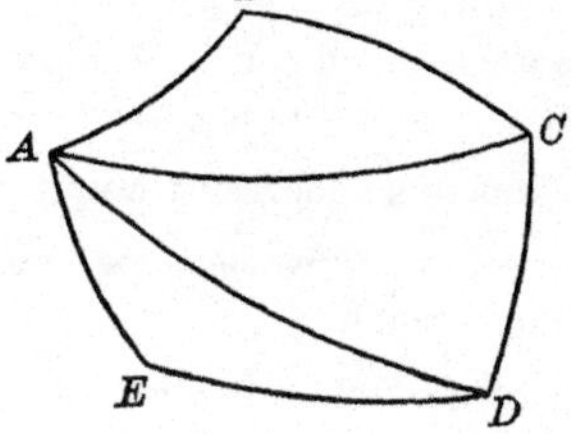

Hyp. $ABCD \cdots$ is a spherical polygon of n sides.

To prove $ABCD \cdots$ is measured by $(A+B+C+\cdots)-2(n-2)$.

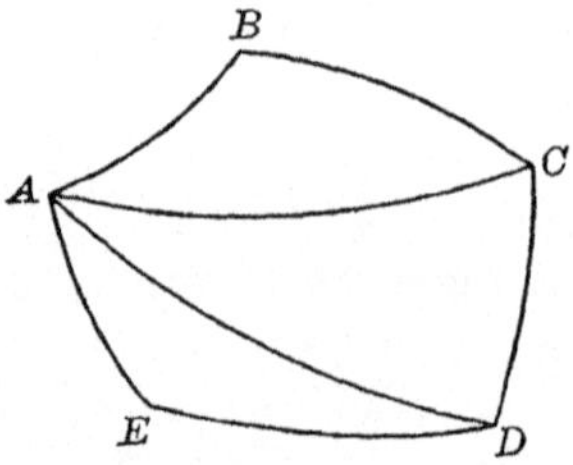

Proof. Draw all diagonals from A, which will divide $ABCD \cdots$ into $n-2$ triangles.

The area of each triangle is equal to the sum of its angles less 2. (751)

Hence the area of the polygon $ABCD \cdots$ is equal to

$$(A + B + C + \cdots) - 2(n-2).$$ Q.E.D.

754. Scholium. Props. XXIV and XXV do not give us the absolute, but the relative size of areas on the sphere.

Ex. 1213. Find the area of an equiangular hexagon whose angle is 160°. (The tri-rectangular triangle taken as unit.)

Ex. 1214. What fraction of the surface of the sphere is covered by the hexagon of the preceding exercise.

Ex. 1215. The angles of a spherical quadrilateral are 90°, 100°, 110°, and 120°. Find the angle of an equivalent equilateral triangle.

Ex. 1216. Find the angle of an equiangular spherical triangle equivalent to the sum of three equilateral triangles whose angle is 70°.

Ex. 1217. The medians of a spherical triangle meet in a point.

Ex. 1218. The intersection of two spheres is a circle whose plane is perpendicular to the line of centers.

Ex. 1219. The line of centers of two spheres is 13. Find the radius of the circle of intersection if the radii of the spheres are 5 and 12 respectively.

Ex. 1220. The volume of a circumscribed polyedron is equivalent to its surface multiplied by one-third of the radius.

PROPOSITION XXVI. THEOREM

755. *The area generated by the revolution of a straight line about an axis in its plane is equal to the projection of the line upon the axis multiplied by the circumference of a circle whose radius is the perpendicular erected at the midpoint of the line and terminating in the axis.*

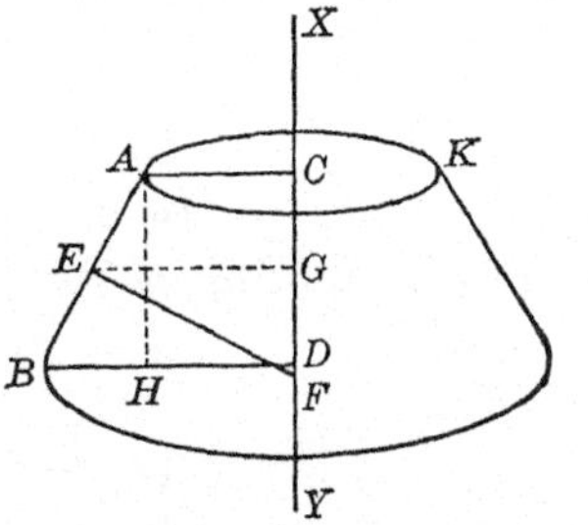

Hyp. *AB* revolving about an axis *XY* in its plane produces surface *ABK*, *CD* is the projection of *AB* upon *XY*, and *EF* is the perpendicular bisector of *AB* terminating in *XY*.

To prove $ABK = CD \times 2\,\pi EF.$

Proof. Draw *EG* ∥ to *BD* and *AH* ∥ to *CF*.

The surface *ABK* is the lateral surface of a cone of revolution.

Hence area $ABK = AB \times 2\,\pi EG.$ (1) (668)

But $\triangle ABH \sim \triangle EFG.$ (291)

$$\therefore \frac{AB}{EF} = \frac{AH}{EG}.$$

Whence $AB \times EG = AH \times EF,$

or $AB \times EG = CD \times EF.$

Multiplying both sides by $2\,\pi$ we have from (1).

$\therefore$ area $ABK = CD \times 2\,\pi EF.$ Q.E.D.

756. DEF. A **zone** is a portion of a spherical surface included between two parallel planes.

757. DEF. The circumferences of the sections made by the planes form the **bases** *of the zone,* and the distance between the two planes is the **altitude** of the zone.

Ex. 1221. In the diagram for Prop. XXVI, find the area ABK, if EF is equal to 10 in., AB is equal to 8 in., and angle BH is one-third of a right angle.

PROPOSITION XXVII. THEOREM

758. *The surface of a sphere is equivalent to four times the area of a great circle.*

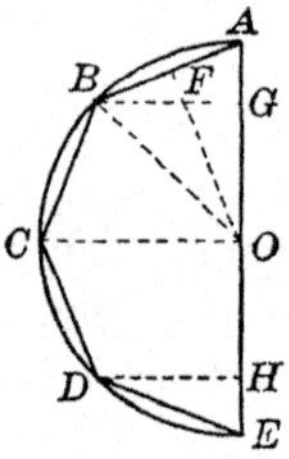

Hyp. S is the area of the surface, R the radius of a sphere generated by revolving semicircle ABE about AE.

To prove $S = 4\,\pi R^2$.

Proof. Inscribe in the semicircle half of a regular polygon of an even number of sides, as $ABCDE$.

Draw BG, CO, and DH perpendicular to AE, and denote the common distance of the chords AB, BC, CD, and DE from O by d.

Then area AB* $= AG \times 2\,\pi d$, (755)

area $BC = GO \times 2\,\pi d$, etc.

* Area AB means area generated by AB.

Adding the equations

$$\text{area } ABCDE = (AG + GO + OH + HE) \times 2\,\pi d$$

or $$\text{area } ABCDE = 2\,R \times 2\,\pi d = 4\,R\pi d.$$

Now let the number of sides of the inscribed polygon be increased indefinitely. Then the area $ABCDE$ approaches S, and d approaches R as a limit.

Hence $$S = 4\,\pi R^2. \qquad (213)$$ Q.E.D.

759. Cor. 1. The areas of the surfaces of two spheres are to each other as the squares of their radii.

760. Cor. 2. The area of a zone is equal to its altitude multiplied by the circumference of a great circle.

The zone generated by the revolution of arc BC is equal to $GO \times 2\,\pi R$. (See diagram for Prop. XXVII.)

761. Cor. 3. Zones on the same or on equal spheres are to each other as their altitudes.

762. Scholium. Prop. XXVII, in connection with Props. XXIV, XXV, and 746, enables us to find the absolute size of spherical surfaces.

Ex. 1222. Find the surface of a sphere whose radius is 10.

Ex. 1223. Find the surface of a zone whose altitude is 5, if the radius of the sphere is 20.

Ex. 1224. Find the number of square feet in the surface of a lune whose angle is 40°, if the radius of the sphere is 20 ft.

Ex. 1225. Find the number of square feet in the surface of a spherical triangle, if its angles are respectively equal to 80°, 90°, and 100°, if the radius of the sphere is 20 ft.

Ex. 1226. Find the number of square feet in a hemispherical dome whose diameter is 40 ft.

Ex. 1227. The surface of a sphere is 100 square meters. Find the radius.

Ex. 1228. Find the angle of an equilateral spherical triangle on a sphere whose radius is 2, if the area of the triangle is equal to π.

SPHERICAL VOLUMES

763. DEF. A **spherical sector** is the solid generated by any plane sector of the semicircle when the semicircle revolves about its diameter.

764. The **base** of the spherical sector is the zone generated by the revolution of the arc.

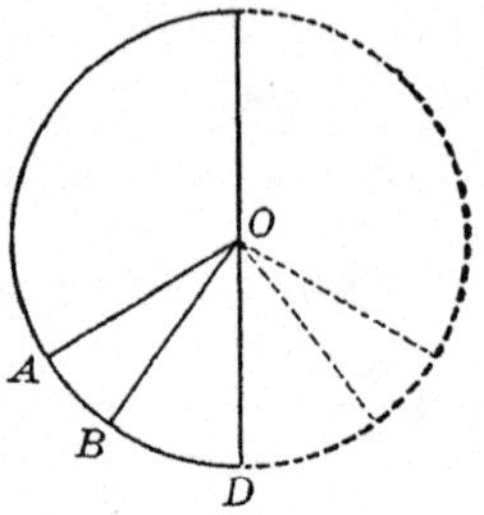

Thus, the solid generated by the revolution of *AOB* about *OD* is a spherical sector.

765. DEF. A **spherical pyramid** is a portion of a sphere bounded by a spherical polygon and the faces of the corresponding polyedral angle.

The spherical polygon is called the base of the pyramid.

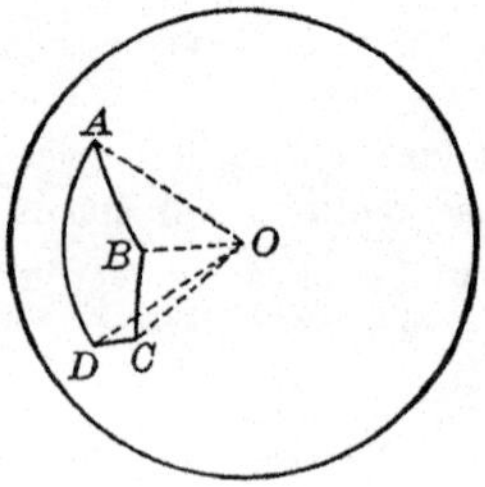

Thus, *OABCD* is a spherical pyramid having base *ABCD*.

766. DEF. A **spherical segment** is a portion of a sphere bounded by two *parallel* planes.

The sections (circles) made by the parallel planes are the **bases**, and the distance between the planes is the *altitude* of the segment.

If one of the bounding planes is tangent to the sphere, the segment is called a **segment of one base.**

767. DEF. A **spherical wedge** (or ungula) is the portion of a sphere bounded by a lune and the planes of its sides.

PROPOSITION XXVIII. THEOREM

768. *The volume of a sphere is equal to the area of its surface multiplied by one-third its radius.*

Hyp. V is the volume of a sphere whose radius is R and whose surface is S.

To prove $V = S \times \frac{1}{3} R$.

Proof. Conceive any polyedron circumscribed about the sphere, and let each vertex of the polyedron be joined to the center of the sphere.

The polyedron is then divided into pyramids, each having one face as a base and R as the common altitude.

Therefore, the volume of the polyedron is equivalent to the sum of the bases multiplied by one-third of R, or it is equivalent to the surface multiplied by one-third of the radius.

But by increasing the number of faces of the polyedron indefinitely, its volume will approach the volume of the sphere and its surface, the surface of the sphere as a limit.

Hence $V = S \times \frac{1}{3} R$. (213) Q.E.D.

769. COR. 1. If V denotes the volume, R the radius, and D the diameter of the sphere,

$$V = \frac{4}{3}\pi R^3 \text{ or } V = \frac{D^3\pi}{6}.$$

770. COR. 2. **The volumes of two spheres are to each other as the cubes of their radii.**

771. Cor. 3. The volume of a spherical pyramid is equal to the area of its base multiplied by one-third the radius of the sphere.

772. Cor. 4. The volume of a spherical sector is equivalent to the area of its base multiplied by one-third the radius of the sphere.

The proof is analogous to the proof of Prop. XXVIII.

773. Cor. 5. If h denotes the altitude of a spherical sector, V its volume, and Z the surface of the base,

$$V = Z \times \frac{R}{3}.$$

But $$Z = 2\pi R \times h. \qquad (760)$$

$$\therefore\ V = \tfrac{2}{3}\pi R^2 h.$$

774. Cor. 6. The volume of a sphere is to the volume of the circumscribed cylinder as two to three.

For the volume of the circumscribed cylinder is $2\pi R^3$.

Ex. 1229. Find the volume of a sphere whose radius is 10.

Ex. 1230. Find the volume of a spherical sector whose base is 20, if the radius of the sphere is 10.

Ex. 1231. Find the radius of a sphere whose volume is 100 cu. ft.

Ex. 1232. Find the volume of a sphere whose surface is equal to 10 square meters.

Ex. 1233. Find the surface of a sphere whose volume is equal to M.

Proposition XXIX. Problem

775. *To find the volume of a spherical segment.*

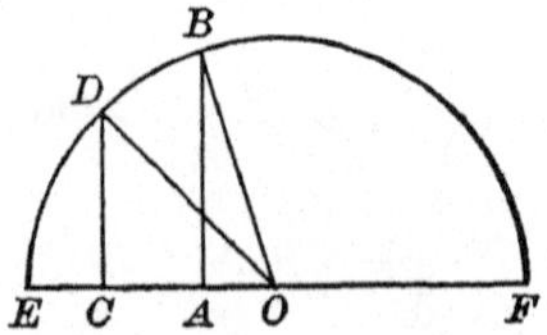

Given. A spherical segment generated by the revolution of $ABCD$ about EF.

Required. To find the volume V of the spherical segment, if $AB = r$, $CD = r_1$, and $CA = h$.

Join OD and OB.

Then the volume generated by $ABCD$ is equal to the spherical sector generated by DBO, plus the cone generated by ODC, minus the cone generated by BAO.

Hence

$$V = \tfrac{2}{3}\pi R^2 h + \tfrac{1}{3}\pi \overline{CD}^2 \cdot CO - \tfrac{1}{3}\pi \overline{AB}^2 \cdot AO,$$

$$= \tfrac{1}{3}\pi[2R^2h + (R^2 - \overline{CO}^2)CO - (R^2 - \overline{AO}^2)AO]$$

$$= \tfrac{1}{3}\pi[2R^2h + R^2(CO - AO) - (\overline{CO}^3 - \overline{AO}^3)].$$

But $CO - AO = h$, and $\overline{CO}^3 - \overline{AO}^3$ can be factored, hence:

$$V = \tfrac{1}{3}\pi[2R^2h + R^2h - h(\overline{CO}^2 + CO \times AO + \overline{AO}^2)]$$

$$V = \frac{\pi h}{3}[3R^2 - (\overline{CO}^2 + CO \times AO + \overline{AO}^2)]. \qquad \cdots (1)$$

But

$$h^2 = (CO - AO)^2 = \overline{CO}^2 - 2\,CO \times AO + \overline{AO}^2,$$

and

$$\overline{CO}^2 + CO \times AO + \overline{AO}^2$$

$$= \tfrac{3}{2}\overline{CO}^2 + \tfrac{3}{2}\overline{AO}^2 - \left(\frac{\overline{CO}^2}{2} - \frac{2\,OC \cdot AO}{2} + \frac{\overline{AO}^2}{2}\right)$$

$$= \tfrac{3}{2}(\overline{CO}^2 + \overline{AO}^2) - \frac{h^2}{2}$$

$$= \tfrac{3}{2}(R^2 - r^2 + R^2 - r_1^2) - \frac{h^2}{2}$$

$$= 3R^2 - \frac{3(r^2 + r_1^2)}{2} - \frac{h^2}{2}.$$

Hence by substituting this value in (1)

$$V = \frac{\pi h}{3}\left[\tfrac{3}{2}(r^2 + r_1^2) + \frac{h^2}{2}\right]$$

$$= \frac{\pi h}{2}(r^2 + r_1^2) + \frac{\pi h^3}{6}.$$

776. Cor. In a spherical segment of one base, $r_1 = 0$.

$$\therefore \quad V = \frac{h\pi r^2}{2} + \frac{\pi h^3}{6}.$$

EXERCISES

Ex. 1234. Find the volume of a spherical segment, the radii of whose bases are 4 and 5 and whose altitude is 1.

Ex. 1235. The volumes of two spheres are to each other as 8 to 125. Find the ratio of their radii.

Ex. 1236. The volumes of two spheres are to each other as 125 to 216. Find the ratio of their surfaces.

Ex. 1237. Find the radius of a sphere whose surface is equivalent to the sum of the surfaces of two spheres whose radii are 3 and 4 respectively.

Ex. 1238. Find the volume of a spherical shell whose exterior radius is 13 and whose thickness is 8.

Ex. 1239. Find the radius of a sphere equivalent to the spherical shell in the preceding exercise.

Ex. 1240. Find the radius of a sphere equivalent to a cube whose edge is equal to a.

Ex. 1241. A cylindrical vessel, 4 in. in diameter, is partly filled with water. Upon immersing a ball the surface of the water rises 1 in. Find the diameter of the ball.

Ex. 1242. A sphere whose radius is 2 in. weighs 32 oz. Find the weight of a sphere of the same material whose radius is 3 in.

Ex. 1243. Find the volume of a spherical pyramid whose base is an equilateral triangle with its angles equal to 80°, if the radius of the sphere is equal to 10.

Ex. 1244. A square whose side is 4 revolves about one of its diagonals. Find the surface and the volume of the generated solid.

Ex. 1245. Find the volume of a spherical segment of one base, if its curved surface is 20 and its altitude is 2.

Ex. 1246. Find the radius of a sphere whose surface is equivalent to the entire surface of a cube whose edge is equal to 4.

Ex. 1247. The edge of a cube is 10 in. Find the diameter of the circumscribed sphere.

Ex. 1248. A lune whose angle is equal to 40° is equivalent to a zone on the same sphere. Find the ratio of the altitude of the zone to the radius of the sphere.

Ex. 1249. The diedral angles of a spherical pyramid of six sides are 140°. Find the volume of the pyramid if the radius is equal to 10.

Ex. 1250. Through a sphere whose diameter is 10 m. a cylindrical hole of 5 m. diameter is bored. Find the volume of the solid if the axis of the cylinder passes through the center of the sphere.

Ex. 1251. The surface of a sphere is equivalent to the lateral surface of the circumscribed cylinder.

Ex. 1252. Two bi-rectangular spherical triangles are equal if the oblique angles are equal.

Ex. 1253. Find the ratio of a sphere to its circumscribed cube.

Ex. 1254. The area of a zone on a sphere is 20, its altitude 4. Find the radius of the sphere.

Ex. 1255. If the diagonals of a spherical quadrilateral bisect each other, the opposite sides are equal.

Ex. 1256. The radius of a sphere is 9 in. Find the volume of a spherical wedge whose angle is equal to 60°.

Ex. 1257. Find the radius of a sphere equivalent to a cone of revolution, the radius of whose base is equal to r and whose altitude is equal to h.

Ex. 1258. The area of a zone is equal to A, its altitude is equal to h. Find the radius of the sphere.

Ex. 1259. The volume of a sphere is numerically equal to one-half its surface. Find the radius.

Ex. 1260. The volume of a cylinder of revolution is equal to one-half the product of its lateral surface by the radius of its base.

Ex. 1261. How many square miles of the surface of the earth can be seen from a point 1000 miles above the surface, if the earth is supposed to be a perfect sphere whose radius is equal to 4000 miles?

Ex. 1262. If from a point without a sphere a tangent and a secant be drawn, the tangent is the mean proportional between the secant and its external segment.

Ex. 1263. If through the line of intersection of two spheres a plane be passed, tangents from a point of the plane to the spheres are equal.

Ex. 1264. The radius of a sphere is r, the area of a small circle a. Find its distance from the center.

Ex. 1265. The volume of a sphere is V. Find the surface of an equilateral spherical triangle whose angle is equal to 100°.

INDEX OF DEFINITIONS

Printed in Dunstable, United Kingdom

91219953R00218